A first course in fluid dynamics

A first course in

fluid dynamics

A. R. PATERSON

Lecturer in Mathematics, University of Bristol

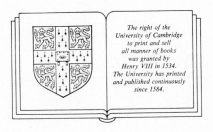

*The right of the
University of Cambridge
to print and sell
all manner of books
was granted by
Henry VIII in 1534.
The University has printed
and published continuously
since 1584.*

CAMBRIDGE UNIVERSITY PRESS

Cambridge

London   New York   New Rochelle

Melbourne   Sydney

Published by the Press Syndicate of the University of Cambridge
The Pitt Building, Trumpington Street, Cambridge CB2 1RP
32 East 57th Street, New York, NY 10022, USA
10 Stamford Road, Oakleigh, Melbourne 3166, Australia

First published 1983
Reprinted 1985

Printed by Nene Litho, Irthlingborough, Northants
Bound by Woolnough Bookbinding, Irthlingborough, Northants

Library of Congress catalogue card number: 82–23437

British Library cataloguing in publication data

Paterson, A. R.
A first course in fluid dynamics
1. Fluid dynamics
I. Title
532′.05   QC151

ISBN 0 521 25416 7 hard covers
ISBN 0 521 27424 9 paperback

TM

Contents

Contents

Preface

An author makes his excuses in a preface, so here are mine. For a number of years at Bristol University we tried to find a suitable text to introduce fluid dynamics to second year mathematics students, and failed. The modern texts with the 'right' attitude to the subject were too hard for a first course, the older texts were dominated by potential theory and unrealistic examples. This text has been tried in draft form for several years on our students, and has been judged 'hard, but interesting'. New work in mathematics is always hard, but I believe that the level chosen here is a suitable one.

I apologise to my colleagues for the gross over-simplification of their work and their subject which is committed in this book. And also for the errors and misapprehensions – students, beware! all texts have mistakes in them. I thank my colleagues for helpful comments and discussions over many years; I also thank a succession of seminar speakers for maintaining my awareness of the full range of fluid dynamics.

Introduction

1. Fluid dynamics

There are two main reasons for studying fluid dynamics. Firstly, understanding (though in some places still only partial) can be gained of a great range of phenomena, many of which are of considerable complexity. And secondly, predictions can be made in many areas of practical importance which involve fluids.

As we shall see later, a 'fluid' is a way of looking at a large collection of particles, so as to avoid dealing with each particle separately. One of the largest examples of such a collection of 'particles' is a galaxy, composed of a vast number of individual stars. A more obvious fluid composes the sun: the particles here are largely electrons and nucleons, and the fluid dynamics here is complicated by electromagnetic forces, nuclear reactions and radiation effects. Astrophysics provides another example of fluid motion in the solar wind, the outflow of (isolated) particles from the sun: this is a fluid in which the particles are very thinly spread, but it is a fluid which interacts importantly with the Earth's magnetic field and the upper layers of the atmosphere. Both atmosphere and magnetic field are much studied examples of fluid dynamics; climate predictions, weather forecasts and studies of local climate are of obvious interest, while the origin of the Earth's magnetic field in the inner motions of the Earth's material is rather less obvious, but no less interesting. You may list for yourself some of the physical phenomena associated with the existence of oceans, rivers, lakes and underground water.

At this sort of scale one can start to see practical considerations coming to the fore. The engineer who designs a hydroelectric system must have a good knowledge of how water will behave and what forces it will exert. As a further example, an aeroplane (or a ship) is built by an interaction of past experience, mathematical calculation and testing of models in wind tunnels (or wave tanks). High speed trains, and suspension bridges, are in constant interaction with the wind. Even a low speed bicycle rides on a few thin films of lubricating oil.

Man himself can be regarded as a collection of fluid dynamic systems (a rather restricted view), and physiological fluid dynamics has recently emerged as an important area of research.

The above examples of areas involving the motion of fluids could easily be added to almost indefinitely: you should, for example, list all the fluid dynamic aspects of a petrol engined car. But these examples will serve to show the range and applicability of fluid dynamics. Moreover, the range and the applications increase, because fluid dynamics is an active field of research, not only in universities but also in industrial research associations and in national research centres.

Fluid dynamics is a branch of applied mathematics; the subject cannot be studied to any depth without a considerable skill in mathematics. This is one of its fascinations for any mathematically inclined person, to see how much of the apparatus of mathematics is needed to describe such a 'simple' problem as the flow of a fluid past an obstacle. In fact, problems in fluid dynamics have caused developments in mathematical techniques; the idea of a boundary layer (Chapter IX, §3) has stimulated the growth of the mathematics of 'singular perturbation theory', to give one example described in this text.

The application of mathematics to problems in the real world of physics and engineering is a skill that is hard to learn. Any real problem has too many aspects for us to hope to describe them all mathematically at once – if a man drops a book in a room, the air flow that is generated must depend on the shape of the room and the position and shape of all the objects in it: how can we solve a problem as complicated as that? The art is to describe the non-essential parts of the problem one by one until a mathematical problem can be formulated that is easy enough to solve, but still contains the essence of the original situation. The fluid dynamics in this text provides many examples of this reduction of reality to a simple model which can be treated mathematically and which shows the nature of the phenomenon under discussion. We may hope that after an intelligent study of these examples you will be ready to try your hand at 'mathe-matical modelling' of this kind.

2. Structure of the text

This text cannot start at the beginning, or it would become far too long. We must assume that you already know something about mathematics, about physics, and about the world around you and how to describe it mathematically. The first three chapters (all pleasantly short) are to remind you about some important things and, if necessary, to introduce you to others. All this material is, of course, important; but perhaps the mathematics for vector fields contained in Chapter I is the most important: you certainly cannot get by without it in later work.

The next two chapters (IV and V) are about the description of the velocities of a fluid. This is 'just' kinematics, there is as yet no discussion of the forces acting or of an equation of motion to predict what will result from a given initial situation. Some of the basic models and concepts are brought in at this stage, and the mathematics of

$$\nabla \cdot \text{ and } \nabla \times$$

or

$$\text{div and curl}$$

is needed.

The next three chapters (VI – VIII) discuss the forces acting in a fluid, and lead up to a full equation of motion for a fluid. All three chapters are concerned with the idea of pressure in a fluid. Chapter VI introduces pressure and deals with the easiest case, that of the pressure in a fluid at rest, and the forces it causes. Chapter VII discusses the possible relations between pressure and density, and outlines some necessary thermo-dynamics. Chapter VIII derives the relation between pressure and accelera-tion in a fluid, and discusses how it might be simplified in commonly occurring situations. Some of the derivations of equations in this part of the text are quite complicated, and need not be mastered at a first reading; but it would be a pity not to look through them, at least.

By the end of Chapter VIII there is nothing left to do except solve the equation of motion! This 'nothing' is a very large amount of work, because the equations of fluid dynamics are impossibly hard in their full generality. So we set off in Chapter IX with some moderately easy flows and using almost the full equations; these are flows in which the viscosity of the fluid is important, and they give valuable ideas on when viscosity must be included, and what some of its effects will be. Chapter X continues this line of thought by discussing flows for which viscosity can mainly be neglected, and Chapter XI is about a smaller class of rather simple flows which have a very easy (relatively!) equation to describe them and which are sometimes realistic. This group of three chapters concludes the most basic course on fluid dynamics: you can pick and choose from other chapters, but you must cover (even if not in all the details) Chapters I–XI.

The two chapters on sound and water waves are about small disturbances to a state of rest, with simple boundaries. They provide interesting preliminary descriptions of a number of obvious phenomena, without being too hard. The mathematics is mainly linear, and so comparatively easy.

The two chapters after that provide an introduction to what happens when disturbances are not small. Chapter XIV is about some of the effects of compressibility, and shows that discontinuities ('shock waves') can appear in the solutions of the apparently appropriate equations. Chapter XV uses some rather similar mathematics to deal with water waves of larger size on shallow water. Because much of the mathematics is similar, these two chapters go together, much as the previous two do.

The last two chapters use some advanced mathematics to discuss an approximate version of the fluid dynamics of aeroplane wings. You should not be unduly put off by the need for complex function theory here – not too much is needed, and you may find it easier than you expect. These two chapters are independent of the previous four.

Generally speaking, the mathematical techniques you need for this text increase as you go through it, and extra methods are not brought in until they are absolutely necessary. This is often done by postponing particular examples until the required methods have been explained, at the point when they are absolutely essential. Brief versions of the new mathematics are given in this text; if you need more, you must seek help in books on mathematical techniques or advanced calculus.

3. Method of working

If you try to study fluid dynamics purely as a branch of mathematics, then you are liable to get answers which do not agree with experiment or observation. This is because there always has to be a careful choice of the mathematical model that is to be used to describe a particular phenomenon: injudicious modelling will retain the wrong terms in the equations, and so not give a description of what you want. The test of the mathematical model must always be against reality.

It is only rarely that a student has the time or the opportunity to carry out serious experiments in fluid dynamics for himself. So a text has to provide some sort of substitute. What is recommended here is the study of films of carefully devised experiments: these give a better idea of the motions involved than any (reasonable) number of still photographs, even though these can be quite revealing. There is a good set of film loops, each lasting only a few minutes and each devoted to a single idea; and there are some longer films available which cover whole areas. You should spend some time with these films or loops as each new area of theory comes along.

Moreover, you should spend time looking around you. Examples of fluid dynamics are everywhere, and you should try to relate what you see to the topics you are studying. Sometimes you *can* do simple experiments, or use the weather maps in newspapers as a source of observational data. But do not expect too much; this is only a first text on fluid dynamics, and so the theories will be rather too simple to give more than a roughly correct answer.

The models we shall use in this text will be as simple as possible, and even so the mathematics may be felt to be rather hard. You must study the models carefully to see not only how the mathematics is operated, but also how the physical reality has been modelled. There are many examples in the text to help you along.

Next comes the real test. Can you do the problems at the end of each chapter? If not, ask yourself what section of the text it is about, and return there for guidance or inspiration. If that fails, try the hints at the back of the text, and start again.

No text book is ever perfect for everyone; you may need other books to give a different explanation of some points, to give some other examples, or to give some photographs for study. Or you may want to find a book to cover some application for which there is no room here. At the end of each chapter you will find some references to help with any of these quests, and also suggestions for books at a higher level for when you feel you are ready. But do not feel guilty if you never look at other texts; you will have quite enough to do working through this one.

Reference

The film loops and films referred to above are fully described in *Illustrated Experiments in Fluid Dynamics*, National Committee for Fluid Mechanics Films, 1972.

They are distributed in Great Britain by Fergus Davidson Associates Ltd, 376 London Road, West Croydon, Surrey CRO 2SU on behalf of Encyclopaedia Brittanica Educational Corp, 425 N. Michigan Ave, Chicago, Ill 60611.

If you have no opportunity to see films or film loops, then you should spend some time looking at photographs in other texts.

1

Mathematical preliminaries

1. Background knowledge

Quite a lot of mathematics is needed for this text, and most of it must be assumed to have been met elsewhere. However, it would be unfair to assume that you have already learned every mathematical technique, so some new methods will be introduced in sufficient detail as they become necessary. If you need to revise any of the topics mentioned below, do it now, before your lack of knowledge interferes with the fluid dynamics that is being expounded.

(a) Vectors

Fluid dynamics is about the motion of fluids, so velocities must come in; that is, the whole subject will be full of work with vectors. You must be confident in your use of the two products

$$\mathbf{A} \cdot \mathbf{B} \text{ and } \mathbf{A} \times \mathbf{B},$$

and in the use of components and unit vectors. These components may be those appropriate to cartesian axes or to polar directions; the important material about polar coordinates and directions is summarised below.

(b) Functions of several variables

The velocities in fluid dynamics will in general depend on position and time, e.g.

$$\mathbf{v}(x, y, z, t).$$

For an example of this you can think of weather maps, which usually show wind velocities at the surface. These velocities are different for different parts of the country and change from day to day. So we shall certainly need to consider functions of several variables; whenever possible we shall reduce from the generality of four variables down to special cases involving only two, but we cannot often use only one variable.

(c) *Vector calculus*

Since the velocities change in space and time, we shall need vector calculus. For example, the divergence of $\mathbf{v}$

$$\nabla \cdot \mathbf{v} \text{ or div } \mathbf{v}$$

and the curl of $\mathbf{v}$

$$\nabla \times \mathbf{v} \text{ or curl } \mathbf{v}$$

are most important. This whole area is vital, and so it is summarised below with special results that are needed in fluid dynamics. If you are not happy with the operations of vector calculus, then some of the work in later chapters will be almost impossible: you should learn now what is needed later, either from the material below or from some other text, such as the ones quoted in the list of reference books.

(d) *Tensor notation*

Some of the manipulations of vector calculus are most easily carried out by using the suffix notation of cartesian tensors. For example, the vector $\mathbf{v}$ is represented by its ith component v_i in many calculations. There are two second order tensors which have to come into later chapters, and so it is necessary to have some background in this area. The material is summarized below, and references are given at the end of this chapter if you need to start from scratch.

(e) *Differential equations*

Because fluid dynamics is a branch of dynamics, we shall have differential equations expressing the relation

$$\text{force} = \text{mass} \times \text{acceleration.}$$

Basic methods for ordinary differential equations should be known before you start. It would be a help if you had met some methods for simple first and second order partial differential equations; the necessary material will be presented in the text, but it is easier to get a working knowledge of the techniques if you have met them some time before you find that you have to apply them.

In particular, the method of characteristics will be important in Chapters XIV and XV; and the method of separation of variables is needed from Chapter XI on. The wave equation

$$c^2 \, \partial^2 u/\partial x^2 = \partial^2 u/\partial t^2$$

forms the basis for much of the work described in Chapters XII and XIII; Laplace's equation

$$\nabla^2 \phi = 0$$

is needed in Chapter XI; and the diffusion equation

$$k \, \partial^2 u/\partial x^2 = \partial u/\partial t$$

occurs in Chapter IX. The more you know about these three equations, the better. For all these methods and equations the necessary material appears in the text, but it would be a considerable advantage if that was not your first meeting with the ideas and methods.

(*f*) *Fourier series*

It is often found to be convenient to represent the solutions of partial differential equations in terms of series based on eigenfunctions of some ordinary differential equation. If this is not to seem a totally strange method to you, then an acquaintance with Fourier series will be needed. All that is needed is the techniques, we shall not need theorems. Other examples of eigenfunction expansions will come in, from Chapter XI on. If you already know about Legendre polynomials and Bessel functions, so much the better. But you should be able to manage by using the explanations given when these functions arise naturally in solving particular problems in fluid dynamics.

(*g*) *Complex numbers and functions*

Some complex function theory is needed in the last two chapters, and this is not described in this text – if you intend to work on these chapters, you must find a text on functions of a complex variable to help you. But for the two chapters (XII and XIII) on small amplitude waves you will need to be familiar with complex numbers and the formula

$$\cos kx + i \sin kx = e^{ikx}.$$

This will be used to replace the rather clumsy cosine and sine functions with the easier exponential function. We assume that you are already familiar with the algebra of complex numbers, including

$$z = |z| e^{i\theta}$$

and

$$z = \mathscr{R}e\, z + i\mathscr{I}m\, z.$$

2. Polar coordinate systems

We start with a careful revision of some results for plane polar coordinates, and then go on to use the same methods for the three-dimensional cylindrical polar and spherical polar systems.

(a) Plane polar coordinates

Plane polar coordinates (r, θ) are related to cartesian coordinates (x, y) by (see fig. I.1)

$$\begin{cases} x = r \cos \theta, \\ y = r \sin \theta, \end{cases}$$

and the unit vectors $\mathbf{i}, \mathbf{j}$ of cartesians are related to those for polars, $\hat{\mathbf{r}}$ and $\hat{\boldsymbol{\theta}}$, by

$$\begin{cases} \hat{\mathbf{r}} = \mathbf{i} \cos \theta + \mathbf{j} \sin \theta, \\ \hat{\boldsymbol{\theta}} = -\mathbf{i} \sin \theta + \mathbf{j} \cos \theta. \end{cases}$$

If you calculate $\partial\hat{\mathbf{r}}/\partial\theta$ and $\partial\hat{\boldsymbol{\theta}}/\partial\theta$ from these formulae, you get

$$\begin{cases} \partial\hat{\mathbf{r}}/\partial\theta = \hat{\boldsymbol{\theta}}, \\ \partial\hat{\boldsymbol{\theta}}/\partial\theta = -\hat{\mathbf{r}}. \end{cases}$$

Clearly also

$$\partial\hat{\mathbf{r}}/\partial r = \partial\hat{\boldsymbol{\theta}}/\partial r = 0.$$

Now suppose that you allow small changes dr and $d\theta$ in the formulae for x and y. The corresponding small changes dx and dy are given by

$$\begin{cases} dx = dr \cos \theta - r \sin \theta \, d\theta, \\ dy = dr \sin \theta + r \cos \theta \, d\theta. \end{cases}$$

Fig. I.1. Plane polar coordinates.

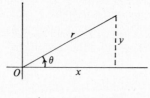

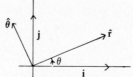

Fig. I.2. Length elements in cartesian and plane polar coordinates.

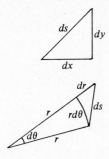

The length element ds is given by

$$ds^2 = dx^2 + dy^2,$$

and substituting for dx and dy gives, after a little work,

$$ds^2 = dr^2 + r^2 d\theta^2$$

(See fig. I.2). That is, the changes dr and $d\theta$ correspond to length changes

$$\begin{cases} dr \\ r d\theta. \end{cases}$$

These length changes are in many ways more important than the changes in the coordinates.

(b) Cylindrical coordinates

Cylindrical polar coordinates will be denoted by (r, θ, z) as shown on the diagram. The coordinate r is now distance (perpendicularly) from the z-axis. There are advantages in using a letter other than r for this distance, and keeping r for distance from an origin; but it is common practice to use r in this context: you *must* check what coordinate system is being used in each example later in the text.

The point P in fig. I.3 has coordinates r, θ, z using $r = 0$ on the z-axis and $\theta = 0$ on the x-axis.

The relations between cartesian and cylindrical polar coordinates are almost exactly as in (a) above:

$$x = r \cos \theta, \ y = r \sin \theta, \ z = z,$$

$$\begin{cases} \hat{\mathbf{r}} = \mathbf{i} \cos \theta + \mathbf{j} \sin \theta, \\ \hat{\boldsymbol{\theta}} = -\mathbf{i} \sin \theta + \mathbf{j} \cos \theta, \\ \mathbf{k} = \mathbf{k}, \end{cases}$$

and

$$ds^2 = dr^2 + r^2 d\theta^2 + dz^2.$$

Fig. I.3. Cylindrical polar coordinates.

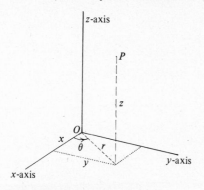

(c) *Spherical coordinates*

Spherical polar coordinates will be denoted by (r, θ, λ) as shown on the diagram. The coordinate r *here* means distance from the origin O, and the coordinate θ *here* means angle measured from the 'polar line' which is the z-axis. The angle λ is angle round the 'equator'; this angle is commonly denoted by ϕ, but we shall need to use ϕ for potential, and it looks silly to have $\partial\phi/\partial\phi$ with the two ϕ having different meanings.

The point P in fig. I.4 has coordinates r, θ, λ with $\theta = 0$ on the z-axis and $\lambda = 0$ on the x-axis. Note that the range of θ is $0 \leqslant \theta \leqslant \pi$, while $0 \leqslant \lambda < 2\pi$.

The relations to cartesian coordinates and unit vectors are

$$\begin{cases} x = r \sin \theta \cos \lambda, \\ y = r \sin \theta \sin \lambda, \\ z = r \cos \theta, \end{cases}$$
$$\begin{cases} \hat{\mathbf{r}} = \mathbf{i} \sin \theta \cos \lambda + \mathbf{j} \sin \theta \sin \lambda + \mathbf{k} \cos \theta, \\ \hat{\boldsymbol{\theta}} = \mathbf{i} \cos \theta \cos \lambda + \mathbf{j} \cos \theta \sin \lambda - \mathbf{k} \sin \theta, \\ \hat{\boldsymbol{\lambda}} = - \mathbf{i} \sin \lambda + \mathbf{j} \cos \lambda. \end{cases}$$

Fig. I.4. Spherical polar coordinates.

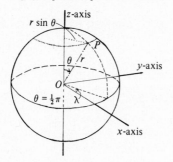

The length elements are

$$dr, rd\theta, r\sin\theta\, d\lambda$$

and

$$ds^2 = dr^2 + r^2 d\theta^2 + r^2 \sin^2\theta\, d\lambda^2.$$

From the formulae above it is easy to calculate derivatives of $\hat{\mathbf{r}}, \hat{\boldsymbol{\theta}}, \hat{\boldsymbol{\lambda}}$. For example

$$\partial\hat{\mathbf{r}}/\partial\lambda = \partial/\partial\lambda(\mathbf{i}\sin\theta\cos\lambda + \mathbf{j}\sin\theta\sin\lambda)$$
$$= -\mathbf{i}\sin\theta\sin\lambda + \mathbf{j}\sin\theta\cos\lambda$$
$$= \hat{\boldsymbol{\lambda}}\sin\theta.$$

In such calculations we use the fact that $\mathbf{i}, \mathbf{j}, \mathbf{k}$ are constant unit vectors, and we show that $\hat{\mathbf{r}}, \hat{\boldsymbol{\theta}}, \hat{\boldsymbol{\lambda}}$, are non-constant unit vectors – because their directions change.

3. The vector derivative, ∇

Let $\phi(x, y, z)$ be a scalar field; then $\nabla\phi$ may be defined as the vector which has cartesian component form

$$\nabla\phi = \mathbf{i}\frac{\partial\phi}{\partial x} + \mathbf{j}\frac{\partial\phi}{\partial y} + \mathbf{k}\frac{\partial\phi}{\partial z}$$

or

$$(\partial\phi/\partial x, \partial\phi/\partial y, \partial\phi/\partial z).$$

In plane polar coordinates $\nabla\phi$ has a different expression. We may proceed in an elementary fashion by transforming $\partial/\partial x, \partial/\partial y, \mathbf{i}, \mathbf{j}$ to r, θ form, as follows

$$\begin{cases} \partial\phi/\partial x = \cos\theta\,\partial\phi/\partial r - r^{-1}\sin\theta\,\partial\phi/\partial\theta, \\ \partial\phi/\partial y = \sin\theta\,\partial\phi/\partial r + r^{-1}\cos\theta\,\partial\phi/\partial\theta, \end{cases}$$

by using the chain rule, and

$$\begin{cases} \mathbf{i} = \hat{\mathbf{r}}\cos\theta - \hat{\boldsymbol{\theta}}\sin\theta, \\ \mathbf{j} = \hat{\mathbf{r}}\sin\theta + \hat{\boldsymbol{\theta}}\cos\theta, \end{cases}$$

from §2 (*a*).

Using these on

$$\nabla\phi = \mathbf{i}\,\partial\phi/\partial x + \mathbf{j}\,\partial\phi/\partial y$$

gives

$$\nabla\phi = \hat{\mathbf{r}}\,\partial\phi/\partial r + \hat{\boldsymbol{\theta}}r^{-1}\,\partial\phi/\partial\theta = (\partial\phi/\partial r, r^{-1}\,\partial\phi/\partial\theta).$$

The similar results in cylindrical and spherical polar systems are

$$\nabla\phi = \hat{\mathbf{r}}\,\partial\phi/\partial r + \hat{\boldsymbol{\theta}}r^{-1}\,\partial\phi/\partial\theta + \hat{\mathbf{k}}\,\partial\phi/\partial z$$

and
$$\nabla\phi = \hat{\mathbf{r}}\,\partial\phi/\partial r + \hat{\boldsymbol{\theta}}r^{-1}\,\partial\phi/\partial\theta + \hat{\boldsymbol{\lambda}}(r\sin\theta)^{-1}\,\partial\phi/\partial\lambda.$$
Note how the length elements
$$\begin{cases} dr, rd\theta, dz \text{ for cylindrical polars,} \\ dr, rd\theta, r\sin\theta\,d\lambda \text{ for spherical polars,} \end{cases}$$
reappear as denominators in components of $\nabla\phi$.

4. Cartesian tensor methods

(a) *Suffix notation for dot and cross products*

The methods of cartesian tensors are very useful in vector calculus and fluid dynamics. Cartesian coordinates x_1, x_2, x_3 are assumed throughout; and the summation convention is used, whereby
$$a_ib_i \text{ means } a_1b_1 + a_2b_2 + a_3b_3$$
and a_jb_j means the same, and three suffixes the same (or more) means that a mistake has been made.

In this notation, instead of dealing with $\nabla\phi$ we deal with $\partial\phi/\partial x_i$, its *i*th component.

The dot and cross products of vector calculus are expressed as follows:
$$\mathbf{A}\cdot\mathbf{B} = A_iB_i$$
and
$$(\mathbf{A}\times\mathbf{B})_i = \varepsilon_{ijk}A_jB_k,$$
which gives the *i*th component of $\mathbf{A}\times\mathbf{B}$ in terms of the alternating tensor ε_{ijk}. This tensor is specially made to express cross products, and has the values
$$\begin{cases} \varepsilon_{ijk} = 0 \text{ if any two of } i, j, k \text{ are the same,} \\ \varepsilon_{ijk} = +1 \text{ if } ijk \text{ is 123, 231 or 312,} \\ \varepsilon_{ijk} = -1 \text{ if } ijk \text{ is 132, 213 or 321.} \end{cases}$$
There is one other very simple tensor that continually comes into the calculations,
$$\delta_{ij},$$
where
$$\begin{cases} \delta_{11} = \delta_{22} = \delta_{33} = 1, \\ \delta_{ij} = 0 \text{ otherwise.} \end{cases}$$
You will see below that there is a theorem relating certain products of two ε to four δ. You may regard δ_{ij} as a unit matrix if you wish, but ε_{ijk} has no matrix representing it.

Now exactly as for $\mathbf{A} \cdot \mathbf{B}$ and $\mathbf{A} \times \mathbf{B}$ we have

$$\nabla \cdot \mathbf{A} = \partial A_i / \partial x_i \text{ (a summation)}$$

and

$$(\nabla \times \mathbf{A})_i = \varepsilon_{ijk} (\partial / \partial x_j) A_k$$

where $\mathbf{A}$ is a vector field.

(b) *Example*

Let us expand $\nabla \cdot (\phi \mathbf{A})$. Use suffix notation to get

$$\partial / \partial x_i (\phi A_i) = \partial \phi / \partial x_i A_i + \phi \, \partial A_i / \partial x_i .$$

Now translate back into vector language:

$$\nabla \cdot (\phi \mathbf{A}) = (\nabla \phi) \cdot \mathbf{A} + \phi \nabla \cdot \mathbf{A}$$
$$= \mathbf{A} \cdot (\nabla \phi) + \phi \nabla \cdot \mathbf{A} .$$

This is exactly what we would expect, using the naive ideas that:

  (i) ∇ is differentiation and $(fg)' = f'g + fg'$,

  (ii) ∇ is a vector and we must keep a dot.

Notice that $\mathbf{A} \cdot (\nabla \phi)$ is $A_i(\partial / \partial x_i)\phi$, which looks as though it could equally well be written as

$$(\mathbf{A} \cdot \nabla)\phi,$$

where $\mathbf{A} \cdot \nabla$ means the scalar operator

$$A_1 \, \partial / \partial x_1 + A_2 \, \partial / \partial x_2 + A_3 \, \partial / \partial x_3 .$$

These are indeed equivalent ways of writing $A_i \, \partial \phi / \partial x_i$ and so we will usually just write $\mathbf{A} \cdot \nabla \phi$ for either. *But*, we must take due care when we are working in any form of polar coordinates and when we have differentiation of a vector field to deal with: see §6 below.

(c) *An important identity*

As a further example let us prove the important identity

$$\mathbf{A} \times (\nabla \times \mathbf{A}) = \nabla (\tfrac{1}{2} A^2) - \mathbf{A} \cdot \nabla \mathbf{A}$$

where $\mathbf{A}^2 = \mathbf{A} \cdot \mathbf{A}$ and $\mathbf{A} \cdot \nabla \mathbf{A}$ is commonly thought of as the scalar operator $\mathbf{A} \cdot \nabla$ acting on the vector field $\mathbf{A}$ – though with due care it may equally be regarded as the scalar product (premultiplying) of $\mathbf{A}$ and the tensor $\nabla \mathbf{A}$.

Now

$$(\nabla \times \mathbf{A})_k = \varepsilon_{klm} \partial A_m / \partial x_l ,$$

and so

$$\{\mathbf{A} \times (\nabla \times \mathbf{A})\}_i = \varepsilon_{ijk} A_j (\nabla \times \mathbf{A})_k$$
$$= \varepsilon_{ijk} \varepsilon_{klm} A_j \partial A_m / \partial x_l .$$

But cyclic permutation is permitted in alternating tensors and so

$$\varepsilon_{ijk}\varepsilon_{klm} = \varepsilon_{kij}\varepsilon_{klm},$$

and a theorem says that this has value

$$\delta_{il}\delta_{jm} - \delta_{im}\delta_{jl}.$$

Hence

$$\{\mathbf{A} \times (\nabla \times \mathbf{A})\}_i = (\delta_{il}\delta_{jm} - \delta_{im}\delta_{jl})A_j\partial A_m/\partial x_l.$$

Now, for example, $\delta_{jm}A_j = A_m$ because $\delta_{jm} = 0$ when $j \neq m$ and has value 1 when $j = m$. Proceeding in a similar fashion we derive

$$\{\mathbf{A} \times (\nabla \times \mathbf{A})\}_i = A_j\partial A_j/\partial x_i - A_m\partial A_i/\partial x_m$$
$$= (\partial/\partial x_i)(\tfrac{1}{2}A_jA_j) - \mathbf{A}\cdot\nabla A_i$$

as required.

(d) Principal axes and isotropic tensors

The material so far presented on tensors has made no mention of what axes have been chosen. This is perfectly reasonable, because the motion of a fluid will not depend on how you choose to describe it – the weather on Earth is made up of the same air motions whether they are measured by ground stations or by satellites. However, there are occasions when a sensible choice of axes makes the calculations a lot easier; for example in the motion of fluid along a pipe it is sensible to choose one axis along the centreline of the pipe, so that a simple model like

$$\mathbf{v} = U(r)\mathbf{k}$$

(in cylindrical coordinates) might be appropriate.

Second order symmetric tensors – those for which

$$a_{ij} = a_{ji}$$

– have a set of axes associated with them for which their form is very easy. If you use 'principal axes' of a symmetric second order tensor, then the tensor has the special form

$$\begin{pmatrix} a_1 & 0 & 0 \\ 0 & a_2 & 0 \\ 0 & 0 & a_3 \end{pmatrix}$$

that is,

$$a_{ij} = 0 \text{ for } i \neq j.$$

The values a_1, a_2, a_3 are the 'eigenvalues' of the tensor, and the principal axes are the 'eigenvectors' of the tensor. These are found, as in matrix theory, by solving

$$|a_{ij} - \lambda \delta_{ij}| = 0$$

to get

$$\lambda = a_1, a_2 \text{ or } a_3;$$

and then solving

$$(a_{ij} - a_n \delta_{ij})b_j = 0$$

for $n = 1, 2, 3$ to get the three eigenvectors b_j. We do not prove this important theorem here, or give examples of such calculations. If you need practice, you must seek one of the texts in the list of references.

There are a few tensors for which the choice of axes is quite irrelevant, because however you rotate the axes you still get the same values for the components of these tensors. These are known as 'isotropic' tensors. The ones important in this course are:

    (i) δ_{ij}, the only constant second order isotropic tensor;

    (ii) ε_{ijk}, the only constant third order tensor which retains its value under rotations;

    (iii) $\alpha \delta_{ij}\delta_{kl} + \beta \delta_{ik}\delta_{jl} + \gamma \delta_{il}\delta_{jk}$ for scalars α, β, γ is the general form of fourth order constant isotropic tensors.

5. Integration formulae

(a) *Derivatives of integrals*

The two fundamental theorems of calculus are

$$\frac{d}{dx} \int_a^x f(t)dt = f(x),$$

$$\int_a^b \frac{df}{dx} dx = f(b) - f(a),$$

for suitable functions f. Fluid dynamics is a three-dimensional subject, so we will need to have the analogues of these theorems ready for occasional use. We shall not attempt to prove any of the theorems, but it will be necessary to note reasonable conditions for their validity. It will also be necessary to manipulate similar expressions into the form of the theorems.

Theorems of the type

'derivative of integral = function'

are:

    (i) $\nabla \int_a^r \mathbf{F}(\mathbf{s}) \cdot d\mathbf{s} = \mathbf{F}(\mathbf{r})$, where $\mathbf{F}$ is a continuous vector field with $\nabla \times \mathbf{F} = 0$ and the integral is along any curve from the fixed point $\mathbf{a}$ to the point $\mathbf{r} = (x, y, z)$.

(ii) $\displaystyle \nabla \int_{p_0}^{p(x,y,z)} F(t)dt = F(p)\nabla p,$

where F is a continuous scalar field and p is a differentiable scalar field.

(b) Integrals of derivatives
The basic theorems of the type

'integral of derivative = boundary values'

are (for reasonably smooth surfaces S and curves l):

(i) $\displaystyle \int_V \nabla \cdot \mathbf{A} dV = \int_S \mathbf{A} \cdot d\mathbf{S},$

if derivatives of $\mathbf{A}$ are continuous in the volume V, S is the surface of V, and $d\mathbf{S}$ is outwards from V. This is usually known as the divergence theorem, and is mainly used in three dimensions in fluid dynamics, though versions in all other numbers of dimensions exist.

(ii) $\displaystyle \int_S \nabla \times \mathbf{A} \cdot d\mathbf{S} = \oint_l \mathbf{A} \cdot d\mathbf{l},$

if derivatives of $\mathbf{A}$ are continuous, l is a closed curve and S is any surface spanning l, and $d\mathbf{S}$ and $d\mathbf{l}$ are suitably oriented (see fig. I.5). This is known as Stokes' theorem.

These theorems may be rather easily generalised. It is, for example, convenient to have a theorem on $\int p d\mathbf{S}$. To obtain this, consider $A_i \int_S p dS_i$, where $\mathbf{A}$ is an arbitrary constant vector and S is a closed surface. This is

Fig. I.5. Stokes' Theorem.

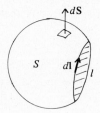

equal to

$$\int_S (\mathbf{A}p) \cdot d\mathbf{S}$$

and using the divergence theorem gives

$$\int_V \nabla \cdot (\mathbf{A}p) dV.$$

Simple manipulations now show that

$$A_i \int_S p dS_i = A_i \int_V (\nabla p)_i dV.$$

Since $\mathbf{A}$ is an arbitrary vector we may choose it to be in turn $\mathbf{i}, \mathbf{j}, \mathbf{k}$ and derive

$$\int_S p dS_i = \int_V (\nabla p)_i dV,$$

for scalar fields p with ∇p continuous. We may also write this as

$$\int_S p d\mathbf{S} = \int_V \nabla p dV.$$

(c) *A useful theorem*

One final theorem on integrals follows.

Let

$$\int_V \phi dV = 0$$

for a continuous scalar field ϕ and *any* volume V within some (larger) volume $\mathbf{V}$. Then $\phi = 0$ inside $\mathbf{V}$.

For if ϕ is *not* zero inside $\mathbf{V}$ then choose a point at which it is not zero, say $\mathbf{r}_0$. Then from the definition of continuity it may be proved that $\phi(\mathbf{r})$ has the sign of $\phi(\mathbf{r}_0)$ in a region surrounding $\mathbf{r}_0$. Hence $\int \phi dV$ over this region is non-zero, which is a contradiction.

This result is often called the Du Bois–Reymond lemma, and it has frequent uses in fluid dynamics.

6. Formulae in polar coordinates

The formulae of fluid dynamics frequently have to be expressed in polar coordinates, so we set out various calculations to show the methods of deriving the required polar forms.

(*a*) $\nabla \cdot \mathbf{A}$ *in plane polar coordinates.*

This may be done in the same way as $\nabla \phi$ was done, by transforming $\partial/\partial x$, $\partial/\partial y$, $\mathbf{i}$, $\mathbf{j}$ into polar forms. But there is a shorter method which is well worth using – though it needs care.

Consider $\nabla \cdot \mathbf{A}$ as 'grad dot $\mathbf{A}$'.

This is

$$\left(\hat{\mathbf{r}} \frac{\partial}{\partial r} + \hat{\boldsymbol{\theta}} \frac{1}{r} \frac{\partial}{\partial \theta} \right) \cdot (A_r \hat{\mathbf{r}} + A_\theta \hat{\boldsymbol{\theta}})$$

in polar components. Now note carefully that $\hat{\mathbf{r}}$ and $\hat{\boldsymbol{\theta}}$ are *not* constant vectors, and so must be differentiated. Hence

$$\nabla \cdot \mathbf{A} = \hat{\mathbf{r}} \cdot \left\{ \frac{\partial}{\partial r} (A_r \hat{\mathbf{r}} + A_\theta \hat{\boldsymbol{\theta}}) \right\} + \hat{\boldsymbol{\theta}} \cdot \left\{ \frac{1}{r} \frac{\partial}{\partial \theta} (A_r \hat{\mathbf{r}} + A_\theta \hat{\boldsymbol{\theta}}) \right\}.$$

Use the rules for differentiating $\hat{\mathbf{r}}$ and $\hat{\boldsymbol{\theta}}$ to get

$$\nabla \cdot \mathbf{A} = \hat{\mathbf{r}} \cdot \left\{ \hat{\mathbf{r}} \frac{\partial A_r}{\partial r} + \hat{\boldsymbol{\theta}} \frac{\partial A_\theta}{\partial r} \right\} + \hat{\boldsymbol{\theta}} \cdot \left\{ \hat{\mathbf{r}} \frac{1}{r} \frac{\partial A_r}{\partial \theta} + \hat{\boldsymbol{\theta}} \frac{1}{r} A_r + \hat{\boldsymbol{\theta}} \frac{1}{r} \frac{\partial A_\theta}{\partial \theta} \right.$$

$$\left. - \hat{\mathbf{r}} \frac{1}{r} A_\theta \right\} = \frac{\partial A_r}{\partial r} + \frac{1}{r} A_r + \frac{1}{r} \frac{\partial A_\theta}{\partial \theta}$$

because $\hat{\mathbf{r}} \cdot \hat{\boldsymbol{\theta}} = 0$. This whole process can become quite quick with practice, and applies to all such differentiations.

Note carefully the term $r^{-1} A_r$. This comes in because the coordinate system is curved, and is the first example of many such 'extra' terms which make a vital difference.

(*b*) $\mathbf{A} \cdot \nabla \mathbf{A}$ *in plane polar coordinates.*

Calculate first the operator $\mathbf{A} \cdot \nabla$. This is

$$(A_r \hat{\mathbf{r}} + A_\theta \hat{\boldsymbol{\theta}}) \cdot \left(\hat{\mathbf{r}} \frac{\partial}{\partial r} + \frac{1}{r} \hat{\boldsymbol{\theta}} \frac{\partial}{\partial \theta} \right) = A_r \frac{\partial}{\partial r} + \frac{1}{r} A_\theta \frac{\partial}{\partial \theta}.$$

Then

$$\mathbf{A} \cdot \nabla \mathbf{A} = \left(A_r \frac{\partial}{\partial r} + \frac{1}{r} A_\theta \frac{\partial}{\partial \theta} \right) (A_r \hat{\mathbf{r}} + A_\theta \hat{\boldsymbol{\theta}})$$

$$= \hat{\mathbf{r}} \left(A_r \frac{\partial A_r}{\partial r} + \frac{1}{r} A_\theta \frac{\partial A_r}{\partial \theta} - \frac{1}{r} A_\theta^2 \right)$$

$$+ \hat{\boldsymbol{\theta}} \left(A_r \frac{\partial A_\theta}{\partial r} + \frac{1}{r} A_\theta \frac{\partial A_\theta}{\partial \theta} + \frac{1}{r} A_r A_\theta \right)$$

on using the methods of (*a*) above.

(c) $\nabla \times \mathbf{A}$ *in polar coordinates*

$\nabla \times \mathbf{A}$ only comes in properly in three dimensions. We may calculate $\nabla \times \mathbf{A}$ in cylindrical polar coordinates by means of the method used in (a) and (b):

$$\nabla \times \mathbf{A} = \left(\hat{\mathbf{r}}\frac{\partial}{\partial r} + \frac{1}{r}\hat{\boldsymbol{\theta}}\frac{\partial}{\partial \theta} + \mathbf{k}\frac{\partial}{\partial z} \right) \times (A_r\hat{\mathbf{r}} + A_\theta\hat{\boldsymbol{\theta}} + A_z\mathbf{k}).$$

The result is found to be equal to

$$\frac{1}{r}\begin{vmatrix} \hat{\mathbf{r}} & r\hat{\boldsymbol{\theta}} & \mathbf{k} \\ \partial/\partial r & \partial/\partial \theta & \partial/\partial z \\ A_r & rA_\theta & A_z \end{vmatrix},$$

which is more convenient both for memory and calculation.

In *spherical* polar coordinates the corresponding formula is

$$\nabla \times \mathbf{A} = \frac{1}{r^2 \sin \theta}\begin{vmatrix} \hat{\mathbf{r}} & r\hat{\boldsymbol{\theta}} & r \sin \theta \, \hat{\boldsymbol{\lambda}} \\ \partial/\partial r & \partial/\partial \theta & \partial/\partial \lambda \\ A_r & rA_\theta & r \sin \theta \, A_\lambda \end{vmatrix}$$

Notice how the length element coefficients

$$1dr, rd\theta, 1dz$$

in cylindrical polars, and

$$1dr, rd\theta, r \sin \theta \, d\lambda$$

in spherical polars, come into the formulae for $\nabla \times \mathbf{A}$. This helps you to remember them.

Another important operator in fluid dynamics is ∇^2, and it may operate either on a scalar field ϕ, or on a vector field $\mathbf{A}$.

(d) $\nabla^2\phi$ *in cylindrical coordinates*

$\nabla^2\phi$ is calculated as $\nabla\cdot(\nabla\phi)$. For example, in cylindrical polar coordinates we have $\nabla\phi = (\partial\phi/\partial r, r^{-1}\partial\phi/\partial\theta, \partial\phi/\partial z)$, and from a slight extension of (a) we have

$$\nabla\cdot\mathbf{A} = \frac{\partial A_r}{\partial r} + \frac{1}{r}A_r + \frac{1}{r}\frac{\partial A_\theta}{\partial\theta} + \frac{\partial A_z}{\partial z}$$

so that

$$\nabla^2\phi = \left(\frac{\partial}{\partial r} + \frac{1}{r}\right)\left(\frac{\partial\phi}{\partial r}\right) + \frac{1}{r}\frac{\partial}{\partial\theta}\left(\frac{1}{r}\frac{\partial\phi}{\partial\theta}\right) + \frac{\partial}{\partial z}\left(\frac{\partial\phi}{\partial z}\right)$$

$$= \frac{\partial^2\phi}{\partial r^2} + \frac{1}{r}\frac{\partial\phi}{\partial r} + \frac{1}{r^2}\frac{\partial^2\phi}{\partial\theta^2} + \frac{\partial^2\phi}{\partial z^2}.$$

(e) $\nabla^2 \mathbf{A}$ *in cylindrical coordinates*

$\nabla^2 \mathbf{A}$ is more difficult. Perhaps the simplest method is to use the identity

$$\nabla \times (\nabla \times \mathbf{A}) = \nabla\nabla\cdot\mathbf{A} - \nabla^2\mathbf{A}$$

and calculate $\nabla^2\mathbf{A}$ by using

$$\nabla^2\mathbf{A} = \nabla\nabla\cdot\mathbf{A} - \nabla \times (\nabla \times \mathbf{A}).$$

The alternative method in the spirit of §6(a) using derivatives of unit vectors, has to be done *very* carefully, as $\nabla\mathbf{A}$ is a tensor whose divergence from the left must be taken to get $\nabla^2\mathbf{A} = \nabla\cdot(\nabla\mathbf{A})$.

Take the easiest case of $\nabla^2\mathbf{A}$ to calculate as an example, that of cylindrical polar coordinates. First we calculate

$$\nabla\nabla\cdot\mathbf{A} = \hat{\mathbf{r}}\frac{\partial}{\partial r}\left\{\frac{\partial A_r}{\partial r} + \frac{A_r}{r} + \frac{1}{r}\frac{\partial A_\theta}{\partial \theta} + \frac{\partial A_z}{\partial z}\right\}$$

$$+ \frac{1}{r}\hat{\boldsymbol{\theta}}\frac{\partial}{\partial \theta}\{\text{same expression}\} + \mathbf{k}\frac{\partial}{\partial z}\{\text{same}\}.$$

Next we need

$$\nabla \times (\nabla \times \mathbf{A}) = \frac{1}{r}\begin{vmatrix} \hat{\mathbf{r}} & r\hat{\boldsymbol{\theta}} & \mathbf{k} \\ \partial/\partial r & \partial/\partial\theta & \partial/\partial z \\ \frac{1}{r}\frac{\partial A_z}{\partial\theta} - \frac{\partial A_\theta}{\partial z} & r\left(\frac{\partial A_r}{\partial z} - \frac{\partial A_z}{\partial r}\right) & \frac{1}{r}\frac{\partial}{\partial r}(rA_\theta) - \frac{1}{r}\frac{\partial A_r}{\partial\theta} \end{vmatrix}$$

where the last line uses the previous formula for the components of $\nabla \times \mathbf{A}$.

If we collect all the terms carefully (a long calculation!) we should get the result

$$\nabla^2\mathbf{A} = \hat{\mathbf{r}}\left\{\nabla^2 A_r - \frac{1}{r^2}A_r - \frac{2}{r^2}\frac{\partial A_\theta}{\partial\theta}\right\} + \hat{\boldsymbol{\theta}}\left\{\nabla^2 A_\theta + \frac{2}{r^2}\frac{\partial A_r}{\partial\theta} - \frac{1}{r^2}A_\theta\right\}$$

$$+ \mathbf{k}\nabla^2 A_z,$$

where, for example, $\nabla^2 A_r$ means 'take the divergence of the gradient of the *r-component* of vector $\mathbf{A}$'.

It is clearly *not* true that $\nabla^2\mathbf{A}$ has components $\nabla^2 A_r, \nabla^2 A_\theta, \nabla^2 A_z$: there are four extra terms which come in because of the curvature of the r, θ part of the coordinate system.

Exercises

The formulae and methods above look a bit fierce to the newcomer to vector analysis, but they must be mastered, as fluid dynamics is based on such formulae. The only way is to study the methods, work at some exercises and finally learn formulae for $\nabla\phi, \nabla\cdot\mathbf{A}, \nabla\times\mathbf{A}$ and $\nabla^2\phi$. Fortunately this course is arranged so that several chapters elapse before much of this material has to be used, and this gap should be used to gain command of vector analysis.

1. For plane or cylindrical polar coordinates express $\mathbf{i}, \mathbf{j}$ in terms of $\hat{\mathbf{r}}, \hat{\boldsymbol{\theta}}$.

2. For cylindrical and spherical polar coordinates prove, either by a sketch (not really a proof!) or by transformations, that ds^2 has the form given.

3. For spherical polar coordinates calculate $\partial\hat{\mathbf{r}}/\partial\theta, \partial\hat{\boldsymbol{\theta}}/\partial\theta, \partial\hat{\boldsymbol{\theta}}/\partial\lambda, \partial\hat{\boldsymbol{\lambda}}/\partial\lambda$; in each case express your answer in terms of the unit vectors $\hat{\mathbf{r}}, \hat{\boldsymbol{\theta}}, \hat{\boldsymbol{\lambda}}$, and not in terms of $\mathbf{i}, \mathbf{j}, \mathbf{k}$.

4. Write down Taylor's theorem for $\phi(x_1 + h_1, x_2 + h_2, x_3 + h_3)$ ending at second degree terms in $\mathbf{h}$. Show that it may be put as

$$\phi(\mathbf{x} + \mathbf{h}) = \phi(\mathbf{x}) + \mathbf{h}\cdot(\nabla\phi)_{\mathbf{x}} + O(\mathbf{h}^2)$$

or as

$$\phi(\mathbf{x} + \mathbf{h}) = \phi(\mathbf{x}) + h_j(\partial\phi/\partial x_j)_{\mathbf{x}} + O(h_k h_k).$$

[Note: $O(h^2)$ means that these terms are no bigger than some constant times h^2 for all small enough values of h.]

5. Show that
   (i) $\nabla\times(\phi\mathbf{A}) = (\nabla\phi)\times\mathbf{A} + \phi\nabla\times\mathbf{A}$,
   (ii) $\nabla\times(\partial\mathbf{A}/\partial t) = \partial/\partial t(\nabla\times\mathbf{A})$,
   (iii) $\nabla\cdot(\phi\nabla\psi) = \nabla\phi\cdot\nabla\psi + \phi\nabla^2\psi$,
   (iv) $\nabla\cdot(\nabla\times\mathbf{A}) = 0$,
   (v) $\nabla\times(\nabla\phi) = 0$,
   (vi) $\nabla\times(\nabla\times\mathbf{A}) = \nabla\nabla\cdot\mathbf{A} - \nabla^2\mathbf{A}$.

6. Show that $\nabla\times(\mathbf{A}\cdot\nabla\mathbf{A})$ may be written as

$$(\nabla\times\mathbf{A})(\nabla\cdot\mathbf{A}) + \mathbf{A}\cdot\nabla(\nabla\times\mathbf{A}) - (\nabla\times\mathbf{A})\cdot\nabla\mathbf{A}.$$

[It is easiest to start with $\mathbf{A}\cdot\nabla\mathbf{A} = \nabla(\frac{1}{2}\mathbf{A}^2) - \mathbf{A}\times(\nabla\times\mathbf{A})$].

7. Show that

$$\int_V (\phi\nabla^2\psi - \psi\nabla^2\phi)dV = \int_S (\phi\nabla\psi - \psi\nabla\phi)\cdot d\mathbf{S}.$$

8. Calculate $\nabla\cdot\mathbf{A}$ and $\nabla\times\mathbf{A}$ in spherical polar coordinates.

9. $\mathbf{A}\cdot\nabla\mathbf{A}$ in spherical polar coordinates may be written as

$$(\mathbf{A}\cdot\nabla A_r, \mathbf{A}\cdot\nabla A_\theta, \mathbf{A}\cdot\nabla A_\lambda) + \text{extra vector}.$$

   Calculate the extra vector.

10. Calculate $\nabla^2\phi$ in spherical polar coordinates.

11. Look up $\nabla^2\mathbf{A}$ in spherical polar coordinates.

References

Most of the mathematical areas quoted in this chapter should have been covered in other courses, so you should already have texts for them, or else course notes. So references are only given to vector analysis and to cartesian tensor texts. There are many such, all covering much the same ground, and it is to some extent a matter of taste which is found to be best. Some suitable ones are:

(a) *Vector and Tensor Methods*, F. Chorlton, Ellis Horwood 1976.
(b) *A Course in Vector Analysis*, L1 G. Chambers, Chapman and Hall 1969.
(c) *Vector Analysis*, N. M. Queen, McGraw-Hill 1967.
(d) *Vector Analysis*, L. Marder, George Allen and Unwin 1970.
(e) *Cartesian Tensors*, H. Jeffreys, C.U.P. 1931.
(f) *Cartesian Tensors in Engineering Science*, L. G. Jaeger, Pergamon 1966.

▌▌
Physical preliminaries

1. Background knowledge

No book like this can go right back to the very beginning, so we must expect you to have some general ideas from earlier study of applied mathematics or physics. In particular it will help considerably if you have some acquaintance with the following.

(i) Newtonian dynamics, including energy, momentum and angular momentum (moment of momentum).

(ii) The physics of any sort of wave motion, including the ideas of standing and travelling waves and reflection at barriers.

(iii) The atomic nature of matter, and especially that a gas is made up of well-separated rapidly moving molecules.

Naturally it will be a help if you have further background, but there will be quite a lot of explanation for everything that is essential. If you find there is not enough description in some areas, you must go to other texts.

2. Mathematical modelling

The activity of modelling (in a mathematical sense) is one that causes unprepared students some trouble. The theories and examples that we deal with in a course like this seem too unreal to be worth the trouble, and unless the reasons behind the theories and the choice of examples are explained, applied mathematics gets a bad name. This happened many years ago in elementary statics and dynamics with interminable questions about ladders leaning against smooth walls and particles sliding down rough wedges with smooth bases.

The problems of the real subject of fluid dynamics are of great complexity, involving *many* physical effects and a considerable set of non-linear partial differential equations. These problems cannot be solved either by advanced techniques or by 'putting them on the computer': the techniques do not exist and the machines are neither powerful enough nor sophisticated enough (to reject spurious solutions). The problems can only be approached by omitting, after much careful thought and perhaps some experiments, a large number of the physical effects; then perhaps a special case can be dealt with analytically, and this will show what sort of calculation the machine must be programmed for in the more general case.

This reduction to a specially simple case followed by a comparison of the predictions from this case with reality is the essence of mathematical modelling. It is this that justifies many of the trivial examples – they are a first try at a complicated reality, to see if the theory has any chance of explaining the reality. Of course, proper mathematical modelling does not stop at the simplest 'explanation'; usually it is only a partial explanation, and a better (more complicated) model is needed to give a fuller story of what is happening. This course is elementary, so we do not usually do more than indicate what effects have been left out, which would have to be included in a better theory.

A good example of this process at work is often seen in a first course about waves on a stretched string. The allowed frequencies (eigenvalues) of transverse vibration for a string stretched between fixed end points are shown to be

$$\omega_n = \begin{cases} n\pi c/l \text{ radians per second,} \\ \frac{1}{2}nc/l \text{ cycles per second, or Hz,} \end{cases}$$

where n is an integer, l is the distance between the fixed points and c, the wave speed, has value

$$c = (T/\rho)^{1/2}$$

for tension T in the string which has mass ρ per unit length. This has assumed many things on the way, including:

(i) the string is perfectly flexible;
(ii) the motion is small;
(iii) there is no air resistance;
(iv) the ends are rigidly fixed.

As an experiment to test whether this model is reasonably accurate, or even precisely accurate to within experimental errors of measurement, we may consider the playing of a note on the piano. This is usually model-led as a first try by solving the initial value problem for the string stretched between fixed points that has:

(1) y, the displacement, zero at $t = 0$ between $x = 0$ and l;

(2) $\partial y/\partial t = v$ at $t = 0$ for a small section of the string near $x = 0$, and $\partial y/\partial t = 0$ for the rest of the string.

This will give a prediction of the relative amplitudes of the various har-monics in the note that is sounded when the piano hammer strikes the string. The results will *not* be exactly correct for a variety of reasons con-nected with the modelling – it is clear that for a real piano string (i) above is false as the string has some stiffness; and (iii) and (iv) are probably also false as the note dies away (though some of the energy loss may go to heating the piano string at non-zero amplitudes of motion); moreover the hammer is usually felt-covered and its equation of motion in contact with the string should be discussed, so that condition (2) above is likely to be inaccurate. In this case, if we are not satisfied by the prediction we have made for the relative amplitudes of the various harmonics, then we must decide which of these many extra effects to discuss next – it will probably be too difficult to deal with them all together.

The discussion of mathematical modelling in the previous paragraph may seem terribly obvious to you. In much fluid dynamics the modelling is not obvious, and it has often taken years of research to find how a given real situation may be realistically modelled. In the early models of flow round bodies, totally incorrect results were obtained by leaving out what was 'obviously' a small effect; early numerical weather forecasting was based on equations which predicted wave motions rather than weather patterns – they were correct equations, but the results were not the ones that were expected; it took years before a model of the North Atlantic was produced that gave even a moderately accurate explanation of the Gulf Stream.

In this course we do not have years available to 'do modelling'. But we must observe it in action, and notice that the results will all have inaccura-cies built into them because of the modelling assumptions that have been made.

3. Properties of fluids

The four states of matter are often said to be solids, liquids, gases and plasmas. This division is oversimple, as rocks and glaciers flow quite well under extreme conditions. But as long as we are looking at reasonable materials under reasonable conditions, we can be sure of what state of matter we are discussing. In this course you may assume that:

(i) solid = wood, glass, metal – deforming little under moderate stresses and returning to its original shape when the stress is removed (stress is force per unit area; if the force is parallel to the area it acts across, then it is shear stress);

(ii) liquid = water, oil, syrup, at usual temperatures – deforming *indefinitely* under small shear stresses, having a surface when put in a container, only slightly compressible;

(iii) gas = air at usual temperatures – having no surface in a container, deformable as in (ii), but also rather compressible.

These materials are constructed from molecules, and many of their properties can be deduced from a consideration of how these molecules interact. In this course we shall stay well above the molecular level, only occasionally looking down to small scales to explain qualitatively what is observed at the larger scales.

In some parts of the course there are no real distinctions to be drawn between liquids and gases, and so in those sections we usually use the word fluid to include both liquid and gas. But of course elsewhere we must use the appropriate word for the material. For example, sound waves occur in fluids; but surface waves only on liquids (and solids).

Note that the particular fluids mentioned in (ii) and (iii) above have rather simple molecular structures. Do not assume that the results in this course all apply to fluids with complex molecules; in particular, fluids containing long-chain polymers behave very differently.

It might seem that there were three easy ways of dividing up fluids into types:

(i) low density gases – much denser liquids;

(ii) compressible gases – much less compressible liquids;

(iii) viscous liquids such as heavy oils and syrup which resist motion of a knife in its own plane – less viscous liquids like water and also the gases which give little resistance to such motion.

This is a false way of making divisions. In many circumstances the density of a fluid is quite irrelevant as both forces *and* mass-accelerations are proportional to density and hence the density does not have an important effect: the flows of air and water down pipes are very similar, if the velocities are the same. Similarly the flow of air round the wing of a glider is very similar to the flow of water round a control vane of a submarine. In neither case does the density affect the pattern of the flow, nor does the compressibility. In the latter case the density *does* affect the resulting forces on the surfaces – the stresses are much larger in water than in air.

The role of viscosity is equally difficult to assess once and for all. When water flows in a narrow tube, then the viscosity of the water is likely to be very important, but for water flowing in the Gulf Stream the exact value of the viscosity makes little difference to the major features of the motion. If one could fill the North Atlantic with oil, there would still be a Gulf Stream!

4. Dimensional reasoning

The functions of mathematics operate on numbers, and the numbers used to describe the physical world depend on the system of measurement used. The number associated with the size of the North Atlantic basin is some 5000, if the basic unit is the kilometre; but it is vastly different if a more fundamental unit of length (such as the atomic radius of hydrogen, or the Earth's radius) is used. To get meaningful formulae in applied mathematics, we must apply the functions to quantities that do *not* change when the unit of measurement changes: these are *dimensionless* quantities. In the example of the piano string above, the shape of the vibration was given by

$$\sin n\pi x/l.$$

Here x/l is dimensionless: its value is the same in all systems of measurements. In the same way, the frequencies of oscillation can be written as

$$\omega_n = \tfrac{1}{2}n(T/\rho l^2)^{1/2} \text{ Hz.}$$

The group $(T/\rho l^2)^{1/2}$ has the dimensions (time)$^{-1}$, i.e. frequency; hence this is a meaningful equation in all systems of measurement.

Thus we can only assess how a physical property of a fluid will affect a flow pattern when we have found the dimensionless grouping that it fits into, and also the functional form of the dependence. This is why we cannot make once-and-for-all judgements on the importance of density or compressibility or viscosity. We must return to this later when we have quantified the idea of viscosity.

Two further physical effects need to be mentioned in this chapter. The first is that for any interface between two fluids there is usually an energy per unit area of surface. This is often expressed as a 'surface tension', which is a force per length (check that energy per area and force per length have the same dimensions). Among its obvious effects are the roundness of small mercury drops on a plate, the break-up of a small stream of water into drops, and the shape of soap bubbles and soap films.

Let us consider the effect of the surface energy γ on surface waves in

deep water. The dimensional quantities that are *likely* to affect this phenomenon are:

   (i) λ, the wavelength, which transforms as a length L;
   (ii) a, the amplitude, which transforms as a length L;
   (iii) ρ, the density, which transforms as ML^{-3};
   (iv) g, gravity, which transforms as LT^{-2}.

The frequency and the wave speed are found not to be independent of
these, just as in the case of the piano wire.

Now the dimensions of γ (i.e. the way it transforms) are MT^{-2}, so it
is easily seen that the grouping

$$\gamma/\rho g$$

has dimensions L^2. So either of the groupings $\gamma/\rho g\lambda^2$ or $\gamma/\rho g a^2$ is dimensionless, but until we examine the situation in more detail below, we cannot
be sure which is the relevant grouping. In fact it is curvature that is important in surface tension effects, and this essentially brings in a term in
$\partial^2/\partial x^2$, which leads to λ^{-2}; thus the grouping

$$\gamma/\rho g\lambda^2$$

is relevant, and if this is small it may be *expected* that surface tension is
unimportant. Now for a *clean* water surface the value of γ is about
$73 \times 10^{-3}\ \mathrm{kg\,s^{-2}}$ and this leads to the useful result that for waves of
length $\gg 3$ mm, surface tension is probably irrelevant.

The second physical effect that should be mentioned here is that of
temperature. The properties of air, water and other fluids can change
significantly over a range of 100 K. For example, the density, viscosity
and thermal conductivity of air all change by around 25% between
273 K and 373 K, and the viscosity of water is halved between 283 K
and 313 K. Only in a few cases will we take any further note of these
effects of temperature changes, though in some applications they can
be important.

Changes in temperature have another effect in providing buoyancy.
Heated air is less dense than unheated, and so tends to rise. Convection
caused by heating at the bottom of a layer of fluid can be a dominant
effect, for example in the atmosphere, but we cannot consider it further
in a first look at fluid motion.

Exercise

One way of estimating the depth of a well is to time the fall of a stone from
the top of the well until the splash is heard. The basic model in this case is

that of free fall under gravity, and the result

$$h_0 = \tfrac{1}{2}gT^2$$

gives the first estimate h_0 of the depth in terms of the observed time T. This is only a first model of the situation, and is good enough in practice, as T will not be measured very accurately and it is probably only idle curiosity that needs to be satisfied. But suppose that T *can* be measured accurately (and it could) and try to refine the model. There are two obvious effects to include:

(1) sound travels back up the well at finite speed c,
(2) air resistance reduces the speed of the stone below that predicted in the basic model.

Deal first with (1) as follows:

(i) Find a dimensionless group involving c, and find its size when $T = 3$ seconds.
(ii) Make and solve a new mathematical model to see how the dimensionless group enters and the approximate change from h_0.

For (2) we must have a model for air resistance to the fall of a stone. Take as a first (rough) approximation that the resistance is $\tfrac{1}{4}\rho v^2 A$, where ρ is air density, v is the speed of the stone and A is the maximum area of the stone's cross-section perpendicular to its motion. Now deal with (b) in a similar manner to (a), making some assumptions about the stone you are using (you will need to look up values for c and ρ).

Are there any other major effects which ought to be included along with (1) and (2)?

Observational preliminaries

1. The continuum model

Fluid dynamics is a (highly mathematical) branch of science. That is, the models which are analysed by mathematics are suggested by observations of and experiments on fluid flows. And the models are, wherever possible, tested by comparing their predictions with the same or new observations and experiments. Models which don't give reasonable answers are usually discarded, though sometimes kept in texts as exercises and warnings!

The importance of observation means that we must spend a little time discussing what can be observed and what cannot. As is often the case, during the course of the mathematics non-observable quantities may be extensively used, but at the beginning and the end of a calculation we should be close to reality.

It is impossible to measure the positions and velocities of the molecules that make up a fluid like air or water. The best we can do is to talk about local averages of the quantities we want to describe, such as density, temperature and velocity. Any measuring instrument does this sort of averaging over the size of the instrument, which is assumed to be rather small compared with the scale of the variations of the quantity being measured. For example, an ordinary mercury thermometer averages over the size of the bulb (and also over the time scale at which it responds), but

Fig. III.1. Variation of average density $\bar{\rho}$(kg m^{-3}) with cube side a (m). The density is poorly defined when a reaches molecular size.

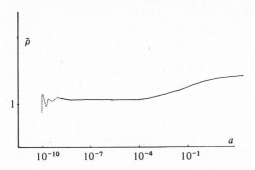

the size of the bulb is not large compared with, say, the size of a patient's mouth or a meteorological recording station.

Let us examine this idea carefully, so as to be sure of the basis on which we are constructing our theories. Suppose we wish to have a measure of the density of the air at a point P. Take a cube centred at P with side a and calculate the average density at P by

$$\bar{\rho} = m/a^3$$

where m is the mass of all the molecules inside the cube at one instant. This quantity $\bar{\rho}$ depends on what size we take for a, as shown in fig. III.1 (roughly). In particular, when a is relatively large, $\bar{\rho}$ varies because of general variations of the density of air in the atmosphere; and when a is very small $\bar{\rho}$ varies because only a small number of molecules will be inside the cube. However, around 10^{-6} m there will still be a very large number of molecules in the cube (more than 10^6) and the scale will still be smaller than the scale of any usual fluid motion in the atmosphere. So in this region $\bar{\rho}$ will have a steady value, independent of a: this is the value that we take as the density at P.

This is an important definition. There is no meaning in the real world to 'the density at P', because P (a mathematical point) is either 'in an atom' when the density is high, or 'not in an atom' when the density is zero. And in the observational world there is no hope of measuring densities on such a fine scale. This definition is the basis of the *continuum model*; we assign a density to each point by this means and then do the mathematics on these values, not on the real fluid. The whole of our subsequent work will be mathematics done on the continuum model, and it is quite possible that our mathematics will give incorrect answers if the continuum model is not a good description of the real situation. For example, if the mathematics predicts a discontinuity (as it does in shock waves), then the reality will be a region of rapid change where the continuum model may not be appropriate. And in the solar wind, particles are so well separated that a sphere of radius 10 m might not contain a large number of them, so that a spacecraft of size 10 m would be below the size at which the continuum model would be useful to describe the flow.

Note at this stage that the 'diameter' of an air molecule is about 10^{-10} m, and that the average distance a molecule travels between collisions (the mean free path) is about 10^{-7} m in the lower atmosphere.

A consequence of using the continuum model is that we must *assume* certain physical properties (such as viscosity and thermal conductivity) which can be predicted on the basis of a detailed molecular theory (the kinetic theory of gases). However, there is as yet no complete theory for liquids, so we would anyway have had to make such assumptions for liquids.

There is no difficulty in defining the local temperature T and local velocity $\mathbf{v}$ in the continuum model by just the same sort of averaging process applied to heat content (or the energy associated with the random motion of the molecules) and momentum.

Note that *because* we have a model in which the variables are defined at every point and every time, we can consider differentiation, which requires, for example, $\delta x \to 0$. Reality cannot be differentiated, because limits do not exist; the model can be differentiated (at least almost always).

2. Fluid velocity and particle paths

The observation and measurement of fluid flows is a large subject, which we shall not go very far into. Modern techniques are sophisticated, expensive and time-consuming; even so they are not always able to measure exactly what is wanted to a very high accuracy, and this means that mathematical models which are of only moderate complexity can match the experimental results to within experimental error.

One of the most primitive ways of observing and measuring a flow is to mark a small piece of fluid with dye and to photograph the subsequent motion of the dye at fixed, short, time intervals. The 'dyes' used have been as various as potassium permanganate, condensed milk, paraffin droplets (smoke), tiny hydrogen bubbles and polystyrene balls. The resulting film shows the path of this 'particle' of fluid and comparisons of adjacent frames can be used to give velocity measurements.

The idea of a fluid particle can be made precise in the continuum model: a fluid particle is a point which moves with the velocity of the fluid at that point. Clearly the dyed blob of fluid is rather larger than a fluid particle, but the two can correspond acceptably over a restricted time if the blob is small (compared with the scales of variation in the fluid).

The path of a particle is, then, an observable; and from many such paths we can measure velocities throughout the fluid. But we can also go the other way, and from the velocity field we can calculate the particle paths. For if the velocity at some point $\mathbf{r}$ at time t is $\mathbf{v}(\mathbf{r}, t)$, then the particle at this point and time moves along this velocity vector, so that

$$d\mathbf{r}/dt = \mathbf{v}(\mathbf{r}, t)$$

gives the path of this particle at other times. Naturally, particles starting at different places or different times will (usually) follow different paths.

If you *do* observe the paths of particles in many real fluid flows, you will find that they are immensely contorted. For example, watch a piece of smoke from a chimney on a windy day: it does not follow a smooth curve, but gets caught up by random gusts and swirls in the wind. This kind of behaviour seems to defy mathematical analysis. It is known as turbulence. Turbulence occurs because many simple fluid motions are unstable, and can lose energy into eddying motions at smaller scales than would seem to be appropriate for the size of the flow as a whole. This is particularly so when the product of the size of the flow and its speed (made dimensionless in a manner that is explained in Chapter IX) is large, so that most flows in atmosphere, ocean or rivers are turbulent. Fortunately it is found that the *average* flow is in many cases much simpler; so when we refer to real flows at moderate or large scale in future, you must assume that it is the average flow that is being referred to.

Examples on particle paths

(1) $\mathbf{v} = (ay, -ax, 0)$

The path of the particle initially at $x = x_0, y = y_0, z = z_0, t = 0$ is given by

$$dx/dt = ay, \, dy/dt = -ax, \, dz/dt = 0.$$

These are easily solved to give

$$\begin{cases} x = y_0 \sin at + x_0 \cos at, \\ y = y_0 \cos at - x_0 \sin at, \\ z = z_0. \end{cases}$$

If we eliminate t, we get

$$\begin{cases} x^2 + y^2 = x_0^2 + y_0^2 = \text{constant}, \\ z = z_0, \end{cases}$$

i.e. a circle in the plane $z = z_0$, as shown in fig. III.2.

Fig. III.2. Particle paths in example (1).

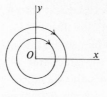

This example looks easier in r, θ, z notation. We have

$$v_r = v_x \cos \theta + v_y \sin \theta = 0$$

and

$$v_\theta = -v_x \sin \theta + v_y \cos \theta$$
$$= -ar \sin \theta \sin \theta - ar \cos \theta \cos \theta$$
$$= -ar.$$

The velocity field is now easily seen to correspond to 'rigid body rotation' about the z-axis, i.e. motion in circles with speed proportional to distance from the axis.

This velocity field is a good first model for the flow in a mug of coffee that has been stirred. Does it also fit weather maps of the centre of a depression?

This example could also have been done by writing

$$dy/dx = dy/dt \div dx/dt = -x/y$$

and solving this simple equation.

(2) $\mathbf{v} = (ay, -a(x - bt), 0)$

The differential equations for the path are similar to those above, and can be reduced to

$$\begin{cases} d^2x/dt^2 + a^2x = a^2bt \\ \qquad ay = dx/dt \\ \qquad dz/dt = 0. \end{cases}$$

Solve the first and use the solution to give y from the second:

$$\begin{cases} x = (y_0 - b/a)\sin at + x_0 \cos at + bt \\ y = (y_0 - b/a)\cos at - x_0 \sin at + b/a \\ z = z_0. \end{cases}$$

This is circular motion of radius $\{(y_0 - b/a)^2 + x_0^2\}^{1/2}$ about the moving centre

$$(bt, b/a, z_0)$$

and so it represents a cycloidal path. It is not clear what this could represent in the real world, because we do not usually see particle paths. There are

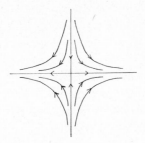

Fig. III.3. Particle paths in example (3).

however several circular patterns which move along fairly steadily, for example whirlpools left by oars in a river, or depressions in the atmosphere.

This example cannot be done by working out dy/dx, as it still has t left in it explicitly

(3) $\mathbf{v} = (a(t)x, -a(t)y, 0)$

(i) Take the general case in which $a(t)$ is not a constant. The equations for the particle paths are

$dx/dt = a(t)x,\ dy/dt = -a(t)y,\ dz/dt = 0.$

It is easiest to calculate dy/dx as in (1):

$dy/dx = -y/x,$

which has solution

$xy = \text{constant}$, and $z = z_0$ as before.

The time dependence of x and y is

$x = x_0 \exp\{A(t)\} \quad y = y_0 \exp\{-A(t)\}$

where $A(0) = 0$ and $dA/dt = a(t)$.

The result is thus a hyperbola for *all* functions $a(t)$, but described at different speeds.

(ii) The special case $a = \text{constant}$ is a good model for either flow against the fixed plane $y = 0$, or flow near the local zero of velocity at $x = y = 0$, as sketched in fig. III.3.

3. Definitions

It is useful to make some definitions based on these last three examples.

(a) A two-dimensional flow

In a suitable system of coordinates we have v_x and v_y independent of z and $v_z = 0$, and other variables such as density and temperature are also independent of z. All three flows in §2 are two-dimensional for the velocity. No real flow is ever exactly two-dimensional, but with careful experimental design it is possible to get quite a close approximation to this state. The mathematical treatment of two-dimensional flows will be much easier in most cases, so they will occur frequently in this elementary text. Some naturally occurring flows have features which are nearly two-dimensional – for example the weather patterns in the atmosphere have velocities mainly parallel to the Earth's surface, and not varying too much over a reasonable height range.

(b) A steady flow

Here velocity, density and so on are independent of t. Flows (1) and (3) (ii) above are steady for the velocity, flows (2) and (3) (i) are not. In the steady cases the positions of the particles depend on t, but this is true for both steady and unsteady flows. Steady flows are again mathematically easier, experimentally attainable in some cases, and represent some aspects of naturally occurring flows.

(c) A stagnation point

This is a point at which $\mathbf{v} = 0$. In examples (1) and (3) there are stagnation points at $x = y = 0$, for all z and all t. In (2) the stagnation point is at $y = 0$, $x = bt$. The stagnation point in (1) is at the centre of a rotary motion, while that in (3) is not. Notice that we should write $\mathbf{v} = \mathbf{0}$, because all components of $\mathbf{v}$ are zero; but it is common practice to be careless and write 0 for $\mathbf{0}$.

(d) Eulerian description

The velocities are given at fixed points in space
$$\mathbf{v}(\mathbf{r}, t)$$
as time varies. This corresponds to the usual experimental arrangement where the measuring devices are fixed and the frame of reference is fixed with them. However it would in some ways be better to follow a particle and see what happens near it as it moves along; in the atmosphere it is the history of a mass of air as it moves along that determines whether it will become a shower, rather than the sequence of air masses that pass a weather station (though they are of course related). This leads to

(e) Lagrangian description

Here quantities are given for a fixed particle at varying times, so that the velocity is
$$\mathbf{v}(\mathbf{r}_0, t)$$

where $\mathbf{r}_0$ was the particle's position at $t = 0$. Unfortunately the mathematics of the Lagrangian description is hard; but it is often useful to consider the particle's life history in order to gain an understanding of a flow. Lagrangian histories can be obtained in the atmosphere from balloon flights, or in the Gulf Stream from just-buoyant devices.

Consider again example (3) with $a = $ constant. The particle path was

$$x = x_0 e^{at}, \, y = y_0 e^{-at}, \, z = z_0$$

and this is in Lagrangian form, as it depends on the initial position. To get the Lagrangian version of the velocity we differentiate keeping x_0, y_0, z_0 constant, so as to stay with the same particle:

$$\mathbf{v} = (\partial \mathbf{r}/\partial t)_{\mathbf{r}_0}$$
$$= (ax_0 e^{at}, \, -ay_0 e^{-at}, \, 0),$$

which is (of course) just the previous form put in terms of $\mathbf{r}_0 = (x_0, y_0, z_0)$ instead of in terms of $\mathbf{r} = (x, y, z)$.

However now consider the acceleration:

(i) Lagrangian $(\partial \mathbf{v}/\partial t)_{\mathbf{r}_0} = (a^2 x_0 e^{at}, a^2 y_0 e^{-at}, 0) = a^2 \mathbf{r}$;

(ii) Eulerian $(\partial \mathbf{v}/\partial t)_{\mathbf{r}} = 0$ because $\mathbf{v} = (ax, -ay, 0)$ does not contain t explicitly.

This can be translated as 'the particles accelerate, the stream does not'. For example, consider a log drifting down a river which has a section of rapids on it: the log accelerates as it enters the rapids, and this is (approximately) the Lagrangian acceleration – following the 'particle'. But an observer on the bank will see that a succession of sticks passing him all do so at the same speed, because the river as a whole is not speeding up – the Eulerian acceleration at a fixed point is zero.

Note that in example (3) (ii) the Lagrangian acceleration is not just a matter of a curved path being followed, the speed of the particle also changes: the speed is

$$v = a(x_0^2 e^{2at} + y_0^2 e^{-2at})^{1/2} = a^2 r$$

which is not constant.

4. Streamlines and streaklines

Suppose that there are many marked particles in the flow, and take two photographs a short time apart, and then superimpose them. An arrow can now be drawn between the first and second position of each particle. This set of arrows (all taken from the same short time interval) will indicate a set of curves, called the streamlines. These are *not* the same as the particle paths, which take a *long* time to trace out, not just a short time interval; though we shall see that the two curves can be identical, for example when the flow is steady.

Let us now calculate the equation for these streamlines. Take a position vector **p** at some point on the curve, and take a parameter s along the curve. Then the tangent vector $d\mathbf{p}/ds$ is parallel to **v**, and so

$$d\mathbf{p}/ds = \lambda\mathbf{v}$$

for some λ, which we can choose to have value 1 by taking it into the parameter s. The equation of the streamlines is then derived by solving

$$d\mathbf{p}/ds = \mathbf{v}(\mathbf{p}, t).$$

This is not the same equation as that for the path of a particle, in general, as t now enters in a different way. There is a different set of streamlines for each different time t, unless **v** is independent of t.

A newspaper weather map provides a set of wind direction arrows from which a streamline pattern (changing day by day) can be guessed.

(a) *Examples on streamlines*

(1) $\mathbf{v} = (ay, -ax, 0)$

This example has **v** independent of t (i.e. the flow is steady), hence the streamline equation

$$d\mathbf{p}/ds = \mathbf{v}(\mathbf{p})$$

is formally the same as the particle path equation

$$d\mathbf{r}/dt = \mathbf{v}(\mathbf{r})$$

with just a change of notation. Hence the streamlines are also circles in the plane $z = z_0$, as were the particle paths.

(2) $\mathbf{v} = (ay, -a(x - bt), 0)$

The streamline equations are

$$\begin{cases} dx/ds = ay, \\ dy/ds = -a(x - bt), \\ dz/ds = 0. \end{cases}$$

They reduce as before to

$$\begin{cases} d^2x/ds^2 + a^2x = a^2bt, \\ ay = dx/ds, \, dz/ds = 0. \end{cases}$$

The solution is, remembering that t is a constant when streamlines are being considered,

$$\begin{cases} x = (x_0 - bt)\cos as + y_0 \sin as + bt, \\ y = -(x_0 - bt)\sin as + y_0 \cos as, \\ z = z_0, \end{cases}$$

where we have chosen to have (x_0, y_0, z_0) on the streamline at $s = 0$.

These are not the particle paths of example (2) in §2. The streamlines are, at each time t, circles (and not cycloids), with radius

$$\{(x_0 - bt)^2 + y_0^2\}^{1/2}$$

and centre

$$(bt, 0, z_0).$$

The relation of the paths to the streamlines is that as the particle moves a short distance parallel to the velocity (shown on the streamline curves) time changes a little, and so the particle now finds a *new* streamline pattern to move parallel to for a short distance. Hence the particle uses information from a succession of streamline patterns to build up its path.

This model velocity field will probably be moderately suitable for a moving depression in the atmosphere provided the constants a and b are chosen sensibly: depressions have moderately circular streamlines and sometimes move eastwards with fairly constant velocity.

The radius of the streamlines is not really changing in time; it is just that the streamline through (x_0, y_0, z_0) has to become a bigger circle as x_0 gets left behind.

(3) $\mathbf{v} = (a(t)x, -a(t)y, 0)$

This is not steady flow, except when $a(t) = $ constant. But the equations of the streamlines

$$\begin{cases} dx/ds = a(t)x, \\ dy/ds = -a(t)y, \\ dz/ds = 0, \end{cases}$$

reduce to

$$dy/dx = -y/x, z = z_0.$$

These are just the particle path equations again; this is because the *direction* of the velocity is independent of time, even though its magnitude varies. The streamlines are the hyperbolas $xy = $ constant, $z = z_0$.

In the examples above we have in effect proved the theorem that follows.

(b) *Theorem*

In steady flows the streamlines and the particle paths are identical, but not conversely.

We have also given examples of the theorem that through each point there is just one streamline, except perhaps through a stagnation point – in (3) the lines $x = 0$, $z = z_0$ and $y = 0$, $z = z_0$ are both streamlines through the point $(0, 0, z_0)$ where $\mathbf{v} = 0$. We do not prove this theorem, which is really a theorem on differential equations. Typical flows near stagnation points are sketched in fig. III.4.

Fig. III.4. Typical flows near stagnation points in the flow and at a wall.

A further observational use of dyed particles is to put a continuous source of dye at a point and observe the 'streakline' that is generated by it in the flow. The calculation of streaklines is somewhat involved as it requires the subsequent position at a given time of particles which at any previous time have passed through the dye source.

Usually it is found that streaklines differ from paths and streamlines, but all three are certainly the same in steady flows.

It should be noted that experimental work in fluid dynamics is often difficult; the measuring devices can be temperamental, and the experiments can refuse to give the same flow patterns and values to different groups of experimenters for no obvious reasons. Partly this is due to a frequent lack of stability of fluid flows: a beginner might think that water would flow down a straight pipe in straight lines, but this tidy solution of the equations is only realised at rather low speeds, and at speeds more usual in plumbing or engineering the flow may be quite chaotic. The tidy mathematical solutions of the theory often require very careful experimental conditions if they are to exist in reality.

Exercises

1. Domestic salt flows quite well out of a container with a hole in the bottom. Does salt match up to the properties quoted for a fluid? What is the size of a typical salt particle? At what size of hole would a continuum model of the salt flow out of the hole be reasonable.

2. Define carefully the velocity at a point in the continuum model of a gas.

3. A depression (a region of low atmospheric pressure p) moves across the British Isles from West to East and fills slightly as it passes across. Use typical values from weather maps (the newspaper ones will do) to estimate

the following rates of change – assume that the depression has its centre moving along the line from Bristol to London.

    (i) $\partial p/\partial t$ as measured by a balloonist travelling at constant height and always in the centre of the depression;

    (ii) $\partial p/\partial t$ as measured by an observer in Bristol;

    (iii) $\partial p/\partial t$ as measured by an observer who leaves Bristol as the centre of the depression passes over, and travels by train to Edinburgh.

4. Calculate and describe particle paths and streamlines for the flow

$$\mathbf{v} = (ay, \, -ax, \, b(t)).$$

What could be modelled by the case $b(t) = $ constant?

5. Sketch streamlines for
    (i) $\mathbf{v} = (a \cos \omega t, \, a \sin \omega t, \, 0)$,
    (ii) $\mathbf{v} = (x - Vt, \, y, \, 0)$,
    (iii) $v_r = r \cos \frac{1}{2}\theta, \, v_\theta = r \sin \frac{1}{2}\theta, \, v_z = 0, \, 0 < \theta < 2\pi$.

6. Find streamlines and particle paths for the two-dimensional flows
    (i) $\mathbf{v} = (xt, \, -yt, \, 0)$,
    (ii) $\mathbf{v} = (xt, \, -y, \, 0)$.

7. The pressure in a fluid (which will be discussed in detail in Chapters VI and VII) is due to the momentum changes of molecules reflected at a boundary. This cannot be defined instantaneously, because there may be no molecules striking a boundary in a short time interval like 10^{-30} seconds. Make a definition, in the spirit of the definition of density in §1, which averages the momentum change over an area a^2 of surface *and* over a time interval τ, so as to allow for a definition of pressure in the continuum model at position $\mathbf{x}$ and time t.

References

In a first text like this you cannot expect to have a full treatment of anything. The references are provided to fill a few of the gaps.

(a) There is a full length 16 mm sound film by the National Committee for Fluid Mechanics Films (NCFMF) which illustrates many of the points in this chapter. It is *Eulerian and Lagrangian Descriptions in Fluid Mechanics* by J. L. Lumley. Film loops from this film are FM-47 and FM-48 (sometimes found as Rank 8 mm Auto Loops Fluid Mechanics Series 28.5062 and 38.5063). These show pathlines and streamlines.

(b) *Physical Fluid Dynamics*, D. J. Tritton, Van Nostrand Reinhold 1977. This book has many excellent photographs of flows, but for this chapter its main supplementary information is on experimental methods (Chapter 23). There are also some interesting (though rather hard) streamline and particle path diagrams for flow around a cylinder (§6.2).

(c) Elementary kinetic theory of gases. The introductory sections to each chapter in *Kinetic Theory of Gases*, L. Loeb, McGraw-Hill 1927, are at the right level. Most texts are too condensed in their explanations. *Elements of the Kinetic*

Theory of Gases, E. Guggenheim, Pergamon 1960, is more recent and rather mathematical, but the elementary material is there.

(d) *Natural Aerodynamics*, R. S. Scorer, Pergamon 1958. This fascinating trip round the atmosphere is an encouragement to anyone interested in fluid motion to use his eyes. Many examples of air motion are explained in terms accessible by half way through this course.

(e) *The Liquid Phase*, D. H. Trevena, Wykeham Publications 1975. Skimming through the easy parts of a book like this will give some useful background to the liquid state.

IV

Mass conservation and stream functions

1. The continuity equation

One of the fundamental laws in Newtonian dynamics is that mass is conserved, and we now convert this statement into a differential equation. Take a fixed smooth surface S in the fluid containing volume V (see fig. IV.1). The rate of flow of mass into V is

$$- \int_S \rho \mathbf{v} \cdot d\mathbf{S},$$

the minus being because $d\mathbf{S}$ is *out* of V, and the density ρ being used to give a mass flux rather than a volume flux. The rate of increase of mass in V is

$$d/dt \int_V \rho dV$$

and because V is a fixed region this is

$$\int_V \partial\rho/\partial t \, dV.$$

Fig. IV.1. Definition diagram for the conservation of mass.

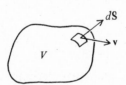

Because mass is conserved we must have

$$\int_V \partial\rho/\partial t \, dV = -\int_S \rho\mathbf{v}\cdot d\mathbf{S}.$$

Using the divergence theorem gives

$$\int_V \{\partial\rho/\partial t + \nabla\cdot(\rho\mathbf{v})\}dV = 0.$$

But this is for any volume V, Hence, by the DuBois–Reymond Lemma in Chapter I §5(*e*), assuming that the integrand is continuous because we rarely observe discontinuities,

$$\partial\rho/\partial t + \nabla\cdot(\rho\mathbf{v}) = 0,$$

at any point in the fluid. This is called the 'continuity equation' or the 'mass-conservation equation'. The method of derivation, using an arbitrary control volume, is important later as well as here. There are two immediate consequences: first if ρ is independent of $\mathbf{r}$ and t, as it would be in a well mixed river, then the equation reduces to

$$\nabla\cdot\mathbf{v} = 0;$$

and if ρ is independent of t, as it would be in a 'steady' experiment, then

$$\nabla\cdot(\rho\mathbf{v}) = 0.$$

Another real situation seems worth pursuing, and that is where the fluid is incompressible (or sufficiently nearly so) but not necessarily all of the same density. This occurs, for example, in water of varying salinity (or temperature): the water can be stratified so that ρ is not independent of $\mathbf{r}$, but as it moves around each particle keeps its density under circumstances where there is little mixing. The resulting form of the continuity equation for incompressible fluid is most easily investigated indirectly, as follows.

2. The convective derivative
(a) Incompressibility

Suppose a particle at position $\mathbf{r}$ at time t has density $\rho(\mathbf{r}, t)$. When it moves on to $\mathbf{r} + \mathbf{v}\delta t$ and $t + \delta t$, the density must become

$$\rho(\mathbf{r} + \mathbf{v}\delta t, t + \delta t).$$

The change in ρ that is experienced by the particle is

$$\rho(\mathbf{r} + \mathbf{v}\delta t, t + \delta t) - \rho(\mathbf{r}, t)$$

which, by Taylor's theorem, is

$$\delta t(\mathbf{v}\cdot\nabla\rho + \partial\rho/\partial t) + O(\delta t^2),$$

where $O(\delta t^2)$ means that this term is less than some constant times δt^2

whenever δt is small enough. Thus the rate of change experienced by the particle is (dividing by δt and taking a limit)

$$\mathbf{v}\cdot\nabla\rho + \partial\rho/\partial t.$$

This is known as the convective derivative or derivative following the fluid or particle (you will find other names in use also). What it does is to give the Lagrangian (following the particle) time rate of change in terms of Eulerian (fixed point) measurements. The usual notation is

$$\frac{D\rho}{Dt} = \frac{\partial\rho}{\partial t} + \mathbf{v}\cdot\nabla\rho$$

or in general

$$\frac{D}{Dt} = \frac{\partial}{\partial t} + \mathbf{v}\cdot\nabla$$

for any fluid property.

If the fluid is incompressible, then the density does not change following the fluid and so

$$D\rho/Dt = 0.$$

However, we already know that

$$\partial\rho/\partial t + \nabla\cdot(\rho\mathbf{v}) = 0,$$

which may be written as

$$\partial\rho/\partial t + \mathbf{v}\cdot\nabla\rho + \rho\nabla\cdot\mathbf{v} = 0$$

or

$$D\rho/Dt + \rho\nabla\cdot\mathbf{v} = 0.$$

So combining these equations gives the important result

$$\nabla\cdot\mathbf{v} = 0$$

for an incompressible fluid.

This result is of wide application, even to flows of gases (for which the compressibility only becomes important as the speed of sound is approached). We will take it as the basic equation of continuity for large parts of this course, though when we come to discuss high speed gas flow or sound waves we must return to a fuller equation of continuity.

(b) Acceleration

The convective derivative D/Dt also applies to velocity. The acceleration of a fluid particle (following the particle) is

$$\partial\mathbf{v}/\partial t + \mathbf{v}\cdot\nabla\mathbf{v}.$$

This explains the difference between Eulerian and Lagrangian accelerations in Chapter III§3, where

$$Dv/Dt = a^2\mathbf{r}$$

and

$$\partial\mathbf{v}/\partial t = 0.$$

The difference must be $\mathbf{v}\cdot\nabla\mathbf{v}$, and in that example

$$\mathbf{v}\cdot\nabla = (ax, -ay, 0)\cdot(\partial/\partial x, \partial/\partial y, \partial/\partial z)$$
$$= ax\,\partial/\partial x - ay\,\partial/\partial y,$$

so that

$$\mathbf{v}\cdot\nabla\mathbf{v} = (ax\,\partial/\partial x - ay\,\partial/\partial y)(ax, -ay, 0)$$
$$= (a^2x, a^2y, 0) = a^2\mathbf{r}.$$

It also explains the acceleration of a log in a steady river: the river has a spatial rate of change, and so the acceleration of the log is $u\,\partial u/\partial x$, taking x as distance along the river.

Finally it explains those examples on $\mathbf{A}\cdot\nabla\mathbf{A}$ in Chapter I: we need $\mathbf{v}\cdot\nabla\mathbf{v}$ in the acceleration and hence in the equation of motion, when we derive it below.

3. The stream function for two-dimensional flows

We go on in the rest of this chapter to investigate consequences of the continuity equation for an incompressible fluid, $\nabla\cdot\mathbf{v} = 0$, in rather simple circumstances. But before starting, we should point out:

(i) that $\nabla\cdot\mathbf{v} = 0$ means, using the divergence theorem, that there is no total volume flow through *any* closed surface—as much flows out as flows in;

(ii) there will be close links with parts of electrostatics, which has $\nabla\cdot\mathbf{E} = 0$ except at charges;

(iii) the time dependence has been left out to make the formulae look easier – really we may have $\mathbf{v}(x, y, t)$ or $\mathbf{v}(r, \theta, t)$ with the results given holding for each value of t.

For the sake of mathematical simplicity we choose to investigate flows in which there are only two non-zero velocity components and two effective coordinates. These are

(i) two-dimensional flows,

$$\mathbf{v} = u(x, y)\mathbf{i} + v(x, y)\mathbf{j}$$

or

$$\mathbf{v} = v_r(r, \theta)\hat{\mathbf{r}} + v_\theta(r, \theta)\hat{\boldsymbol{\theta}},$$

where plane cartesian or plane polar components are used;

(ii) axisymmetric flows,

$$\mathbf{v} = v_r(r, z)\hat{\mathbf{r}} + v_z(r, z)\mathbf{k},$$

where cylindrical polar coordinates are used, $\mathbf{v}$ is independent of θ and there is no $\hat{\theta}$ velocity (no swirl).

When we have such flows, we have two velocity components and one differential equation $\nabla \cdot \mathbf{v} = 0$ connecting them. We can use this differential equation to eliminate one velocity component, and hence reduce the mathematics to the consideration of only *one* function. It is, in fact, more convenient not to do exactly this, but we see here the reason why such flows are particularly easy.

Two-dimensional flows are easier to visualise, so we spend some time discussing them; after that we largely repeat the discussion for axisymmetric flows.

(*a*) *Existence of a stream function*

In a two-dimensional flow of incompressible fluid we have

$$\begin{cases} \mathbf{v} = u(x, y)\mathbf{i} + v(x, y)\mathbf{j}, \\ \partial u/\partial x + \partial v/\partial y = 0. \end{cases}$$

The second equation is automatically satisfied if we put

$$u = \partial\psi/\partial y, \; v = -\partial\psi/\partial x$$

for any suitably differentiable function $\psi(x, y)$. This is rather like what you do in dynamics, where a conservative force

$$F_1(x, y)\mathbf{i} + F_2(x, y)\mathbf{j},$$

which has $\partial F_2/\partial x - \partial F_1/\partial y = 0$ because it is conservative, may be derived from a potential $\phi(x, y)$ such that

$$F_1 = -\partial\phi/\partial x, \; F_2 = -\partial\phi/\partial y.$$

Change a few signs and letters, and there you are!

Let us demonstrate that such a function $\psi(x, y)$ exists, and find its value in terms of u and v. We wish to have

$$\partial\psi/\partial y = u(x, y).$$

This may be achieved by taking

$$\psi(x, y) = \int_b^y u(x, \eta)d\eta + \alpha(x)$$

for any function $\alpha(x)$ and any constant b, from the fundamental theorem of calculus. Then

$$\partial\psi/\partial x = \int_b^y \partial u/\partial x \, d\eta + \alpha'(x),$$

and using the equation of continuity in the form $\partial u/\partial x + \partial v/\partial\eta = 0$, because x and η are the variables in the integration, gives

$$\partial\psi/\partial x = -\int_b^y \partial v/\partial\eta \, d\eta + \alpha'(x).$$

This may be integrated to give

$$\partial\psi/\partial x = -v(x, y) + v(x, b) + \alpha'(x),$$

and we get what we set out to find if

$$\alpha'(x) = -v(x, b),$$

i.e.

$$\alpha(x) = -\int_a^x v(\xi, b)d\xi,$$

for any constant a.

We have, therefore, calculated $\psi(x, y)$ such that

$$\begin{cases} \partial\psi/\partial y = u(x, y), \\ -\partial\psi/\partial x = v(x, y) \end{cases}$$

as required. It is given by

$$\psi(x, y) = \int_b^y u(x, \eta)d\eta - \int_a^x v(\xi, b)d\xi,$$

for any constants a and b.

This result may be put into a general context as follows. If we write

$$\mathbf{A}(x, y) = \psi(x, y)\mathbf{k}$$

then it is easily shown that

$$\begin{aligned} \nabla \times \mathbf{A} &= \nabla \times (\psi\mathbf{k}) \\ &= (\nabla\psi) \times \mathbf{k} \quad \text{because } \mathbf{k} \text{ is constant} \\ &= (\partial\psi/\partial y, \, -\partial\psi/\partial x, 0) \\ &= \mathbf{v}. \end{aligned}$$

It is generally true, but is not proved here, that if $\nabla \cdot \mathbf{B} = 0$, then an $\mathbf{A}$ exists such that

$$\mathbf{B} = \nabla \times \mathbf{A};$$

such an $\mathbf{A}$ is called a 'vector potential' for $\mathbf{B}$. In the present case the vector potential $\psi\mathbf{k}$ is particularly useful as it has only one component. This component $\psi(x, y)$ is called the 'stream function' of the flow.

(b) Properties of the stream function

The name stream function suggests that ψ has something to do with streamlines. And simple analysis shows that it has. For

$$\mathbf{v} = \nabla\psi \times \mathbf{k}$$

Fig. IV.2. Velocity **v** and $\nabla\psi$ for a streamline $\psi = $ constant.

shows that **v** is perpendicular to $\nabla\psi$, and it is a general property that $\nabla\psi$ is perpendicular to the curves (surfaces if you think of it three-dimensionally)

$\psi = $ constant.

Hence **v** is parallel to the curves

$\psi = $ constant

at every point, and so these curves are the streamlines. Fig. IV.2 shows a streamline and the directions of **v** and $\nabla\psi$.

Take two streamlines $\psi = a$ and $\psi = b$ in the x, y plane, and a unit height of fluid above them, as in fig. IV.3. The flow *through* the rectangle $ABCD$ is

$$\int_S \mathbf{v} \cdot d\mathbf{S},$$

where S is the rectangle with normal in the direction of flow. But this is

$$\int_S \nabla \times (\psi \mathbf{k}) \cdot d\mathbf{S},$$

which by Stokes' theorem is

$$\oint_l \psi \mathbf{k} \cdot d\mathbf{l},$$

where the direction round the rectangle $ABCD$ is marked by arrows. This line integral is easily calculated because

$$\begin{cases} \text{on } AB \quad \psi = b \text{ and } \mathbf{k} \cdot d\mathbf{l} = dl, \\ \text{on } BC \text{ and } DA \quad \mathbf{k} \cdot d\mathbf{l} = 0, \\ \text{on } CD \quad \psi = a \text{ and } \mathbf{k} \cdot d\mathbf{l} = -dl, \end{cases}$$

Fig. IV.3. Flow through a rectangle of unit height.

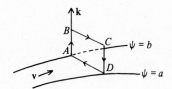

where $d\mathbf{l}$ is a vector length element on the contour $ABCD$. Hence

$$\int_S \mathbf{v} \cdot d\mathbf{S} = b - a$$

because the rectangle has unit height. We have thus proved that the flow between streamlines $\psi = b$ and $\psi = a$ is at rate $b - a$ per unit height. Consequently, if the streamlines are close together the flow must be fast, and streamlines with different values of ψ can only meet at a singularity of the flow.

This simple result makes it worth returning to the general form for ψ and trying to interpret it as a rate of flow. Consider the (curvy) 'triangle' in fig. IV.4. The flow rate per unit height through the base AB will be

$$-\int_a^x v(\xi, b) d\xi.$$

The flow rate out through the side BP will be

$$\int_b^y u(x, \eta) d\eta.$$

Hence the total flow *out* of the figure is just $\psi(x, y)$, where we note that we have defined $\psi(a, b)$ to be zero – this arbitrary choice of a zero for ψ is of no consequence as only derivatives of ψ or differences of ψ are observable.

Now the curve from (a, b) to (x, y) was arbitrary, and the flow out of the 'triangle' must be balanced by a flow in; so we have shown that the flow rate per height through any curve from (a, b) to (x, y) is

$$\int_b^y u(x, \eta) d\eta - \int_a^x v(\xi, b) d\xi = \psi(x, y).$$

Now we have pointed out that $\psi(a, b) = 0$ from our definitions. By a shift of origin we may easily show that the flow rate through any curve from (c, d) to (x, y) is

$$\psi(x, y) - \psi(c, d)$$

per unit height. This is the most revealing interpretation of the stream function ψ.

Fig. IV.4. Flow through an arbitrary curve, seen from above.

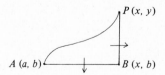

4. Some basic stream functions

We will now take some rather simple forms for ψ in an attempt to describe some of the easy real flows that can be produced in the laboratory or that are seen in reality. This modelling is of the primitive kind, where we do not try to *explain* the flows, only to describe what they look like.

(a) Flows parallel to the x-axis

A uniform stream of speed U parallel to the x-axis has

$$\begin{cases} u = U = \partial\psi/\partial y, \\ v = 0 = -\partial\psi/\partial x \end{cases}$$

and so

$$\psi = Uy$$
$$= U\,r\sin\theta$$

in polars. We do not need a constant of integration here: as explained above it makes no difference.

A stream parallel to the x-axis, whose speed is proportional to distance from the x-axis has

$$\begin{cases} u = \beta y = \partial\psi/\partial y, \\ v = 0 = -\partial\psi/\partial x, \end{cases}$$

and so

$$\psi = \tfrac{1}{2}\beta y^2 = \tfrac{1}{2}\beta r^2 \sin^2\theta.$$

A 'uniform shear flow' like this is an (excessively) simple model for some real flows, such as the wind at low levels in the atmosphere which increases in strength as you go up from ground level. The variation of velocity (across the stream) is called the 'velocity profile', and is shown in fig. IV.5.

Flows like this one with

$$\mathbf{v} = u(y)\mathbf{i}, \quad \psi = \int_a^y u(\eta)d\eta$$

are moderately realistic in many cases, when the development of the flow

Fig. IV.5. The velocity profile $u = \beta y$.

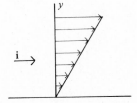

in the downstream x-direction is slow enough to be ignored, and also any change in time; we are also, of course, ignoring changes in the cross-stream z-direction. This can be useful as a model in a wide, flat-bedded river, or as said above, for some aspects of the atmosphere. The particular velocity profile $u(y)$ that is relevant may have many different functional forms, depending on the detailed dynamical processes at work; a logarithmic profile can be suitable in the fast flow near a wall in a channel,

$$u(y) = U \ln(y/a).$$

One particular model worth mentioning here is the 'transition layer' between a fast stream and a slow one:

$$u(y) = U_0 + \tfrac{1}{2}(U_1 - U_0)\{1 + \tanh(y/a)\},$$

whose velocity profile is sketched in fig. IV.6. Because of the properties of the tanh function, $u(y)$ is approximately U_0 for $y/a \leqslant -2$, and approximately U_1 for $y/a \geqslant 2$. The width of the transition layer is determined by the size of a: if you take $a = 20$ cm, almost all the transition has taken place inside 100 cm.

(b) Flow radially outwards

A simple source of fluid at the origin can be produced (not exactly) in a water tank by using a hose with multiple perforations; water then leaves the vicinity of the hose radially with roughly the same speed at all angles θ:

$$v_r = f(r), \quad v_\theta = 0.$$

Clearly this will be easiest to deal with in polar coordinates. Now

$$\mathbf{v} = (\nabla \psi) \times \mathbf{k}$$
$$= (\partial \psi/\partial r, \ r^{-1} \partial \psi/\partial \theta, \ 0) \times \mathbf{k}$$
$$= (r^{-1} \partial \psi/\partial \theta, \ -\partial \psi/\partial r, \ 0).$$

Fig. IV.6. The transition layer velocity profile.

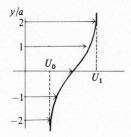

That is, the relation between ψ and $\mathbf{v}$ in plane polars is

$$\begin{cases} v_r = r^{-1}\partial\psi/\partial\theta, \\ v_\theta = -\partial\psi/\partial r. \end{cases}$$

Thus for the simple source we have

$$\begin{cases} \partial\psi/\partial r = 0, \\ r^{-1}\partial\psi/\partial\theta = f(r), \end{cases}$$

for which the only solution is

$$\psi = A\theta$$

because ψ must be independent of r and $\partial\psi/\partial\theta$ must be independent of θ. If ψ is to be a properly defined function on the plane, it must have a single value at each point, and so we restrict θ by $-\pi < \theta \leqslant \pi$.

This stream function gives

$$v_r = A/r, \quad v_\theta = 0,$$

which satisfies $\nabla \cdot \mathbf{v} = 0$ except at $r = 0$. The total outflow from $r = 0$ is the flow through a circle surrounding $r = 0$, and is called the source strength. Its value is

$$\int_0^{2\pi} v_r \times r d\theta$$

$$= 2\pi A,$$

which is just velocity times circumference, the same for any size of circle.

(c) A dipole along the x-axis

A source and an equal sink (a negative source) placed parallel and close together form an approximate dipole, for which the flow pattern is shown in Fig. IV. 7. A true dipole is a useful model of flows of this shape, in which the sources are infinitesimal in size, the distance apart tends to zero and the strength-times-the-distance remains at a constant value. The

Fig. IV.7. Streamlines for a source and sink.

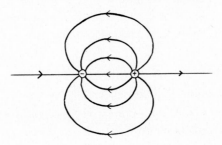

Fig. IV.8. Definition diagram for §4(c).

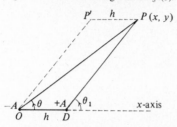

limiting operation can be most easily achieved by differentiation; this is essentially because we are using the *difference* of two source stream functions (+ for the source and − for the sink), and we have to divide by the distance apart so as to get the (constant) source strength times distance as a multiplier, and then finally take the limit as the distance tends to zero. This time we will carry out the full argument by limits; later we shall just differentiate.

Take then, a sink of constant − A at the origin O, and a source whose constant is A at the point D which is a distance h along the x-axis, as in fig. IV.8. The combined stream function is

$$\psi(P) = -A\theta + A\theta_1$$
$$= Ah\{(-\theta + \theta_1)/h\},$$

where we have isolated Ah which will be the (constant) strength μ of the dipole after we have taken the limit. Thus

$$\psi_{\text{dipole}}(P) = \mu \lim_{h \to 0} \{(-\theta + \theta_1)/h\}.$$

Now θ_1 is not only the angle between DP and the x-axis; it is also the angle between OP' and the axis, where P' is the point $(x - h, y)$. So the dipole stream function can be rewritten as

$$-\mu \lim_{h \to 0} \frac{1}{h} \{\theta(P) - \theta(P')\}.$$

This is now obviously in the form of a differentiation, because P' is just P moved by h. It looks as though the signs are wrong, in that we have θ at $(x - h, y)$ rather than at $(x + h, y)$, but h may tend to zero from either side in the definition of differentiation, so it makes no difference.

The result then is that

$$\psi_{\text{dipole}}(P) = -\mu \, \partial\theta/\partial x.$$

Notice the minus sign; and notice that more generally we have

'result for dipole = − constant × derivative along the dipole
                              direction of the result for a source'.

In plane polar coordinates we need to calculate

$$- \mu \partial/\partial x \{\tan^{-1}(y/x)\}$$

and we find that

$$\psi_{\text{dipole}} = (\mu/r)\sin\theta.$$

We could calculate this in another way by expressing the general statement above as

$$\psi_{\text{dipole}} = - \boldsymbol{\mu} \cdot \nabla\theta \text{ with } \boldsymbol{\mu} = \mu\mathbf{i},$$

and then using

$$\nabla\theta = (\partial/\partial r, \, r^{-1}\,\partial/\partial\theta)\theta$$
$$= r^{-1}\hat{\boldsymbol{\theta}},$$

with, finally, $\mathbf{i} \cdot \hat{\boldsymbol{\theta}} = -\sin\theta$.

As a further example, you may show that a dipole of strength μ at angle θ_0 to the x-axis has stream function

$$(\mu/r)\sin(\theta - \theta_0).$$

(d) Flow in circles

Another obvious kind of motion is motion in circles round an axis. For this

$$v_r = 0 \text{ and } v_\theta = \text{some function of } r,$$

and so $\psi = F(r)$, any function of r. We will find later, and we observe in reality, that two special forms of $F(r)$ are most important; these correspond to

$$v_\theta = C/r$$

and

$$v_\theta = Dr.$$

The latter is rigid body rotation, which we have met before; the former is called a 'line vortex', and is another singularity of the flow field rather like a line source. It cannot actually exist, but many real flows are rather like it except for small r. For this line vortex we have

$$v_\theta = - \partial\psi/\partial r = C/r$$

and so

$$\psi = - C\ln(r/a),$$

where a is some constant length.

The quantity analogous to the strength of a line source is the 'circulation' of a line vortex, which is measured by

$$\oint_l \mathbf{v} \cdot d\mathbf{l}$$

where l is a circle round the origin. Its value, often called κ, is

$$\kappa = \int_0^{2\pi} v_\theta r d\theta = 2\pi C,$$

the same for any size of circle. Note that in terms of the circulation κ, a line vortex has stream function

$$\psi = -(\kappa/2\pi)\ln(r/a).$$

All these simple forms for ψ have analogues in electrostatics and magnetostatics. For example, a line charge has radial lines of force just like the streamlines from a line source; and a line current has circular magnetic field lines just like the streamlines for a line vortex.

5. Some flow models and the method of images

Quite interesting flows can be built up using these elementary stream functions. What we do is to take two or more of these 'building blocks' and add them together to produce a new stream function. We may find streamlines of the new flow by sketching (or calculating) curves $\psi = $ constant. And the new flow will still possess the singularities of its parts, so will still have source, sink, dipole or vortex character near these singularities. Notice carefully that streamlines are *not* usually preserved when you add two stream functions – you must start again in your sketching of the flow.

(a) Flow round a cylinder

Start with a stream and a dipole along opposite directions. The new stream function is

$$\psi = Ur\sin\theta - \mu\sin\theta/r.$$

We note three immediate results:

(i) for large r, $\psi \sim Ur\sin\theta$, a uniform stream (where the sign $\sim$ means that less important terms have been omitted);

(ii) for small r, $\psi \sim -(\mu/r)\sin\theta$, a dipole;

(iii) the streamline $\psi = 0$ is the circle

$$r = (\mu/U)^{1/2} = a, \text{ say},$$

together with the lines $\theta = 0, \pi$.

This is really enough to sketch the flow; you get something like fig. IV.9;

Fig. IV.9. Streamlines for example 5(a).

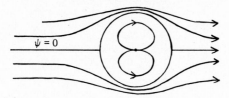

you may wish to try to sketch more accurately the streamlines $\psi = \pm (U\mu)^{1/2}$ as well. Now there is no flow across the streamline $r = a$, which may therefore be taken as a solid boundary in this simple descriptive modelling of flows. So *one* flow past the solid cylinder of radius a has stream function

$$\psi = Ur\sin\theta - Ua^2\sin\theta/r.$$

Notice that the streamlines are closer together above and below the cylinder, which corresponds to higher speeds there. Notice also the symmetry upstream and downstream of the centre plane $x = 0$; this symmetry is enough to suggest that this is not too good a model. For you would expect a downstream force on the cylinder, and this force would remove momentum from the flow, and so make it unsymmetric. We consider this point again later. However this does provide a good model for the initial relative motion of a cylinder starting from rest, or of a cylinder oscillating at small amplitudes, in a fluid at rest at large distances.

The problem of the flow of a uniform stream past a fixed circular cylinder has been central in fluid dynamics for many years, and it is only recently that a reasonable understanding of it seems to have been reached. A series of photographs and calculated streamline patterns in *Physical Fluid Dynamics* by D. J. Tritton is well worth examining just to see the variety of flow patterns which result from such a simple experiment, and how much has to be explained.

(b) *The image of a line vortex in a wall*

Take next a pair of equal line vortices of opposite sign, as shown in fig. IV.10. The stream function at P has value

$$\psi(P) = -C\ln(r_1/a) + C\ln(r_2/a)$$
$$= C\ln(r_2/r_1).$$

Clearly the x-axis has $\psi = 0$, because any point on it is at equal distances from the two vortices. Hence the x-axis has no flow across it, and could be taken as a solid surface. So we have also found the stream function for

Fig. IV.10. Definition diagram for §5(*b*).

Fig. IV.10. Definition diagram for §5(*b*).

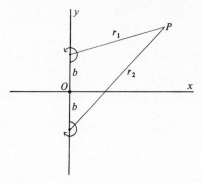

a vortex near a solid wall: it is derived by using an 'image' vortex of equal and opposite strength.

The 'method of images' is useful in fluid dynamics, and also in electro-magnetic theory. It is used to find solutions outside simple geometrical shapes by choosing 'images' of suitable strengths at suitable image points so that the boundary condition on the solid body is satisfied by the combined field of singularity and images; this combined field can then be used outside the body. The simple geometrical shapes are almost always planes, cylinders or spheres, just those objects in which simple optical images form by reflection.

The above stream function is not in fact as easy as it seems: the vortex moves under the influence of its image in a fashion we shall investigate later. All we can get is the instantaneous flow pattern; later on the whole flow (image as well) will have moved elsewhere, because a vortex is hard to fix.

(*c*) *The image of a line source in a cylinder*

Let us use the method of images to find the stream function for a line source outside a cylinder. In fig. IV.11 the source is at S and its 'image point' I is on the line from the centre O to S at distance a^2/c from O, where the cylinder has radius a and the line source is at a distance c from the axis

Fig. IV.11. Definition diagram for §5(*c*).

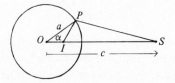

of the cylinder. We require a stream function which has a singularity for a source at S, is constant all round the circle and has no more singularities in the flow. The method of images suggests that we can fix this up by introducing an image source of strength B at I; if we do this we must also have a matching sink $-B$ inside the cylinder, or there will be a net outflow across it – the centre seems an obvious place to put this sink, because of the symmetry of this point.

Before we go on with the calculation, we should ask why the image of a source is going to be in terms of sources. At this stage we have no real answer, because we have not yet considered the *dynamics* of the fluid; when we do, we shall find that there are some uniqueness theorems, and these will put the method of images on a sounder footing. At present it is plausible that a source has images that are sources (or sinks) and not anything else, but that is all we can say. When you look in *any* shape of mirror, you see something related to the original object.

We take then the system of sources shown in fig. IV.12 and calculate $\psi(P)$ for it, and show that this is independent of α, for a special choice of B, and hence is a constant on the circle of radius a. The total stream function must be

$$\psi(P) = A\theta + B\beta - B\alpha,$$

using the angles at the appropriate points to find the various contributions to ψ. Now the triangles POI and SOP are similar triangles, because α is common to each and the sides adjacent to this angle have the same ratio in each triangle:

$$a^2/c \div a = a \div c,$$

$$(IO \div OP = PO \div OS).$$

Hence the angles in the triangles are the same, so that angle $OPI =$ angle $OSP = \pi - \theta$. That is

$$\beta - \alpha = \pi - \theta,$$

$$\theta + \beta - \alpha = \pi.$$

Hence $\psi(P)$ is a constant for P anywhere on the cylinder provided that $B = A$.

This 'solution' for a source outside a cylinder suffers from the same

Fig. IV.12. Image system for a line source outside a circular cylinder.

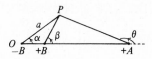

troubles as the 'solution' for a uniform stream past a cylinder, which will be discussed later. But it has some interest and use.

6. The (Stokes) stream function for axisymmetric flows

The other type of flow that was said in §3(ii) to be rather simple was the axisymmetric (swirl free) flow

$$\mathbf{v} = v_r(r, z)\hat{\mathbf{r}} + v_z(r, z)\mathbf{k},$$

where cylindrical polar coordinates are used. The development of the argument is very similar to that for two-dimensional flows, so not all the steps in each demonstration will be given.

(a) Existence of the stream function

The continuity equation $\nabla \cdot \mathbf{v} = 0$ may be put as

$$\partial/\partial r(rv_r) + \partial/\partial z(rv_z) = 0,$$

where the 'extra' r in each bracket allows for the 'extra' term in the form for $\nabla \cdot \mathbf{v}$ in cylindrical polars. It follows immediately that $r\mathbf{v}$ can be derived from a stream function, just as $\mathbf{v}$ could in two-dimensional flow, as the equations are the same. Thus

$$\begin{cases} rv_z = \partial\Psi/\partial r, \\ rv_r = -\partial\Psi/\partial z, \end{cases}$$

where we use a capital for the stream function to emphasize the difference; this is called Stokes' stream function Ψ (or the Stokes stream function).

You may verify that

$$\Psi(r, z) = \int_a^r sv_z(s, z)ds - \int_c^z av_r(a, \zeta)d\zeta,$$

and that the vector potential form is (in cylindrical polars)

$$\mathbf{v} = \nabla \times (r^{-1}\Psi\hat{\boldsymbol{\theta}}).$$

From this vector potential form it is easy to calculate the formulae for spherical polar coordinates – this is the first time we have changed from cylindrical to spherical coordinates, so note carefully what changes have to be made: in spherical coordinates

$$\begin{cases} r \text{ now means distance from } O, \\ \theta \text{ now means angle } from \text{ (not round) the } z\text{-axis.} \end{cases}$$

So we need to calculate $\nabla \times \mathbf{A}$ in spherical polars, where

$$\mathbf{A} = \frac{1}{r \sin \theta}\Psi\hat{\lambda}$$

Fig. IV.13. Relation between cylindrical and spherical coordinates.

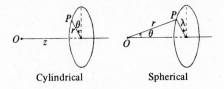

Cylindrical   Spherical

because in spherical polars the distance from the axis is not r but $r \sin \theta$, and the vector round the axis is not $\hat{\theta}$ but $\hat{\lambda}$. These changes are shown in fig. IV.13.
So

$$\mathbf{v} = \frac{1}{r^2 \sin \theta} \begin{vmatrix} \hat{\mathbf{r}} & r\hat{\theta} & r \sin \theta \, \hat{\lambda} \\ \partial/\partial r & \partial/\partial \theta & 0 \\ 0 & 0 & \Psi \end{vmatrix}$$

or

$$v_r = \frac{1}{r^2 \sin \theta} \frac{\partial \Psi}{\partial \theta}, \quad v_\theta = -\frac{1}{r \sin \theta} \frac{\partial \Psi}{\partial r}.$$

(b) Properties of the stream function Ψ

As before Ψ is constant on the streamlines, but because of the axial symmetry it is now more sensible to talk about 'stream tubes': all the streamlines through a circle about the axis of symmetry form a stream tube which, as in fig. IV.14, is a surface of revolution about the axis of symmetry.

The important flux properties of the stream function are derived in the same fashion as before, but using rather more complicated arguments. Take two stream tubes $\Psi = a$ and $\Psi = b$, sketched in fig. IV.15; the rate

Fig. IV.14. A stream tube in an axisymmetric flow.

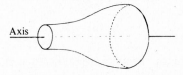

Axis

Fig. IV.15. The stream tubes for flow through an annulus S.

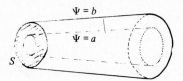

$\Psi = b$

$\Psi = a$

S

Fig. IV.16. The contour for use in Stokes' theorem.

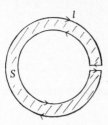

of flow of volume through the annulus shown is

$$\int_S \mathbf{v} \cdot d\mathbf{S},$$

which in terms of Ψ is

$$\int_S \nabla \times (r^{-1}\Psi\hat{\boldsymbol{\theta}}) \cdot d\mathbf{S},$$

and using Stokes' theorem we get

$$\oint_l r^{-1}\Psi\hat{\boldsymbol{\theta}} \cdot d\mathbf{l},$$

where S spans l. We must take l as shown in fig. IV.16, with the integrals along the two sides of the link between the two circles:

(i) being zero because perpendicular to $\hat{\boldsymbol{\theta}}$;

(ii) cancelling out if not perpendicular to $\hat{\boldsymbol{\theta}}$.

[Did we use the analogue of (i) or (ii) in the two-dimensional case?] Now on the outer circle (θ does not increase in the direction you might expect; see the diagram of cylindrical coordinates above)

$$\hat{\boldsymbol{\theta}} \cdot d\mathbf{l} = r d\theta$$

and on the inner circle $\hat{\boldsymbol{\theta}} \cdot d\mathbf{l} = -r d\theta$. Hence we derive

$$\int_S \mathbf{v} \cdot d\mathbf{S} = 2\pi(b - a),$$

the flux through an annulus is equal to 2π times the difference of stream function values.

Finally, as before, the integral for Ψ represents a flux. This proof is left to the exercises.

7. Models using the Stokes stream function

We will now calculate some easy examples of Ψ, and go on to use these to build up some interesting and realistic flows. Note carefully that in some places we use cylindrical coordinates, and in some spherical coordinates.

(a) Basic stream functions

(i) A uniform stream along the z-axis has

$$\mathbf{v} = U\mathbf{k}$$

and so

$$\partial\Psi/\partial r = rU, \ \partial\Psi/\partial z = 0.$$

The stream function is (in cylindrical polars)

$$\Psi = \tfrac{1}{2}Ur^2$$

choosing $\Psi = 0$ on $r = 0$ for convenience. In spherical polars it is $\tfrac{1}{2}Ur^2 \sin^2 \theta$.

   The non-uniform stream $U(r)\mathbf{k}$ is easily shown to have stream function

$$\Psi = \int_0^r sU(s)ds.$$

(ii) The point source at the origin is best treated in spherical co-ordinates, when

$$v_r = m/4\pi r^2$$

is the only velocity component – the $(4\pi r^2)^{-1}$ ensures that the volume flow out of a sphere of radius r and surface $4\pi r^2$ is just m, the 'strength' of the source. We therefore have

$$m/4\pi r^2 = (r^2 \sin \theta)^{-1}\partial\Psi/\partial\theta$$

and we may take

$$\Psi = (m/4\pi)(1 - \cos \theta)$$

to satisfy this and also $\Psi = 0$ along the axis $\theta = 0$.

(iii) To derive the stream function for a dipole along the axis of symmetry, we must differentiate along this direction, and so we start by putting the source stream function into cylindrical coordinates,

$$\Psi_{\text{source}} = (m/4\pi)\{1 - z(z^2 + r^2)^{-1/2}\}.$$

Now differentiate with respect to z, and change the sign, to get

$$\Psi_{\text{dipole}} = (m/4\pi)r^2(z^2 + r^2)^{-3/2}$$

and return to spherical coordinates to get the standard dipole stream function

$$\Psi = \mu \sin^2 \theta / r.$$

(b) Flow past a semi-infinite body

Let us now take a source at the origin and a stream together, and use cylindrical coordinates:

$$\Psi = \tfrac{1}{2}Ur^2 + (m/4\pi)\{1 - z(z^2 + r^2)^{-1/2}\}.$$

The stream tubes $\Psi =$ constant can, of course, be plotted in detail; but it is always better to get some feeling for the general shape of the curves in the r, z plane before trying to plot them.

(i) The curve $\Psi = 0$ is $r = 0$, $z > 0$: note that when $z < 0$ we do not get $\Psi = 0$ but $\Psi = m/2\pi$, and this agrees with half of the fluid from the source setting out to the left and half to the right (the mathematical reason is that $(z^2)^{1/2} = |z|$ and so for $z < 0$, $(z^2)^{1/2} = -z$).

(ii) When z is large and positive we may approximate Ψ by the binomial theorem:

$$\Psi = \tfrac{1}{2}Ur^2 + (m/4\pi)\{1 - (1 + r^2/z^2)^{-1/2}\}$$
$$\doteqdot \tfrac{1}{2}Ur^2 + (m/4\pi)\{1 - 1 + \tfrac{1}{2}r^2/z^2\}.$$

Hence for large z and moderate r we have $\Psi = \tfrac{1}{2}Ur^2$ and so the surface $\Psi = \Psi_0$ has radius

$$(2\Psi_0/U)^{1/2},$$

just as it would in a uniform stream without a source – the flow is uniform far downstream.

(iii) The stagnation points, if any, occur where

$$v_r = v_z = 0,$$

i.e.

$$r^{-1}\partial\Psi/\partial z = r^{-1}\partial\Psi/\partial r = 0.$$

This is easily shown to need $r = 0$ (on the axis of symmetry) at z given by

$$U + (mz/4\pi)(z^2 + r^2)^{-3/2} = 0.$$

The only solution is for $z < 0$, when (with $r = 0$) we get

$$z = -(m/4\pi U)^{1/2}.$$

[Note that the only length scale that can be formed from the parameters m and U of this problem is $(m/U)^{1/2}$, so an answer of this general form is inevitable.]

Fig. IV.17. Streamlines near the axis in §7(*b*).

Fig. IV.18. A sketch of the streamlines, showing the dividing streamline.

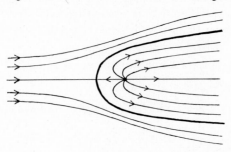

We have reached the stage shown in fig. IV.17: the stream surface $\Psi = m/2\pi$ must divide at the stagnation point, and the stream fluid flows around outside the source fluid – which is what can be observed to happen. And from (ii) the radius of the stream surface $\Psi = m/2\pi$ at $z \to \infty$ is $(m/\pi U)^{1/2}$. Thus we sketch the flow as in fig. IV.18. The dividing stream surface has, in fact, a simple equation in spherical coordinates: it is

$$\tfrac{1}{2}m/\pi = \tfrac{1}{2}Ur^2 \sin^2\theta + (m/4\pi)(1 - \cos\theta),$$

or

$$r = (m/4\pi U)^{1/2} \operatorname{cosec} \tfrac{1}{2}\theta.$$

In this particular case we could have derived the results of (i) – (iii) more easily from this formula; we chose to show a general method which can be used in other cases.

(iv) This stream function has provided the flow round a blunt ended semi-infinite body. We may find flows corresponding to other semi-infinite bodies by taking some other grouping of sources on the axis, rather than just a single source. If the total strength of these sources is zero, then the cylinder is closed at the downstream end also. This method of finding flows round axisymmetric bodies by putting sources on the axis is known as Rankine's method.

(*c*) *Flows past a sphere*

The easiest closed body will be that due to a stream and an opposing dipole. For this arrangement

$$\Psi = \tfrac{1}{2}Ur^2 \sin^2\theta - A\sin^2\theta/r$$

and the stream surface $\Psi = 0$ is

$$\theta = 0, \pi \ \text{or} \ r = (2A/U)^{1/3}.$$

Hence we have a model for the flow of a uniform stream past a sphere. As for the flow past a cylinder, we have a symmetric flow (upstream–downstream as well as axially), so this cannot be the whole answer to flow past a sphere; but it is a useful model for flow past a small rising bubble as well as for the initial relative motion of a sphere moving from rest and for an oscillating sphere in fluid at rest at large distances.

We can see that this is not the whole story by considering

$$\Psi = \tfrac{1}{4}Ua^2(2r^2/a^2 - 3r/a + a/r)\sin^2\theta.$$

This too satisfies $\Psi = 0$ on $r = a$ and behaves like a uniform stream as $r \to \infty$ because

$$\Psi \sim \tfrac{1}{2}Ur^2\sin^2\theta \ \text{as} \ r \to \infty.$$

But it gives a different flow near the sphere, with an extra term in $r\sin^2\theta$. This is in fact an appropriate model for very slow streams moving round a sphere, but this can only be shown by solving a problem in dynamics – it is too advanced for this book.

The sphere solution derived from stream and dipole can be regarded as an image solution, when you notice that a uniform stream can be regarded as a 'dipole at infinity': take two very strong sources of opposite sign and very far apart and you get an almost uniform flow near the axis half way between them. The image of this 'dipole at infinity' in a sphere is then an image dipole at the centre, with a strength chosen to give $\Psi =$ constant on the sphere.

There are image examples in three-dimensional axisymmetric flow, just as there were in two-dimensional flow. We shall not do them here.

Exercises

1. Take a volume V of the North Sea with surface S, and let the density of fish be $n(\mathbf{r}, t)$ – this is taken to be defined as for a continuum model. Set up a differential equation that expresses the same ideas as the continuity equation in fluid dynamics. Fish are, of course, not conserved.

2. Let $\mathbf{v} = mr^{-2}\hat{\mathbf{r}}$ in a spherical system of coordinates; show that $\nabla\cdot\mathbf{v} = 0$ except at O. Let S be *any* smooth surface surrounding O: show that volume flows through S at rate $4\pi m$. What is the corresponding result if O lies on S?

3. (i) Calculate $D\mathbf{v}/Dt$ for the steady two-dimensional circular flow $\mathbf{v} = f(r)\hat{\boldsymbol{\theta}}$. Does your result fit in with particle dynamics?

(ii) Water flows along a pipe whose area of cross-section $A(x)$ varies slowly with the coordinate x along the pipe. Use conservation of mass to calculate the mean velocity along the pipe at x, and calculate the acceleration of a particle moving with this mean velocity. [Take the mean velocity in the pipe to be along the pipe and depending only on x.]

4. The velocity profile in a narrow two-dimensional jet of incompressible fluid is given to be

$$u = U\beta x^{-1/3} \operatorname{sech}^2(\alpha yx^{-2/3}), \quad x \neq 0,$$

where α, β, U are constants. Find a stream function for this flow, such that $\psi = 0$ when $y = 0$. Calculate the velocity $v(x, y)\mathbf{j}$ in this flow, where

$$\mathbf{v} = u\mathbf{i} + v\mathbf{j}.$$

5. The two-dimensional velocity field $\mathbf{v}$ has

$$\mathbf{v} = \nabla \times (\psi\mathbf{k}).$$

Calculate $\nabla \times \mathbf{v}$, $\nabla^2\mathbf{v}$ and $\mathbf{v} \cdot \nabla\mathbf{v}$ in terms of ψ.

6. Sketch streamlines for the stream function

$$\psi = r^{1/2} \sin \tfrac{1}{2}\theta.$$

What flow might be modelled by this?

7. Show that $\psi = r^{-1}a^2V(t) \sin \theta$ is a suitable stream function for a cylinder moving at $V(t)$ in a liquid at rest at large distances, provided r and θ are suitably measured.

8. The stream function

$$\psi = rV(\theta \cos \theta + \tfrac{1}{2}\pi\theta \sin \theta - \sin \theta)$$

represents a flow for $0 < \theta < \tfrac{1}{2}\pi$. Calculate velocity components on $\theta = 0$, $\tfrac{1}{2}\pi$, and hence describe what this flow might model. Calculate $\nabla^2\psi$ and show that $\nabla^4\psi$ (i.e. $\nabla^2\nabla^2\psi$) is zero.

9. Show, by using Stokes' theorem, that the circulation for $\psi = -C \ln(r/a)$ is the same for any simple curve once round the origin. What is the result if the curve does not enclose the origin, or goes twice round it?

10. Consider the streamlines

$$Ur \sin \theta - Ua^2 \sin \theta/r = \pm nUa$$

for the model of flow round a cylinder. Calculate the separation of adjacent pairs of these streamlines at $\theta = \pm\tfrac{1}{2}\pi$ and as $x \to \pm\infty$. Verify that 'velocity times spacing' is approximately constant from your calculations.

11. A source is held at distance c from a fixed wall; use an image method to find where the fluid has largest speed along the wall
    (i) in two dimensions,
    (ii) in three dimensions.

12. What image system would you expect to need for the following?

    (i) A two-dimensional dipole pointing at angle α to a rigid wall.

    (ii) A line vortex outside a cylinder.

    (iii) A line vortex inside a cylinder.

    (iv) A source outside a sphere – you will almost certainly be wrong about this one: look up images in spheres in a larger text.

13. Verify the integral form for Ψ, and show that it represents a flux.

14. A source and a sink of equal and opposite strengths $\pm m$ lie at $r = 0$, $z = \mp a$ in a cylindrical system and a uniform stream U flows parallel to the z-axis at infinity. Write down the stream function Ψ and find the equation for the dividing streamline. Find equations for the length and width of the Rankine body so formed and examine the limit as $a \to 0$ with ma held constant.

15. Let
$$\Psi_1 = -Ar^2(a^2 - r^2 - z^2) \quad \text{for } r^2 + z^2 < a^2$$
and
$$\Psi_2 = \tfrac{1}{2}Ur^2 - \tfrac{1}{2}Ua^3r^2(r^2 + z^2)^{-3/2} \quad \text{for } r^2 + z^2 > a^2.$$

Sketch stream surfaces for both stream functions. Find the value of A that gives a continuous velocity (vector) for all r – put it into spherical polars for this part.

16. In §5(*b*) above the stream function for a line vortex and its image in a wall is given. Use polar coordinates r, θ based at O to express ψ in terms of r, θ, b. Now find the approximate form of ψ when $r \gg b$ by using suitable Taylor series.

    Show that the same result can be obtained by differentiating the stream function for a line vortex.

References

(*a*) A number of film loops give useful illustrations of some of the flows discussed in this chapter. Loops 39.5117 and 39.5118 on Hele–Shaw analogues of potential flows show sources, sinks, dipoles and streams; they also show images in a wall, and Rankine bodies constructed from sources and sinks. Loop 28.5074 on the sink-vortex shows a combination of a sink and a vortex, having a smooth transition between two regions, with $v_\theta \propto 1/r$ away from the centre and $v_\theta \propto r$ near the centre (approximately). Loop 38.5048 on tornadoes covers rather similar ground, with some spectacular shots of real flows. Loop 28.5052 on instabilities in circular Couette flow shows that the simple flows of this chapter are in fact unsuitable as models at higher speeds. (These loops are numbered in the Rank numbering. In the NCFML numbering they are FM-80, 81, 70, 26, 31.)

(*b*) *Theoretical Hydrodynamics*, L. M. Milne-Thomson, Macmillan 1949, contains a large amount of material on stream functions and images in Chapters 4 and 15.

V
Vorticity

---

1. Analysis of the motion near a point

The two easiest derivatives of a vector field $\mathbf{v}$ which have tensor character are the divergence $\nabla \cdot \mathbf{v}$ and the curl $\nabla \times \mathbf{v}$. If both of these have given values, say

$$\nabla \cdot \mathbf{v} = f(\mathbf{r}),$$

$$\nabla \times \mathbf{v} = \mathbf{g}(\mathbf{r}),$$

then we have three differential equations for the three components of $\mathbf{v}$ (not four, because the components of $\mathbf{g}$ are not independent, needing to satisfy $\nabla \cdot \mathbf{g} = 0$ everywhere). It is plausible that these three equations will be enough to determine $\mathbf{v}$, if boundary conditions are also given in a suitable fashion. And indeed there is a theorem – Helmholtz' theorem – which says essentially this; this theorem may be looked up in books on vector analysis, it need not delay us further here, as all we want is the idea contained in it, that the vorticity $\nabla \times \mathbf{v}$ is an important part of a fluid motion.

(a) Physical meaning of vorticity

The easiest flow with vorticity is the two-dimensional shear flow of Chapter IV §4(a),

$$\begin{cases} u = \beta y, \ v = 0, \\ \psi = \tfrac{1}{2}\beta y^2. \end{cases}$$

For this flow $\nabla \times \mathbf{v} = -\beta\mathbf{k}$.

Fig. V.1. 'Vorticity rollers' for the profile $u = \beta y$.

Fig. V.2. Distortion of a cross marked in the flow $\mathbf{v} = \beta y \mathbf{i}$ after a time δt.

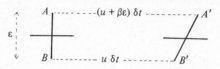

We may look at this in two quite easy ways. Firstly, adjacent sheets of fluid slide over each other, and 'need rollers' to do this – the rotation rate of these rollers is an angular velocity along the direction of the vorticity. Such rollers are shown in fig. V.1. Secondly let us ask how much a small cross marked in the fluid will rotate in time δt. Take one arm along the y-axis and of length ε, the other along the x-axis. The velocity at B in fig. V.2 is taken to be $u = \beta y_B$, and the velocity at A is

$$\beta y_A = \beta(y_B + \varepsilon) = u + \beta\varepsilon.$$

Hence in time δt, B moves to B' as shown while A moves an extra distance

$$\beta\varepsilon\delta t.$$

Thus AB has rotated through an angle

$$\beta\delta t.$$

Naturally the other marked line does not rotate, so the average angular velocity of the two lines is $\frac{1}{2}\beta$.

This kind of rotation can be observed by floating a crossed pair of vanes (sketched in fig. V.3) in a water channel: the rotation of the cross detects the local vertical component of vorticity.

Fig. V.3. A crude vorticity meter.

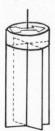

(b) *Local analysis of motion*

We now go on to a thorough analysis of the motion of a short line in a velocity field $\mathbf{v}(\mathbf{r}, t)$. Let one end be at $\mathbf{x}$ initially, the other at $\mathbf{x} + \boldsymbol{\xi}$, and calculate where they get to in time δt as the velocity field moves them on. To first order, $\mathbf{x}$ will be moved by the velocity field at $\mathbf{x}$, t and arrive at

$$\mathbf{x} + \mathbf{v}(\mathbf{x}, t)\delta t.$$

Similarly the end $\mathbf{x} + \boldsymbol{\xi}$ will be moved by the velocity field at $\mathbf{x} + \boldsymbol{\xi}$, t to

$$\mathbf{x} + \boldsymbol{\xi} + \mathbf{v}(\mathbf{x} + \boldsymbol{\xi}, t)\delta t.$$

Expand this by Taylor's theorem to get

$$\mathbf{x} + \boldsymbol{\xi} + \mathbf{v}(\mathbf{x}, t)\delta t + \boldsymbol{\xi}\cdot\nabla\mathbf{v}\delta t + O(\xi^2).$$

Both ends of the line element have been carried along a distance $\mathbf{v}(\mathbf{x}, t)\delta t$, which is a rigid translation of the whole element. But there is a relative motion of the two ends of amount

$$\boldsymbol{\xi}\cdot\nabla\mathbf{v}\delta t \quad \text{or} \quad \xi_j \partial v_i/\partial x_j \delta t,$$

which needs further analysis.

(c) *The antisymmetric tensor and* $\nabla \times \mathbf{v}$

Any tensor A_{ij} can be expressed in terms of a symmetric tensor (which stays the same on interchange of suffixes) and an antisymmetric tensor (which changes sign on interchange of suffixes), by means of the identity

$$A_{ij} = \tfrac{1}{2}(A_{ij} + A_{ji}) + \tfrac{1}{2}(A_{ij} - A_{ji}).$$

$$\text{symmetric} \quad \text{antisymmetric}$$

Apply this to the tensor $\partial v_i/\partial x_j$:

$$\frac{\partial v_i}{\partial x_j} = \frac{1}{2}\left(\frac{\partial v_i}{\partial x_j} + \frac{\partial v_j}{\partial x_i}\right) + \frac{1}{2}\left(\frac{\partial v_i}{\partial x_j} - \frac{\partial v_j}{\partial x_i}\right)$$

$$= e_{ij} + r_{ij},$$

say. Analyse the antisymmetric part r_{ij} first: such a tensor must have the array

$$\begin{pmatrix} 0 & -r_{21} & r_{13} \\ r_{21} & 0 & -r_{32} \\ -r_{13} & r_{32} & 0 \end{pmatrix}$$

in order to be antisymmetric, and we may write this as

$$\begin{pmatrix} 0 & -R_3 & R_2 \\ R_3 & 0 & -R_1 \\ -R_2 & R_1 & 0 \end{pmatrix}.$$

Now since we have seen that the relative motion is given by

$$\xi_j \delta t \partial v_i / \partial x_j$$
$$= \xi_j \delta t (e_{ij} + r_{ij}),$$

this antisymmetric part of the tensor transforms ξ_i to $\xi_i + \xi_j r_{ij} \delta t$ or

$$\xi_i + \tfrac{1}{2} \xi_j (\partial v_i / \partial x_j - \partial v_j / \partial x_i) \delta t,$$

i.e.

$$\boldsymbol{\xi} \mapsto \boldsymbol{\xi} + (R_2 \xi_3 - R_3 \xi_2, R_3 \xi_1 - R_1 \xi_3, R_1 \xi_2 - R_2 \xi_1) \delta t,$$

which is just

$$\boldsymbol{\xi} \mapsto \boldsymbol{\xi} + (\mathbf{R} \times \boldsymbol{\xi}) \delta t.$$

In other words, this part of the motion is an angular velocity $\mathbf{R}$, a rigid rotation of the line element $\boldsymbol{\xi}$. Moreover, if you look at the components of $\mathbf{R}$, you see that

$$\begin{cases} R_1 = r_{32} = \dfrac{1}{2} \left(\dfrac{\partial v_3}{\partial x_2} - \dfrac{\partial v_2}{\partial x_3} \right) = \tfrac{1}{2} (\nabla \times \mathbf{v})_1, \\[3mm] R_2 = r_{13} = \dfrac{1}{2} \left(\dfrac{\partial v_1}{\partial x_3} - \dfrac{\partial v_3}{\partial x_1} \right) = \tfrac{1}{2} (\nabla \times \mathbf{v})_2, \\[3mm] R_3 = r_{21} = \dfrac{1}{2} \left(\dfrac{\partial v_2}{\partial x_1} - \dfrac{\partial v_1}{\partial x_2} \right) = \tfrac{1}{2} (\nabla \times \mathbf{v})_3. \end{cases}$$

Hence the angular velocity of the line element is in general $\tfrac{1}{2} \nabla \times \mathbf{v}$, which agrees with the value found for the rotation rate of the cross in (a) above.

(d) The symmetric tensor and $\nabla \cdot \mathbf{v}$

We are left with the symmetric part of the tensor to analyse, e_{ij}. We have already taken care of translation and rotation by means of $\mathbf{v}$ and $\tfrac{1}{2} \nabla \times \mathbf{v}$, so all there is left ought to be distortion.

Any symmetric tensor e_{ij} has a system of perpendicular axes with respect to which it has diagonal form, so we use these axes to discuss e_{ij}, as no particular axes have been specified earlier. Let

$$e_{ij} = \begin{pmatrix} e_1 & 0 & 0 \\ 0 & e_2 & 0 \\ 0 & 0 & e_3 \end{pmatrix},$$

say, in these axes. Now e_{ij} transforms ξ_i by

$$\xi_i \mapsto \xi_i + e_{ij} \xi_j \delta t$$

or

$$\begin{cases} \xi_1 \mapsto \xi_1 + e_1\xi_1\delta t, \\ \xi_2 \mapsto \xi_2 + e_2\xi_2\delta t, \\ \xi_3 \mapsto \xi_3 + e_3\xi_3\delta t. \end{cases}$$

These are three stretches by factors

$$(1 + e_1\delta t), (1 + e_2\delta t), (1 + e_3\delta t)$$

along the three axes. Naturally some of the e_i may be negative and correspond to compressions.

It is interesting to note that a little block of fluid, of sides ξ_1, ξ_2, ξ_3, has its volume $\xi_1\xi_2\xi_3$ changed to

$$(1 + e_1\delta t)(1 + e_2\delta t)(1 + e_3\delta t)\xi_1\xi_2\xi_3$$

or

$$\{1 + (e_1 + e_2 + e_3)\delta t\}\xi_1\xi_2\xi_3,$$

neglecting terms in δt^2. If you go back through the definitions, you find that

$$e_1 + e_2 + e_3 = e_{11} + e_{22} + e_{33},$$

because this diagonal sum is invariant (by a tensor or matrix theorem) under changes of axes, and so

$$e_1 + e_2 + e_3 = \partial v_1/\partial x_1 + \partial v_2/\partial x_2 + \partial v_3/\partial x_3 = \nabla \cdot \mathbf{v}.$$

Hence the rate of change of volume locally is $\nabla \cdot \mathbf{v}$, which agrees with $\nabla \cdot \mathbf{v} = 0$ meaning incompressibility. Naturally, $\nabla \cdot \mathbf{v} = 0$ requires some positive e_i and some negative (except in the trivial case when all are zero), in other words extension in some directions and compression in others.

(e) Examples

The 'rate of strain' tensor e_{ij} is in this chapter only for completeness. But it is important later, so we now calculate it for two simple flows.

Take the simple two-dimensional shear flow of (a)

$$\begin{cases} u = \beta y, v = 0, \\ \nabla \times \mathbf{v} = -\beta\mathbf{k}. \end{cases}$$

The tensor we need is $\partial v_i/\partial x_j$, which in this case has array

$$\begin{pmatrix} 0 & \beta & 0 \\ 0 & 0 & 0 \\ 0 & 0 & 0 \end{pmatrix}.$$

Hence e_{ij} has array

$$\begin{pmatrix} 0 & \frac{1}{2}\beta & 0 \\ \frac{1}{2}\beta & 0 & 0 \\ 0 & 0 & 0 \end{pmatrix}$$

and r_{ij} has array

$$\begin{pmatrix} 0 & \frac{1}{2}\beta & 0 \\ -\frac{1}{2}\beta & 0 & 0 \\ 0 & 0 & 0 \end{pmatrix}.$$

The angular velocity is

$$\mathbf{R} = (0, 0, -\tfrac{1}{2}\beta)$$

from (c), i.e. the angular velocity of the fluid is (everywhere) $-\frac{1}{2}\beta\mathbf{k}$, or half of the vorticity.

The tensor e_{ij} has eigenvalues (principal rates of strain) given by

$$\text{determinant } (e_{ij} - e\delta_{ij}) = 0$$

or

$$\begin{vmatrix} -e & \frac{1}{2}\beta & 0 \\ \frac{1}{2}\beta & -e & 0 \\ 0 & 0 & -e \end{vmatrix} = 0$$

with solutions $e = \frac{1}{2}\beta, -\frac{1}{2}\beta, 0$.

The corresponding normalised eigenvectors (principal axes) are

$$\begin{aligned} e &= \tfrac{1}{2}\beta &&: 2^{-1/2}(1, 1, 0) && \text{or} && 2^{-1/2}(\mathbf{i} + \mathbf{j}), \\ e &= -\tfrac{1}{2}\beta &&: 2^{-1/2}(-1, +1, 0) && \text{or} && 2^{-1/2}(-\mathbf{i} + \mathbf{j}), \\ e &= 0 &&: (0, 0, 1) && \text{or} && \mathbf{k}. \end{aligned}$$

With these directions as new axes, the rate of strain tensor has the array

$$\begin{pmatrix} \frac{1}{2}\beta & 0 & 0 \\ 0 & -\frac{1}{2}\beta & 0 \\ 0 & 0 & 0 \end{pmatrix}.$$

Thus the distortion of fluid elements is represented by

$$\begin{cases} \text{an extension at rate } \tfrac{1}{2}\beta \text{ along } \mathbf{i} + \mathbf{j}, \\ \text{a compression at rate } -\tfrac{1}{2}\beta \text{ along } -\mathbf{i} + \mathbf{j}, \\ \text{no overall change of volume.} \end{cases}$$

For example, a small circular patch of fluid will be distorted into an ellipse whose longer axis is along $\mathbf{i} + \mathbf{j}$, with the (small) ellipse becoming narrower as time goes on.

As a second example, consider flow along a pipe. For modest speeds and narrow cylindrical pipes the velocity of flow straight along the pipes is (see later for the derivation)

$$u = v = 0, w = b(a^2 - x^2 - y^2)$$

for constant b and pipe radius a. In this case

$$\nabla \times \mathbf{v} = (-2by, 2bx, 0),$$

and e_{ij} has array

$$\begin{pmatrix} 0 & 0 & -bx \\ 0 & 0 & -by \\ -bx & -by & 0 \end{pmatrix}.$$

It is not hard to calculate the principal rates of strain and the principal axes. The main point here is that e_{ij} (and so the principal rates and axes) depends on the position in the flow, and so there is no *single* transformation to principal axes which is valid for the whole flow.

2. Simple model flows

(a) Vorticity and stream functions

In the last chapter we derived many flows from stream functions ψ or Ψ. It is useful to ask how the vorticity ω is related to these stream functions.

Let $\mathbf{v} = \nabla \times (\psi \mathbf{k})$ be a two-dimensional flow. Then

$$\omega = \nabla \times \mathbf{v}$$
$$= \nabla \times \{\nabla \times (\psi \mathbf{k})\}$$
$$= (\nabla \nabla \cdot - \nabla^2)(\psi \mathbf{k})$$

from a formula of vector analysis. Now

$$\nabla \cdot (\psi \mathbf{k}) = \partial/\partial z \, \psi(x, y)$$
$$= 0$$

and $\nabla^2(\psi \mathbf{k}) = (\nabla^2 \psi)\mathbf{k}$ because $\mathbf{k}$ is constant, and so

$$\omega = -\nabla^2 \psi \, \mathbf{k}.$$

Let

$$\mathbf{v} = \left(-\frac{1}{r} \frac{\partial \Psi}{\partial z}, 0, \frac{1}{r} \frac{\partial \Psi}{\partial r} \right)$$

be an axisymmetric flow given in cylindrical polars. Then $\omega = \nabla \times \mathbf{v}$ is given by

$$\omega = \frac{1}{r} \begin{vmatrix} \hat{\mathbf{r}} & r\hat{\boldsymbol{\theta}} & \mathbf{k} \\ \partial/\partial r & \partial/\partial\theta & \partial/\partial z \\ -r^{-1}\partial\Psi/\partial z & 0 & r^{-1}\partial\Psi/\partial r \end{vmatrix}$$

$$= \left(0, \; -\frac{1}{r}\frac{\partial^2\Psi}{\partial z^2} - \frac{\partial}{\partial r}\left(\frac{1}{r}\frac{\partial\Psi}{\partial r}\right), \; 0 \right).$$

This is *not* $-r^{-1}(\nabla^2\Psi)\hat{\boldsymbol{\theta}}$, unfortunately. It is sometimes given the symbol

$$-r^{-1}D^2\Psi\hat{\boldsymbol{\theta}},$$

where D^2 is the operator

$$D^2 = \partial^2/\partial z^2 + r\partial/\partial r(r^{-1}\partial/\partial r).$$

Notice that in this case also the vorticity is in the third perpendicular direction, the $\hat{\boldsymbol{\theta}}$ direction, when the velocities and their variations are to do with r and z.

(b) Examples

Most of the two-dimensional flows in Chapter IV have $\omega = 0$, and we shall see the reason for this later; of those that have a non-zero vorticity, we have already examined one, the simple shear flow $u = \beta y$. The other shear flows

$$\mathbf{v} = u(y)\mathbf{i}$$

all have vorticity in the $\mathbf{k}$ direction

$$\omega = \nabla \times \mathbf{v} = -u'(y)\mathbf{k}.$$

This vorticity is generally not constant. Take for example the transition layer of Chapter IV §4(a)

$$u(y) = U_0 + \tfrac{1}{2}(U_1 - U_0)\{1 + \tanh(y/a)\}.$$

Here we have

$$\omega = -\tfrac{1}{2}a^{-1}(U_1 - U_0)\operatorname{sech}^2(y/a)\,\mathbf{k},$$

and the vorticity is all concentrated into the region near $y = 0$, because $\operatorname{sech}^2 \to 0$ rapidly away from the origin. A rather useful model is derived from this flow by letting $a \to 0$; the transition layer then has zero thickness, and all the vorticity is concentrated along the plane $y = 0$. For example, in the flow of the wind over a wall (sketched in fig. V.4), there is a region of

Fig. V.4. Flow of wind over a wall, leaving a region of almost calm air below a vortex sheet at $y = 0$.

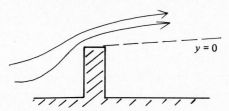

nearly still air behind the wall, and a sharp transition to rapid movement near the level of the wall, taken as $y = 0$. The transition layer is then called a 'vortex sheet'. There is a finite total amount of vorticity concentrated into effectively zero height difference. Such a vortex sheet can be described mathematically by the 'Dirac delta function' or 'generalised function'

$$\omega = -(U_1 - U_0)\delta(y)\,\mathbf{k}.$$

The velocity uses the 'Heaviside step function' $H(y)$ which has value 1 for $y > 0$ and zero for $y < 0$,

$$u(y) = U_0 + (U_1 - U_0)H(y).$$

The remaining two-dimensional flow from Chapter IV is the one in circles round the z-axis, as in fig. V.5,

$$v_r = 0,\ v_\theta = f(r).$$

Here

$$\omega = \nabla \times \mathbf{v}$$

$$= \frac{1}{r}\frac{d}{dr}\{rf(r)\}\mathbf{k}.$$

In the two most realistic cases we have

(i) Rigid rotation,

$$v_\theta = Dr,$$

and so

$$\omega = 2D\mathbf{k},$$

Fig. V.5. Streamlines of a flow in circles.

which gives us back the result that the angular velocity $D\mathbf{k}$ is half the vorticity.

(ii) The line vortex,

$$v_\theta = C/r.$$

In this case $\omega = 0$, because $rf(r) = C$, and so a line vortex appears to have no vorticity! This apparent contradiction is resolved below by taking a closer look at what must happen near $r = 0$, where the model $v_\theta = C/r$ is clearly not very sensible.

The axisymmetric flows of Chapter IV are also largely 'irrotational' (i.e. have zero vorticity). The only two with non-zero vorticity are the non-uniform stream $U(r)\mathbf{k}$, an example of which we have discussed briefly in §1(e) above; and the model for slow flow round a sphere in Chapter IV §7(c). If we calculate ω from

$$\Psi = \tfrac{1}{4}Ua^2(2r^2/a^2 - 3r/a + a/r)\sin^2\theta$$

we get (being careful because D^2 is in terms of cylindrical coordinates and Ψ is in terms of spherical coordinates)

$$- (3Ua\sin\theta/2r^2)\hat{\lambda},$$

which is a rather simple form. It is good practice at vector analysis to show that $\nabla^2\omega = 0$ in this example, which is the model from which this stream function is derived.

3. Models for vortices

(a) Two-dimensional vortex: Rankine's vortex

A more realistic model of the line vortex (due to Rankine) can be found by smoothing out the singularity that the elementary model has at $r = 0$. So take as new model

$$\begin{cases} \omega = \Omega\mathbf{k} & \text{for } r < a, \text{ where } \Omega \text{ is a constant,} \\ \omega = 0 & \text{for } r > a. \end{cases}$$

This preserves the outer flow – which is observed quite easily to be irrotational – while giving a rigid rotation at the centre, which has no singularity, and which is also observable.

We solve

$$\begin{cases} \nabla^2\psi = -\Omega, r < a, \\ \nabla^2\psi = 0, r > a, \end{cases}$$

and impose various boundary conditions later. Taking $\psi = \psi(r)$ (because we need no θ variation) leads to $\nabla^2\psi = r^{-1}d/dr(r\psi')$ and

$$\begin{cases} \psi = -\tfrac{1}{4}\Omega r^2 + A \ln r + B, r < a, \\ \psi = C \ln r + D, r > a. \end{cases}$$

Choose to have $\psi = 0$ on $r = a$, and reject the singularity at $r = 0$ by choosing $A = 0$, to get

$$\begin{cases} \psi = \tfrac{1}{4}\Omega(a^2 - r^2), r < a, \\ \psi = C \ln(r/a), r > a. \end{cases}$$

If we choose also to match v_θ at $r = a$, which seems very reasonable, we end up with

$$\begin{cases} \psi = \tfrac{1}{4}\Omega(a^2 - r^2), r < a, \\ \psi = -\tfrac{1}{2}\Omega a^2 \ln(r/a), r > a. \end{cases}$$

This model is now reasonably satisfactory. The outer flow is the irrotational flow

$$v_\theta = C/r$$

where the value of C is related to the inner flow, and the inner flow is smooth. We will find later that this model predicts the shape of the water surface above an isolated vortex (e.g. from an oar in a river, or near the plug hole in a bath) quite well. But it is still unsatisfactory in that the vorticity (an observable quantity) is assumed to be discontinuous at $r = a$. A top class model would allow for smooth variation of the vorticity, and this will be within our reach later in the course.

The relation of Rankine's vortex to the line vortex singularity that was put forward as a model in Chapter IV §4(d) is quite simple. We just have to let

$$a \to 0 \quad \text{and} \quad \Omega \to \infty$$

in Rankine's vortex in such a way that $\tfrac{1}{2}\Omega a^2 = C$ (a constant). The inner region shrinks to nothing, and in the outer region

$$\psi = -C \ln(r/a)$$

as in Chapter IV §4(d). So a line vortex is a very thin Rankine vortex.

(b) Axisymmetric vortex: Hill's vortex

An axisymmetric region of vorticity within a generally irrotational flow can also be constructed. We use cylindrical polar coordinates and take

$$\omega = -Br\hat{\theta} \quad \text{for} \quad r^2 + z^2 < a^2.$$

Fig. V.6. Sketch of streamlines for Hill's spherical vortex inside the sphere $r^2 + z^2 = a^2$.

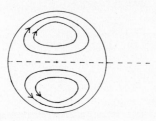

We do not take a constant vorticity because, by symmetry, ω must vanish on the axis, and this is the easiest form for ω that does this.

If you now solve

$$-r^{-1}D^2\Psi = -Br$$

with $\Psi = 0$ when $r^2 + z^2 = a^2$ and no singularity inside this sphere, you get

$$\Psi = -Br^2(a^2 - r^2 - z^2)/10,$$

whose streamlines are sketched in fig. V.6.

Outside the sphere we need a solution of $D^2\Psi = 0$, i.e. one with no vorticity; and if it is to match smoothly to our inner solution, then it will need the same dependence round the sphere, which is (spherical coordinates) $\sin^2 \theta$. We already have *two* basic solutions in spherical coordinates which have no vorticity and $\sin^2 \theta$ dependence; these are:

(i) a uniform stream, $\Psi = \frac{1}{2}Ur^2 \sin^2 \theta$;

(ii) a dipole, $\Psi = A \sin^2 \theta/r$.

We have already seen a combination of them which has $\Psi = 0$ on the sphere, in Chapter IV §7(c),

$$\Psi = \tfrac{1}{2}Ur^2 \sin^2 \theta - \tfrac{1}{2}Ua^3 \sin^2 \theta/r.$$

This then is taken as the stream function for $r > a$, a stream plus a dipole. It only remains to choose B so that the velocities round the sphere match, and this needs (spherical coordinates)

$$(\partial\Psi/\partial r)_{\text{inner}} = (\partial\Psi/\partial r)_{\text{outer}} \quad \text{as} \quad r \to a,$$

which leads to $B = 15U/2a^2$.

This flow, known as Hill's spherical vortex, is less easily set up than the corresponding line vortex flow. It is somewhat similar to a smoke ring flow, except that a smoke ring usually has the region along the axis free of vorticity (and so is more like a piece of a line vortex bent round to join onto itself). It is certainly like the smoke ring in that if it is left by itself with *no* outer stream it will move along, to the left if the vorticity constant B is positive.

The outer dipole flow is interesting; it seems that such a dipole flow can be set up (except near $r = 0$) by giving the fluid near $r = 0$ the general circulatory motion of a Hill's spherical vortex, for example by moving a disc perpendicular to its plane, or by applying a force to the fluid in any other axisymmetric way. A case in which the Hill's type of flow inside a sphere has been observed with reasonable accuracy is that of a spherical drop of glycerine falling through castor oil.

4. Definitions and theorems for vorticity

(a) *Vortex lines, surfaces and tubes*

We have seen in the last section that a spherical vortex of Hill's type, whose vorticity is in loops around the z-axis, will move by itself along the z-axis. This motion of vorticity is not confined to Hill's vortex and the smoke ring type of vortex: if a line vortex is set up near a wall, it and its image will move along parallel to the wall. We will now investigate the motion of vorticity in a fluid; this will require one dynamical theorem, whose proof will follow later.

To start with, we make some definitions.

(i) A 'vortex line' is a curve in the fluid which has its tangent everywhere parallel to ω. This is exactly analogous to a streamline, which was based on $\mathbf{v}$. Do not confuse this with a line vortex, which is a particular two-dimensional singularity. As an example of a vortex line, take any circle round the axis and inside the spherical vortex in §3(*b*).

(ii) A 'vortex surface' is a surface which is composed of all the vortex lines which pass through some curve. Again, this is just like a stream surface. If we took the line $r = \frac{1}{2}a, -\frac{1}{2}a \leqslant z \leqslant \frac{1}{2}a, \theta = 0$ as the given curve, and the vorticity in Hill's vortex, then the vortex surface is (part of) a cylinder of radius $\frac{1}{2}a$ round the z-axis, as shown in fig. V.7.

Fig. V.7. A particular vortex surface in Hill's vortex.

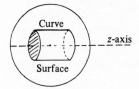

Fig. V.8. A particular vortex tube, which is a torus, in Hill's vortex.

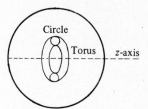

(iii) A 'vortex tube' is the vortex surface formed when the given curve is closed. This is just like a stream tube. Take any circle in the plane $\theta = 0$ and inside Hill's vortex and you will get a torus as the vortex tube (see fig. V.8).

(iv) The 'strength of a vortex tube' is defined to be

$$\int_S \omega \cdot d\mathbf{S} = \int_S \nabla \times \mathbf{v} \cdot d\mathbf{S}$$

$$= \int_C \mathbf{v} \cdot d\mathbf{l},$$

i.e. it is the circulation round C. In this definition S spans the vortex tube, and C is a curve round the intersection of S and the tube. In the Hill's vortex example, choose the closed curve C to be the small circle shown, and take S to be in the plane of fig. V.9. The strength of the tube through this curve is

$$\int_S \omega dS$$

because ω and $d\mathbf{S}$ are both perpendicular to the plane of the diagram, and so the strength is

$$-B \int_S r dS$$

which could be calculated for a particular choice of tube.

Fig. V.9. Definition diagram for §4(a) (iv).

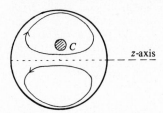

Fig. V.10. Vortex lines forming a vortex tube.

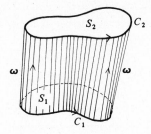

The strength of a tube, as so far defined, seems to depend on the choice of the closed curve C and the surface S. It is easy to prove that it is a sensible definition, independent of C and S. Take a *closed* surface S bounded by a piece of vortex tube with ends S_1 and S_2 spanning curves C_1 and C_2 round the tube, as in fig. V.10. Then

$$\int_S \omega \cdot d\mathbf{S} = \int_V \nabla \cdot \omega \, dV$$

by the divergence theorem, and $\nabla \cdot \omega = \nabla \cdot (\nabla \times \mathbf{v}) = 0$, automatically. Hence

$$\int_S \omega \cdot d\mathbf{S} = 0.$$

Now over a piece of vortex tube, $d\mathbf{S}$ is perpendicular to ω (the lines are everywhere parallel to ω), and so $\int_{\text{sides}} \omega \cdot d\mathbf{S} = 0$. This leaves

$$\int_{S_1} \omega \cdot d\mathbf{S} + \int_{S_2} \omega \cdot d\mathbf{S} = 0$$

and when we make allowance for the direction of the normals – so far they are all *out* of V, and so differently oriented for S_1 and S_2 – we find that the strength is the same for both choices of S and C; hence it is independent of the choice of C and S in the definition.

(b) Kelvin's theorem on circulation

The circulation round a closed circuit

$$\Gamma = \int_C \mathbf{v} \cdot d\mathbf{l}$$

satisfies a most important dynamical theorem called Kelvin's theorem. This is that, under certain approximations which are often adequately closely satisfied,

$$D\Gamma/Dt = 0.$$

In words: if you follow the particles of fluid that make up the circuit C, then the circulation round C is always the same. The proof of the theorem, and the conditions under which it is sufficiently nearly true, will follow later. We take it early so as to get a good idea of how vorticity behaves before we meet the dynamical equations. If you do not build up a little physical and intuitive feeling for vorticity at this stage, the later dynamical equations for vorticity are very hard to understand.

There is an immediate consequence of this. Suppose that the state of strain near C is such that the circuit C is shrinking because there is a stretch perpendicular to C. Now

$$\Gamma = \int_S \omega \cdot d\mathbf{S}$$

is constant, where S spans C; and the only way to achieve this is by an increase in ω. Hence we have the interesting result that vorticity is increased by a stretching motion parallel to the vorticity. You may think of this in terms of angular momentum of the fluid in a vortex; if the vortex becomes thinner, its speed must increase to conserve angular momentum.

We can give an example of these two results based on Hill's vortex. Take a vortex tube based on a small circle C in the plane $\theta = 0$, in the position shown in fig. V.11. As the fluid moves round, so the small circle shrinks until it reaches C'. Because the cross-sectional area of the tube has decreased, the vorticity must have increased to keep the tube's strength constant. And indeed we know that in this flow the vorticity is pro-

Fig. V.11. Definition diagram for §4(*b*).

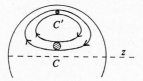

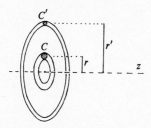

Fig. V.12. Change of area for a vortex tube in Hill's vortex.

portional to distance from the axis. Let us calculate these changes approximately. The vortex tube through C is a torus with cross-sectional area A and radius r, that through C' has area A' and radius r'. These tubes are illustrated in fig. V.12. Since volume is conserved, we have

$$A2\pi r = A'2\pi r'.$$

But ω is proportional to r, and so

$$A\omega = A'\omega',$$

or

$$\Gamma = \Gamma'.$$

Here we see the stretching out of the material in the vortex tube being accompanied by an increase in vorticity so that the circulation round the tube remains constant.

In this calculation we have assumed that a small circle C becomes approximately a small circle C', which is probably not exactly true; we have also used Pappus' formula for the volume of a torus; but we have also assumed without comment that the particles which formed the vortex tube at C will still form a vortex tube when they move round to C'. This is a consequence of Kelvin's theorem, and we proceed to demonstrate it.

Take any vortex surface, and take any circuit C lying entirely in the surface (see fig. V.13). Then

$$\Gamma_C = \int_C \mathbf{v} \cdot d\mathbf{l}$$

$$= \int_S \mathbf{\omega} \cdot d\mathbf{S} = 0$$

because $d\mathbf{S}$ is perpendicular to the surface which C lies in. But by Kelvin's theorem Γ_C remains zero, and so, as it is true for *any* C, C remains in a vortex surface and the whole surface moves together. If the fluid particles which initially made up the vortex surface had, at a later time, a place

Fig. V.13. Vortex lines forming a vortex surface.

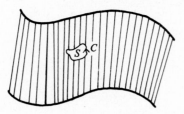

where $\omega \cdot d\mathbf{S}$ was non-zero, we could find a C with $\Gamma_c \neq 0$, contrary to Kelvin's theorem.

In particular, this shows

- (i) that a vortex tube moves with the fluid, as it is a particular vortex surface;
- (ii) that a vortex line moves with the fluid, as it is the intersection of two surfaces;
- (iii) that a line vortex or vortex singularity moves with the fluid, as it is a limiting case of a vortex tube.

Finally, we note that a vortex tube cannot end at any ordinary point in the fluid. The strength is constant along the tube, and if the tube shrank to zero radius we would need to have $\omega \to \infty$; this limit is the line vortex – we cannot go beyond this to the total disappearance of the vorticity, unless there is a singularity or a boundary. For consider an isolated vortex tube that shrinks to a filament and then ends, as in fig. V.14: surround it with a surface S in the fluid and calculate

$$\int_S \omega \cdot d\mathbf{S}.$$

This is equal to

$$\int_V \nabla \cdot \omega dV = \int_V \nabla \cdot (\nabla \times \mathbf{v})dV = 0$$

provided that S and ω are smooth. But we have assumed

$$\int_S \omega \cdot d\mathbf{S} \neq 0,$$

as a single tube enters, and none leaves. This is a contradiction.

This whole discussion has been based on Kelvin's theorem, and in fluid regions where the assumptions of Kelvin's theorem are not true, we can indeed get the strengths of vortex tubes changing with time. We shall return to this matter later (in Chapter X) when the appropriate dynamical equations are ready for use.

Fig. V.14. A vortex tube ending in the fluid.

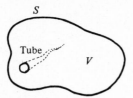

5. Examples of vortex lines and motions

Let us have some simple examples of the motion of vortex tubes and filaments.

(a) Vortices behind aeroplanes

An aeroplane as it flies along sheds vortices from the ends of its wings. These are in fact a necessary part of the dynamics, that enable the aircraft to stay up, and their strength is related to the weight of the aircraft. They trail behind the aircraft as shown in fig. V.15 and the circulations around them cause the air between to move down; each vortex is also caused to travel downwards by the velocity field due to the other. In principle these trailing vortices are permanent, because vortex tubes cannot end except at a boundary or a singularity – so they should stretch from one airport to another. In practice, Kelvin's theorem is not quite true, and they decay slowly. But they are permanent enough to be a menace to any light aircraft that passes through one: the circulation can be quite sufficient to flip a small aeroplane over. These trailing vortices can cause particular problems at airports, where they may upset the landing of the next aircraft.

When water vapour conditions are suitable, these trailing vortices may be seen as condensation trails.

(b) Downwash behind a chimney

The lower layers of the atmosphere in a steady wind have a velocity profile which may be modelled very roughly by

$$u = \beta y,$$

which has vortex lines parallel to the surface and perpendicular to the wind. We consider what happens when a tall, power station chimney interrupts the flow. The vortex lines reach the chimney and cannot pass through it. So they stretch out behind it as they are dragged along by the

Fig. V.15. Trailing vortices behind an aircraft.

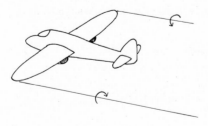

Fig. V.16. Trailing vortices behind a tall chimney. In this diagram and in fig. V.15 each vortex moves with the stream and because of the other vortices.

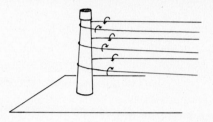

wind, the stretching intensifies them, and they cause a 'downwash' behind the chimney, as sketched in fig. V.16. This can capture some of the emitted smoke, and bring pollution down towards ground level from even quite a tall chimney. It would seem that as more and more vortex lines arrive at the chimney, so the downwash must become more intense. In fact it does not, because Kelvin's theorem is not exact, and the circulations decay, especially near the chimney. So a steady state is actually reached.

(c) A vortex pair in a stream

There are many easy examples of calculation of vortex motion for two-dimensional line vortices. As an example we consider two vortices in a stream, as shown in fig. V.17. The upper vortex has velocity $U - C/2a$ to the right (due to the stream and the effect of the other vortex), the lower has velocity $U - C/2a$ to the right also; so they move as a pair, or remain at rest if $U = C/2a$.

The flow pattern in this example is not hard to calculate as the stream function is

$$\psi = Uy + C \ln\{x^2 + (y-a)^2\}^{1/2} - C \ln\{x^2 + (y+a)^2\}^{1/2},$$

when the origin is taken midway between the vortices (these are *fixed* axes taken at the point which is instantaneously midway between).

The velocity at the origin is

$$U - 2C/a$$

to the right. If this is negative, then there is a region of fluid moving round

Fig. V.17. Definition diagram for §5(c).

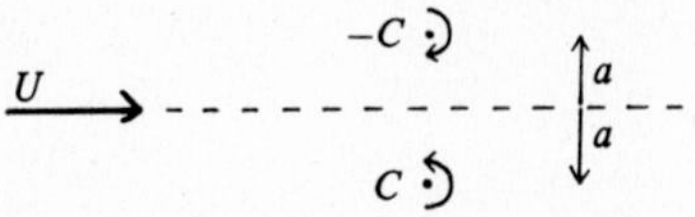

Fig. V.18. Streamlines for two vortices in a slow stream.

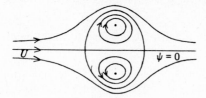

Fig. V.19. Streamlines for two vortices in a fast stream.

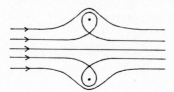

the vortices and staying with them. The dividing streamline $\psi = 0$ has an oval shape as shown in fig. V.18.

On the other hand if $U - 2C/a > 0$, then the stream fluid penetrates between the vortices, which carry round small regions of fluid with each vortex. The stagnation points will now be off the axis, and between the vortices (see fig. V.19). There is naturally a limiting case when $U = 2C/a$, and the stagnation points coincide on the axis.

This piece of theory also applies to the image system of one vortex in a wall, with a stream superimposed. This can often be seen in a river, when a vortex is shed from some obstruction on an otherwise plane wall.

(d) A vortex pair behind a cylinder in a stream

At moderate speeds the flow of a uniform stream past a circular cylinder has two regions of intense vorticity just behind the cylinder, at fixed points. This can be modelled by two line vortices as shown in fig. V.20,

Fig. V.20. Definition diagram for §5(d).

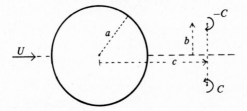

provided that the constants C, c, b can be chosen so that each vortex is at rest under the combined influences of

(i) the stream,

(ii) the dipole representing the disturbance due to the cylinder,

(iii) the other vortex,

(iv) the images of both vortices in the cylinder.

As you can see, this will be a messy calculation to perform, and it is left as an exercise.

The stream function for this model will still be asymptotically $Ur \sin \theta$ as $r \to \infty$. Each vortex stream function is asymptotically smaller than this, and the pair is yet smaller because at large distances they appear to cancel out (they are a 'vortex dipole').

This is another example of the variety of flows which occur round a cylinder in a uniform stream.

Exercises

1. Can you describe the flow

   $$\mathbf{v} = (3z + 4x, \ -5y, \ -2x + z)?$$

   Now calculate the vorticity, the rate of strain tensor and principal rates of strain and axes for this flow, and see if this makes it easier to describe the flow.

2. Poiseuille flow in a pipe has velocity components

   $$u = v = 0, \ w = b(a^2 - x^2 - y^2).$$

   Find the principal rates of strain and axes at any point. Find the vorticity in cylindrical polars and discuss the direction of the vorticity in terms of the slipping of layers of fluid over each other.

3. In Poiseuille flow the rate of strain tensor may be written as

   $$- bx(\mathbf{ik} + \mathbf{ki}) - by(\mathbf{jk} + \mathbf{kj}).$$

   Use the transformation laws from cartesian to cylindrical systems to find the rate of strain tensor in cylindrical polar coordinates.

4. Calculate the vorticity for Chapter IV Exercises 4 and 6.

5. Calculate the vorticity in terms of Ψ using spherical coordinates.

6. It is possible to set up two-dimensional flows like Hill's vortex. The one that follows is easy to calculate, but not real for a reason to do with dynamics, which is considered in Q8 below, i.e. in the exercise numbered 8 in this set of exercises.

   Take $\omega = - Ar \sin \theta$ for $r < a$, and zero outside $r = a$. Then calculate a suitable ψ for $r < a$ that has $\psi = 0$ on $r = a$ and no singularities inside

$r = a$. Finally, match on a suitable exterior flow with the same angular dependence.

7. Flow down a cylindrical pipe with a swirl velocity has a velocity field

$$\mathbf{v} = u(r)\hat{\boldsymbol{\theta}} + w(r)\mathbf{k}.$$

Calculate the vorticity, and describe some vortex lines, surfaces and tubes of this flow.

8. In a two-dimensional flow the vortex tubes are all perpendicular to the flow. If the fluid has constant uniform density, show that the area of cross-section of a tube cannot change, and hence that the vorticity in a tube cannot change. Hence criticise the model in Q6 above.

9. Sketch some streamlines for the limiting case $U = 2C/a$ in §5(c).

10. The corner $0 < \theta < \frac{1}{2}\pi$ is full of fluid and there are rigid walls at $\theta = 0$ and $\theta = \frac{1}{2}\pi$. A line vortex (parallel to the z-axis) passes through the point (x_0, y_0) in the fluid at time $t = 0$. Find an image system and hence the motion of the line vortex.

11. At certain speeds of flow round a circular cylinder there is a regular shedding of vortices, with an opposite circulation being set up round the cylinder. Simplify to a vortex outside a cylinder with no stream and no circulation and calculate the motion of the vortex. Calculate the image system in the cylinder for the vortex when there is an equal and opposite circulation round the cylinder, and hence find the motion of the vortex in this case too.

12. Add the stream into Q11.

13. Try the calculation needed in §5(d).

14. This question uses the technique of 'separation of variables'; if this is unfamiliar, leave this problem until after the technique has been met in Chapter XI.

    When there is a flow past a notch in a wall, a circulatory flow may be set up in the notch. This is modelled as follows. Two rigid planes $\theta = \pm\pi/8$ make an angle of $\frac{1}{4}\pi$, and inside this angle $\omega = $ constant $\times \mathbf{k}$. We find ψ in the corner by solving

$$\nabla^2\psi = -\Omega$$

for $r > 0$ and $-\pi/8 < \theta < \pi/8$. This is done by finding a particular solution depending on r only, and using separation of variables for the rest of the solution. The boundary conditions are $\psi = 0$ on $\theta = \pm\pi/8$, and no singularity at $r = 0$. Finally we reject all the solution except the *least* singular part as $r \to \infty$.

15. In your solution to Q14 there is a place at which the angle of the notch matters. What is the critical angle, for which the solution of Q14 does not apply? [The flow for the critical angle or a larger angle was not calculated until 1979 – it is quite a hard 'easy problem'!]

References

(a) The film loops 28.5042 and 38.5043 on visualisation of vorticity with a vorticity meter are very useful; to some extent they repeat material from the last chapter. (NCFMF numbers 14A and B. There is also a full length film on vorticity, better left till after Chapter X.)

(b) There are excellent photographs of some particular flows with vorticity in *An Introduction to Fluid Dynamics*, G. K. Batchelor, C.U.P. 1967, between pages 352 and 353. These photographs are well worth careful study. This book also has a full discussion of vortices and vorticity: § §5.1–5.4 and § §7.1–7.4 and §7.8 go into far more detail on matters briefly discussed in this chapter and later.

(c) *Theoretical Hydrodynamics*, L. M. Milne-Thomson, Macmillan 1949 has full sections on vortex motion and vorticity, but it is rather short on the reality of the motion. However his §13.7 on the Kármán vortex street is worth looking at, as yet another type of flow due to a circular cylinder in a stream, and an appropriate mathematical model of it.

VI
Hydrostatics

1. Body forces

The fluid modelling we have done so far has been descriptive; in order to construct predictive models we must discuss the forces that act on elements of the fluid, and in this chapter we do this largely in terms of fluid statics for an incompressible fluid, i.e. hydrostatics. This is a subject with important applications, for example in the design of dams and ships; and it is distinctly easier than hydrodynamics.

There is a force $\rho\mathbf{g}$ per volume on every part of a fluid, its weight. This is the main example of a 'body force' which acts on all particles. There are others: a conducting liquid like mercury or molten sodium will undergo a force if it is carrying a current through a magnetic field, and this sort of force is also important in geophysical applications. And in rotating systems for which the coordinate system also rotates there are the 'inertia forces',

Fig. VI.1. Definition sketch for 'centrifugal force' on a volume element dV.

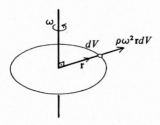

usually called 'Coriolis force' and 'centrifugal force', which arise because a non-inertial frame of reference is being used; the Coriolis force is vitally important in meteorology and the oceanic circulations. We shall mainly confine ourselves to gravity and centrifugal force in this part of the course; that is, to fluids which are at rest relative to either an inertial or a steadily rotating frame of reference. The forces on volume dV are $\rho \mathbf{g}\,dV$ and $\rho\omega^2\mathbf{r}$ dV, where $\mathbf{r}$ is a position vector perpendicular to the axis of rotation (as in fig. VI.1) and ω is the angular velocity of rotation. Both these forces can be derived from potentials per unit mass:

$$\begin{cases} gz \text{ for gravity, } z \text{ vertically upwards,} \\ -\tfrac{1}{2}\omega^2 r^2 \text{ for centrifugal force.} \end{cases}$$

2. The stress tensor

(a) The force across a surface: stress

Take any surface S in the fluid. For molecular reasons the fluid on one side exerts a force on the fluid on the other side of S: this is really a momentum transfer rate between molecules, but by Newton's law this is equivalent to a force and we are not trying to construct a molecular theory. Because the origin of the force is molecular and so local, its magnitude is proportional to the area of surface, so we call the force contribution

$$d\mathbf{F} = \mathbf{\Sigma}dS,$$

where $\mathbf{\Sigma}$ is a force per unit area, i.e. a stress. To get agreement on notation we label the sides of the surface 1 and 2, take the normal to dS *from* side 2 to side 1 to be $\mathbf{n}$, and take $\mathbf{\Sigma}dS$ to be the force *on* side 2 due to side 1. This notation is shown in fig. VI.2. With this convention

$$\mathbf{\Sigma}\cdot\mathbf{n} > 0$$

means that side 1 is pulling on side 2, a tension (this would rarely occur in a fluid); and $\mathbf{\Sigma}\cdot\mathbf{n} < 0$ as in the diagram is a push on side 2, a pressure.

Fig. VI.2. Definition sketch for stress $\mathbf{\Sigma}$ across an element dS.

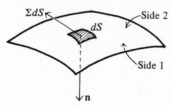

Now by Newton's third law, the force on side 1 due to side 2 must be just

$$- \Sigma dS;$$

in other words, if we take the normal in the opposite direction, the force is changed in sign. This is an example of the general result, that the force on a surface element dS will depend (unless there are special reasons to the contrary) on the orientation of dS. So that we write the force element as

$$d\mathbf{F} = \Sigma(\mathbf{n}, P)dS$$

to show its dependence on orientation $\mathbf{n}$ and position P, and have demonstrated that

$$\Sigma(-\mathbf{n}, P) = -\Sigma(\mathbf{n}, P).$$

(*b*) *How the stress depends on the direction of the surface*

We now need to find exactly how Σ depends on $\mathbf{n}$. To do this we consider a small tetrahedron of fluid, three of whose faces lie on coordinate planes, as in fig. VI.3. We let the areas of these three faces be

$$BPC: \delta A_1,$$
$$CPA: \delta A_2,$$
$$APB: \delta A_3,$$

and let the face ABC have area δA. The unit normals (out of the tetrahedron) to these faces are

$$\begin{cases} BPC: -\mathbf{i}, \\ CPA: -\mathbf{j}, \\ APB: -\mathbf{k}, \\ ABC: \mathbf{n}. \end{cases}$$

Fig. VI.3. An elementary tetrahedron.

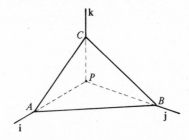

These areas and normals are related by

$$\begin{cases} \delta A_1 = \mathbf{i} \cdot \mathbf{n} \delta A, \\ \delta A_2 = \mathbf{j} \cdot \mathbf{n} \delta A, \\ \delta A_3 = \mathbf{k} \cdot \mathbf{n} \delta A, \end{cases}$$

(these formulae are most easily derived by using vector products).

Now consider the force on the face *BPC from* the material outside, *on* the material inside. This is

$$\int_{BPC} \Sigma(-\mathbf{i}, \mathbf{r}) dS,$$

where $\mathbf{r}$ is a point on the face. Approximate this integral by using Taylor's theorem on the function Σ:

$$\Sigma(-\mathbf{i}, \mathbf{r}) = \Sigma(-\mathbf{i}, P) + O(\delta \mathbf{r} \cdot \nabla \Sigma),$$

where $\delta \mathbf{r}$ is a displacement in this face. Hence the force on this face is

$$\Sigma(-\mathbf{i}, P)\delta A_1 + O(\delta A_1 \delta \mathbf{r} \cdot \nabla \Sigma).$$

Similar calculations show that the forces on the various faces may be approximated by

$$\begin{cases} BPC : \Sigma(-\mathbf{i}, P)\delta A_1, \\ CPA : \Sigma(-\mathbf{j}, P)\delta A_2, \\ APB : \Sigma(-\mathbf{k}, P)\delta A_3, \\ ABC : \Sigma(\mathbf{n}, P)\delta A, \end{cases}$$

with correction terms which are an order of magnitude smaller than δA and which multiply derivatives of Σ.

The equation of motion of this tetrahedron of fluid is

$$\text{density} \times \text{acceleration} \times \delta V = \text{body force} \times \delta V$$
$$+ \Sigma(\mathbf{n}, P)\delta A + \Sigma(-\mathbf{i}, P)\delta A_1 + \Sigma(-\mathbf{j}, P)\delta A_2$$
$$+ \Sigma(-\mathbf{k}, P)\delta A_3 + \text{correction terms.}$$

When we divide by δA and take the limit as all three sides PA, PB, PC tend to zero at the same rate, we get

$$0 = 0 + \Sigma(\mathbf{n}) + \Sigma(-\mathbf{i})\mathbf{n} \cdot \mathbf{i} + \Sigma(-\mathbf{j})\mathbf{n} \cdot \mathbf{j} + \Sigma(-\mathbf{k})\mathbf{n} \cdot \mathbf{k} + 0.$$

Here we have dropped the reference to P, as it is true for any P, and have used the form for each δA_i in terms of δA.

Next we use $\Sigma(-\mathbf{i}) = -\Sigma(\mathbf{i})$ and two similar results to get

$$\Sigma(\mathbf{n}) = \Sigma(\mathbf{i})\mathbf{n} \cdot \mathbf{i} + \Sigma(\mathbf{j})\mathbf{n} \cdot \mathbf{j} + \Sigma(\mathbf{k})\mathbf{n} \cdot \mathbf{k}.$$

This is easier to deal with after a change of notation. Put

$$\begin{cases} \sigma_{rn} = r\text{th component of stress vector } \Sigma(\mathbf{n}), \\ \sigma_{r1} = r\text{th component of stress vector } \Sigma(\mathbf{i}), \\ \sigma_{r2} = r\text{th component of stress vector } \Sigma(\mathbf{j}), \\ \sigma_{r3} = r\text{th component of stress vector } \Sigma(\mathbf{k}). \end{cases}$$

(The quantities σ_{rs} can be interpreted directly; σ_{rs} is the component in the rth direction of the force per unit area on an element of area whose normal is in the sth direction.) Then the result we have derived is

$$\sigma_{rn} = \sigma_{r1}n_1 + \sigma_{r2}n_2 + \sigma_{r3}n_3, \text{ for } r = 1, 2, 3,$$

on using the components of $\mathbf{n}$. This may be put into words as

'the stress for direction $\mathbf{n}$ is obtained by
resolving the stresses for directions $\mathbf{i}, \mathbf{j}, \mathbf{k}$
and adding the results, so that $dF_i = \sigma_{ij}dS_j$.'

What it shows is that σ_{rs} is a second order tensor, because the stress vector $\Sigma(\mathbf{n})$ can always be obtained from the array of nine quantities σ_{rs} by taking a scalar product with the direction vector $\mathbf{n}$: a form of the quotient theorem for tensors then ensures the result.

(c) Stress on a boundary

So far we have discussed forces between two parts of the fluid; there will also be forces between the fluid and any solid in contact with it, such as a container, or an immersed body. Fortunately we do not need another lengthy analysis for this – the results are just the same whether the interaction is between two lots of fluid molecules, or between molecules of fluid on one side and molecules of a solid on the other side. The only detail of the working that has to be noted is that the face ABC of the tetrahedron in (b) is now part of the solid surface.

The result for a solid surface in a fluid is easily stated:

'take normal $\mathbf{n}$ from solid into fluid, and the force
on the surface area dS of the solid is given by
$dF_i = \sigma_{ij}n_j dS = \sigma_{ij}dS_j$.'

The tensor σ_{ij} depends on position in the fluid (and also on the time in general). We set out to find further general properties of this tensor.

3. The form of the stress tensor

(a) Symmetry of the stress tensor

A rather similar calculation to that above can be made for moments of forces. Take some origin O in a small region V of fluid. The surface forces due to the fluid outside V have moment

$$\int_S \mathbf{r} \times d\mathbf{F},$$

where $d\mathbf{F}$ is the surface force on the piece of surface dS. The ith component of this is (writing $d\mathbf{S} = \mathbf{n}dS$ and $dF_k = \sigma_{kl}n_l dS$)

$$\int_S \varepsilon_{ijk}x_j\sigma_{kl}dS_l,$$

if we write $\mathbf{r} = (x_1, x_2, x_3)$. The divergence theorem (extended for use with tensors) can be applied here to give

$$\int_V \partial/\partial x_l \{\varepsilon_{ijk}x_j\sigma_{kl}\}dV$$

and, using $\partial x_j/\partial x_l = \delta_{jl}$, we get

$$\int_V \{\varepsilon_{ijk}\sigma_{kj} + \varepsilon_{ijk}x_j\partial\sigma_{kl}/\partial x_l\}dV.$$

If the average radius of V is α, the first term here has size about that of

$$\sigma\alpha^3,$$

where σ is the size of the elements of the stress tensor (say the largest one); and the second term has size about

$$\sigma\alpha^4/L,$$

where L is the scale on which σ changes appreciably. We will in due course divide by α^3 and let $\alpha \to 0$: the second term then vanishes.

We derive an equation involving

$$\int_S \mathbf{r} \times d\mathbf{F}$$

by taking the moment about O of the forces acting on V and equating it to the rate of change of angular momentum about O. All the terms in this equation are of size α^4, having an $\mathbf{r} \times$ and a dV in them, except for the term

$$\int_V \varepsilon_{ijk}\sigma_{kj}dV$$

which we have derived above – there may be another term of size α^3 in electromagnetic applications when a 'body moment per volume' arises. So on dividing by α^3 and letting $\alpha \to 0$ we must in general be left with

$$\varepsilon_{ijk}\sigma_{kj} = 0.$$

This can only be the case if σ_{kj} is a symmetric tensor, i.e.

$$\sigma_{kj} = \sigma_{jk}.$$

Henceforth, therefore, we assume that the stress tensor σ_{kj} is symmetric.

(b) *Isotropic pressure in a fluid at rest*

Use principal axes for the symmetric tensor σ_{kj}; the tensor then has the array

$$\begin{pmatrix} \sigma_1 & 0 & 0 \\ 0 & \sigma_2 & 0 \\ 0 & 0 & \sigma_3 \end{pmatrix}.$$

We may rewrite this into an isotropic part and a remainder by the simple splitting

$$\begin{pmatrix} \tfrac{1}{3}(\sigma_1 + \sigma_2 + \sigma_3) & 0 & 0 \\ 0 & \tfrac{1}{3}(\sigma_1 + \sigma_2 + \sigma_3) & 0 \\ 0 & 0 & \tfrac{1}{3}(\sigma_1 + \sigma_2 + \sigma_3) \end{pmatrix}$$

$$+ \begin{pmatrix} \sigma_1 - \tfrac{1}{3}(\sigma_1 + \sigma_2 + \sigma_3) & 0 & 0 \\ 0 & \sigma_2 - \tfrac{1}{3}(\sigma_1 + \sigma_2 + \sigma_3) & 0 \\ 0 & 0 & \sigma_3 - \tfrac{1}{3}(\sigma_1 + \sigma_2 + \sigma_3) \end{pmatrix}.$$

Remember that the sum of the diagonal elements of a tensor is an invariant; we may therefore write the stress tensor as

$$\tfrac{1}{3}\sigma_{ii}\delta_{kj} + \begin{pmatrix} \sigma_1 - \tfrac{1}{3}\sigma_{ii} & 0 & 0 \\ 0 & \sigma_2 - \tfrac{1}{3}\sigma_{ii} & 0 \\ 0 & 0 & \sigma_3 - \tfrac{1}{3}\sigma_{ii} \end{pmatrix}.$$

The parts are called the isotropic stress tensor and the 'deviatoric stress' tensor.

When σ_{ii} is negative, the isotropic term is an equal pressure in all directions; this is the usual case, as fluids can withstand pressures easily, but are less good at tolerating tensions – a gas can never be under tension, and a liquid can only withstand a tension under careful experimental conditions, usually just splitting apart under even small tensions.

The deviatoric term must contain both positive and negative elements as its diagonal sum is zero: we can at this stage say little else about the general form of this tensor. However in the *special* case of a fluid at rest we can find the precise form of the deviatoric stress tensor. Consider a small spherical piece of a fluid which is at rest: it can certainly withstand a uniform pressure, such as that supplied by the isotropic part of the stress tensor, perhaps by compressing somewhat. But the deviatoric part is exerting a compressive stress in at least one direction and a tensile stress in at least one direction, and a fluid cannot resist such stresses, but must move appropriately, converting the sphere of fluid to an ellipsoid. Such motion is not taking place in a fluid at rest, so the deviatoric stress tensor *must* be zero.

There are two important conclusions here:

    (i) In a fluid at rest

$$\sigma_{ij} = -p\delta_{ij}$$

    where p is the fluid pressure.

    (ii) In a fluid in motion the deviatoric stress tensor depends on the motion. This conclusion is rather weak, and we return to it later. This part of the stress tensor is concerned with viscosity, and shearing forces; neither are of any concern in a fluid at rest.

4. Hydrostatic pressure and forces

(a) Relation of pressure to body force

As a *first* model, the oceans, the atmosphere, lakes, the Earth's interior, the matter in a star are all static; there is a balance of fluid pressure against the body forces acting. Naturally, the motions in these situations are also of great interest, but we must get the easiest model done first.

Let the body force be $\rho\mathbf{F}$ per volume. Consider a volume V bounded by a surface S: the surface force due to fluid outside S has ith component

$$\int_S \sigma_{ij}dS_j,$$

and the body force is similarly

$$\int_V \rho F_i dV.$$

Use the divergence theorem on the former, and the static equilibrium condition that the total force is zero, to get, with $\sigma_{ij} = -p\delta_{ij}$,

$$\int_V (\rho F_i - \partial p/\partial x_i)dV = 0.$$

This is true for any volume V, and the integrand is assumed to be smooth, and so

$$\rho F_i - \partial p/\partial x_i = 0$$

or

$$\rho\mathbf{F} = \nabla p.$$

Note that ρ is not in general constant in any large scale situation. Even in a lake there is very often thermal stratification: we would expect this stratification to be in horizontal layers, and this is easily proved from the above formula.

Let **F** be derived from a potential Φ,

$$\mathbf{F} = -\nabla\Phi.$$

Then

$$\nabla p = -\rho\nabla\Phi$$

and taking $\nabla \times$ gives

$$0 = \nabla\rho \times \nabla\Phi$$

because $\nabla \times \nabla = 0$. Hence $\nabla\rho$ is parallel to $\nabla\Phi$ (assuming neither is zero) and so the surfaces $\rho = $ constant (perpendicular to the vector $\nabla\rho$) are parallel to the surfaces $\Phi = $ constant. When $\Phi = gz$, the usual gravitational potential (with z measured upwards), we must have that the surfaces $\rho = $ constant are horizontal.

We may also deduce that the surfaces $p = $ constant are horizontal here, because ∇p is parallel to $\nabla\Phi$. In the atmosphere it is usually found that the surfaces of constant p, constant ρ and constant potential Φ are *not* parallel; the differences are associated with the motion of the atmosphere.

The easiest case of the above theory is when we take $\rho = $ constant and $\Phi = gz$, certainly the case in a swimming pool and reasonably true in the sea on a calm day. Then

$$\nabla p = -\rho g\mathbf{k}$$

and so

$$p = p_0 - \rho gz,$$

where p_0 is the (atmospheric) pressure at the surface $z = 0$. Pressure increases at constant rate as you descend and z becomes more negative.

If you assume $\rho = \rho(z)$ in a lake, then it is easily seen that the equilibrium equation has solution

$$p = p_0 - g\int_0^z \rho(\zeta)d\zeta:$$

the pressure below the surface equals the surface pressure plus the weight of water per unit area above the observation point – remember that $z < 0$ below the surface, and so we should perhaps write

$$p = p_0 + g\int_z^0 \rho(\zeta)d\zeta.$$

(b) Uniformly rotating liquid

Another interesting case of hydrostatics is that of fluid rotating steadily like a rigid body; the fluid is at rest when viewed in a frame of reference rotating with the fluid, and we may compensate for this steady rotation by adding in a centrifugal force $\rho\omega^2 \mathbf{r}$ (as above) with a potential $-\frac{1}{2}\omega^2 r^2$.

For example, consider a cylindrical can containing water and rotating about its axis, which is vertical. The water will eventually rotate steadily with the can. We will then have, as equilibrium equation,

$$\nabla p = -\rho\nabla(gz - \tfrac{1}{2}\omega^2 r^2)$$

where cylindrical coordinates are used. This may be written as

$$\begin{cases} \partial p/\partial z = -\rho g \\ \partial p/\partial r = \rho\omega^2 r \\ \partial p/\partial \theta = 0, \end{cases}$$

with solution

$$p = p_0 + \tfrac{1}{2}\rho\omega^2 r^2 - \rho g z.$$

We may take the surface to be $p = p_0$, as atmospheric pressure will change little over the scale of such an experiment. Thus the surface has equation (with origin chosen on the axis at the surface)

$$0 = \tfrac{1}{2}\rho\omega^2 r^2 - \rho g z$$

or

$$gz = \tfrac{1}{2}\omega^2 r^2.$$

Fig. VI.4. The steady water surface in a rotating vertical can is a paraboloid.

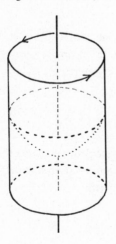

This is a paraboloid of revolution, as shown in fig. VI.4. It may be observed easily in a well-stirred coffee mug, to a fair approximation.

The most interesting rotating system around is the Earth. For it, the relative sizes of g and $\omega^2 r$ are, at 45°N latitude,

$$\begin{cases} g = 10 \text{ m s}^{-2}, \\ \omega^2 r = 0.02 \text{ m s}^{-2} \end{cases}$$

approximately. This centrifugal force is therefore a small effect of rotation. However, a rotating (non-inertial) frame of reference also brings in a 'Coriolis force' when the fluid is not at rest in this frame of reference, and this is more important in connection with the Earth's rotation.

(c) Force due to hydrostatic pressure

Most of the subject of hydrostatics is concerned with the detailed calculation of forces on bodies immersed (wholly or partially) in liquids. This has its place, but it is not really in this course, where we do not have time for many such calculations.

The force on any piece of surface in a fluid at rest is perpendicular to the surface and proportional to its area and to the local pressure. This follows from the hydrostatic stress tensor

$$\sigma_{ij} = -p\delta_{ij},$$

and the relation of force to stress tensor

$$dF_i = \sigma_{ij} dS_j$$
$$= -p dS_i.$$

As an example of this we prove Archimedes' theorem for a fluid of uniform density. The force on V (see fig. VI.5) is

$$\mathbf{F} = -\int_S p d\mathbf{S}$$

$$= -\int_V \nabla p \, dV$$

by the divergence theorem, where p is now the pressure field that would

Fig. VI.5. Definition sketch for Archimedes' theorem.

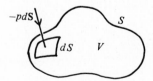

have been there in the absence of the body. But for this field,

$$\nabla p = \rho \mathbf{g},$$

and so

$$\mathbf{F} = -\rho \mathbf{g} \int_V dV$$
$$= -\rho \mathbf{g} V.$$

But $\rho \mathbf{g} V$ is the weight of fluid displaced, so we have the theorem:

'the force on the body is an upthrust equal
to the weight of fluid displaced'.

Should the body intersect the surface of a liquid, then the integration must be done in two parts, but the result is the same.

(d) Examples

A lock gate in a canal (shown roughly in fig. VI.6) is a structure that has to withstand considerable forces from a static fluid. Consider either gate, as shown in fig. VI.7. The force on the deep side is (neglecting

Fig. VI.6. Sketch of lock gates in a canal.

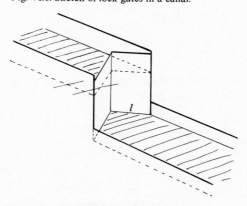

Fig. VI.7. Water levels at the lock gate.

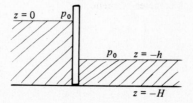

the part above the water on both sides)

$$\int_{-H}^{0} (-\rho g z + p_0) dS,$$

where dS is a horizontal strip of height dz and width l. This force has value

$$p_0 H l + \tfrac{1}{2}\rho g H^2 l.$$

The force on the shallow side is similarly

$$p_0 h l + \int_{-H}^{-h} \{-\rho g(z+h) + p_0\} dS$$
$$= p_0 h l + p_0 (H-h)l + \tfrac{1}{2}\rho g(H-h)^2 l,$$

where the first term comes from air pressure. Thus the net force on the gate due to the water is

$$F = \tfrac{1}{2}\rho g l\{H^2 - (H-h)^2\},$$

in a direction perpendicular to the gate. Notice that p_0 cancels out, as it acts on both sides.

To specify a force system, it is not enough just to give the resultant force, you must also give the resultant moment about some line. So we calculate the moments on each side much as before: on the left side we get (for moments about the base of the gate, and omitting p_0)

$$\int_{-H}^{0} l(-\rho g z)(H+z) dz$$
$$= \tfrac{1}{6}\rho g l H^3 ;$$

and on the right side

$$\int_{-(H-h)}^{0} l(-\rho g z)(H-h+z) dz$$
$$= \tfrac{1}{6}\rho g l(H-h)^3.$$

So the force system on this gate may be specified as a total force

$$\tfrac{1}{2}\rho g l\{H^2 - (H-h)^2\}$$

acting with a total moment

$$\tfrac{1}{6}\rho g l\{H^3 - (H-h)^3\}.$$

This can be achieved by a single force acting on the middle vertical of the gate (by symmetry) at a height

$$\zeta = \tfrac{1}{6}\{H^3 - (H-h)^3\} \div \tfrac{1}{2}\{H^2 - (H-h)^2\}$$

above the gate's base, as shown in fig. VI.8. This point of action of the

Fig. VI.8. The single equivalent force due to water pressure on the lock gate.

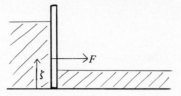

Fig. VI.9. Possible floating positions for a piece of wood.

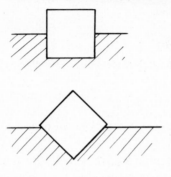

single equivalent force is known as the 'centre of pressure'. In practice, of course, the balance forces are exerted at pivots and where the gates meet: for a large lock gate the total force may easily reach 10^6 N.

The total force and total moment do not tell the whole story. A piece of timber of square section *ought* to be able to float as shown in the first diagram of fig. VI.9, as forces and moments are just in balance with weight. But the wood often prefers to float diagonally (try soft wood, density just less than half that of water). The fluid pressures are again in static equilibrium with the weight. But this time it is a stable equilibrium, whereas the previous position is unstable. In general such a stability problem depends on the shape of the section and also the density ratio.

A boat or a ship must be designed so that, under any reasonable loading, it floats upright in stable equilibrium. Extra requirements will be imposed when the dynamic interaction of water and ship are considered, which provides some very hard mathematical problems.

Exercises

1. A frame of reference which is accelerating (with respect to inertial frames) is used to describe an experiment. If the acceleration has the constant value f in the $\mathbf{i}$ direction, show that an 'inertial force' $-\rho f \mathbf{i}$ acts on a fluid (per

volume), and that a potential fx may be used. Hence find the equilibrium water surface in an accelerated tank of water.

2. Show that $\delta A_1 = \mathbf{n} \cdot \mathbf{i} \delta A$, in the notation of §2.

3. Show that the normal component of the surface force vector is
   $\sigma_{ij} n_j n_i$ (summed over both i and j)
   per area, and find an expression for the tangential force on area dS (i.e. the force parallel to the surface).

4. Prove that $\int_S A_{il} dS_l = \int_V \partial A_{il}/\partial x_l \, dV$, where A_{il} is a tensor function of position.

5. A mercury barometer is fixed vertically in a frame of reference which has acceleration $f\mathbf{i}$. What is the reading on a scale fixed with the tube of the barometer? What would be the reading if the barometer hung from a fixed point in the accelerating frame?

6. A cylindrical can of height a and radius a is filled to $3a/4$ with water, and then set rotating at angular velocity ω about its axis. What is the maximum value of ω for which the water remains in the can? What is the minimum depth of water in the can at this maximum value of ω?

7. The most primitive theory of the tides is that the ocean surface is an equipotential of the potentials
   (i) Earth's gravity,
   (ii) centrifugal force,
   (iii) Moon's gravity,
   (iv) Sun's gravity.
   Estimate the maximum height of a tide on this theory. What is the maximum observed? [Tides are in fact dynamic effects: high tide does not occur as the moon passes overhead.]

8. A gutter is in the form of half a cylinder (whose cross-section is sketched in fig. VI.10), and is full of water. Prove, by integrating surface forces, that the total force on the gutter is equal to the weight of water in the gutter. Calculate the moment, about the lowest level of the gutter, of the surface forces on the half of the gutter on one side of this lowest line. Calculate also the force on one half of the gutter.

9. Water half fills a hemispherical container which has its plane face vertical. Find the single forces that are equivalent to the surface pressures on
   (i) the plane face,
   (ii) the curved face.
   Where are the centres of pressure in the two cases?

Fig. VI.10. A gutter full of water.

Fig. VI.11. A conical container.

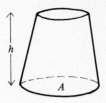

Fig. VI.12. Definition sketch for Q11.

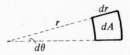

10. A container shown in fig. VI.11 in the shape of part of a cone is full of water. The height of the container is h and the area of the base is A, so that the water pressure on the base is ρgh and the total force on the base is ρghA, which exceeds the weight of water in the container. Resolve, in detail, this apparent paradox.

11. Water is flowing in circles (e.g. in a cylindrical can) so that

$$\mathbf{v} = v(r)\hat{\boldsymbol{\theta}}$$

and

$$p = P(r) - \rho gz.$$

Consider the pressures on the faces of a small volume whose cross-section is shown in fig. VI.12, and equate the resultant force to a mass-acceleration, to show that

$$dP/dr = \rho v^2/r,$$

provided that no other forces act.

References

This chapter has covered very little of the well developed subject of hydrostatics. A brief look over a text would give an idea of the range of problems that can be solved. There are no modern texts just on hydrostatics. Most engineering texts on mechanics of fluids have an early section on hydrostatics. Otherwise:

(a) *Hydrostatics*, A. S. Ramsey, C.U.P. 1947;
(b) *Statics* (3rd edn.), H. Lamb, C.U.P. 1949.

VII

Thermodynamics

1. Basic ideas and equations of state

(*a*) *Functions of state*

In our discussions so far we have introduced five functions to describe the motion of a fluid:

(i) density ρ;

(ii) velocity $\mathbf{v}$, three components;

(iii) pressure p.

So far there are only four equations in sight for these functions, the equation of mass-conservation, and the equation of momentum change (three components). We clearly need something like an energy equation, which must include energy of compression of a gas; and it is well known that compressing a gas heats it (try pumping a bicycle tyre), so that temperature will come in as well, as a related quantity to heat energy. In order to discuss these ideas clearly, we must set up some, but not too much, of the theory of thermodynamics. By the end of the chapter we will have three standard mathematical models to work with, one for liquids and two for gases under reasonable conditions; but we should also have a good idea of when and why these models are adequate – a model that is not understood is a model that will be used in the wrong way.

A new area of physical theory will have new observables, and the mathematical theory will bring in new functions which are not observable and yet which are the best ones for framing the theory. Our early stages in fluid dynamics are helped by the fact that density and velocity are common measurable concepts in other forms of mechanics; and the idea of a stress tensor, which though not directly observable is extremely useful, is not too unlikely a generalisation from a force vector. But thermodynamics brings in ideas which may be quite new to those without much background in physics, so the early stage of thermodynamics needs careful attention, to note where axioms based on experiments are being brought in, and where new definitions are being made.

We start with an example to clarify some ideas. A cyclist in a road race is observed at a certain time to be travelling at a certain speed in the centre of Bristol; later he is observed at the top of Shap Fell in the Lake District. Write these as

$$\begin{cases} \text{Bristol, time } t_1, \text{ position } \mathbf{r}_1, \text{ velocity } \mathbf{v}_1; \\ \text{Shap, time } t_2, \text{ position } \mathbf{r}_2, \text{ velocity } \mathbf{v}_2. \end{cases}$$

In other words 'state 1' 'state 2'.

Now there are some things we know at once when we know his state – we immediately know his kinetic energy $\frac{1}{2}mv^2$ and his gravitational potential energy mgz: these are 'functions of state', given directly by the state of the system. And we know the work done against gravity in the *whole* trip: it is just

$$mg(\text{height of Shap} - \text{height of Bristol}).$$

Similarly the work done to change kinetic energy in the whole trip is known:

$$\tfrac{1}{2}m(\mathbf{v}_2^2 - \mathbf{v}_1^2).$$

But we do *not* know the total work done in the whole trip, as this involves many local effects on the way – wind, passing lorries, lubrication, route taken, rider's posture and so on. The total work done is not a function of the current state of the rider, but depends on the whole detailed history of the journey.

Changes between neighbouring states of a function of state are denoted by a differential:

$$d(mgz) = mgdz,$$
$$d(\tfrac{1}{2}mv^2) = m\mathbf{v}\cdot d\mathbf{v}.$$

But the total work done between neighbouring states is *not* the differential of a function of state, and so some other notation is needed – here we use Δ, so that the work done between neighbouring states is ΔW.

This distinction between functions of state and process dependent quantities is important in thermodynamics, and is not usually emphasised in mechanics, where models involving potential energy are most frequent.

(b) Equilibrium thermodynamics and the equation of state

Elementary thermodynamics is very concerned with slow changes between equilibrium states, and it would seem that fluid dynamics must (except for fluid statics) be concerned with rather faster changes, and non-equilibrium states. But we find that the molecular time scales are *so* short that a 'slow' change can be completed in what seems to us a short time, and effective equilibrium can be maintained all the time.

Consider a gas, whose molecules are at a separation such that on average a molecule travels 10^{-7} m between collisions. The average speed of such a molecule is around 5×10^2 m s^{-1}, so that the time between collisions is about 2×10^{-10} s. Thus five collisions occur for each molecule in about 10^{-9} s, and this is about enough to randomise the motion of any molecule which starts with a special velocity. The time scale on which local equilibrium for quite a large group of molecules is reached would be some 10^{-6} s, which is *far* less than any important time scale in most flows.

The above argument may not be totally convincing. But in practice the use of equilibrium thermodynamics seems to cause no errors.

Thermodynamics, like fluid dynamics, is about matter viewed at larger than the molecular scale. However, it is often very revealing to dip down to the kinetic theory of gases to understand what the meaning of the larger scale phenomena is.

For example, the density ρ of a fluid has been defined in terms of the number of molecules in a small region of space. And the pressure p, which was introduced in the last chapter as the isotropic part of the stress tensor, can be understood in terms of the recoil force on a surface (see fig. VII. 1 for direct reflection) when molecules bombard it – a molecule reflected at speed u requires a momentum change $2mu$, and many such collisions with a wall lead to a pressure on the wall.

It has been found that the 'hotness' of a material can be described by

Fig. VII.1. Impact of a molecule on a wall.

(i)   (ii)

a function of state, the temperature T. This is sometimes called the zeroth law of thermodynamics, and is an axiom derived from observation. The temperature can be interpreted as a direct measure of the average kinetic energy of molecules, which is why it, like density and pressure, is a function of state; the history of molecule only goes back at most 10^{-9} s, because successive collisions 'cause it to forget' its more distant past.

Since we have interpretations of all three of ρ, p and T in terms of just the two quantities density and molecular speed (molecular motion in a gas is locally isotropic, so we do not need three velocity components here), we might suspect that there should be a relation between ρ, p and T. This has in fact been found to be generally true, that the state of a simple fluid (for which no other descriptive parameter is obviously needed) can be specified by just two functions of state, for example, ρ and T. This is another axiom, and as a consequence the third of these physical quantities must be a function of the other two, $p = g(\rho, T)$; or in more general form

$$f(p, \rho, T) = 0.$$

This is known as an 'equation of state'.

Real materials have complex equations of state. Some simple model equations follow.

(i) $p = R\rho T$ for a 'perfect gas', where R is a constant for any particular mass of gas, and is a universal constant if ρ is measured in mol m^{-3} and T is in K (degrees Kelvin).

(ii) $(p + a\rho^2)(\rho^{-1} - b) = RT$, Van der Waal's equation; this is more realistic than (i), but (i) is good enough in this course for any usual gas.

(iii) $\rho = \rho_0$, a constant, is a useful model equation of state for a liquid.

(iv) $p = (p_0 + B)(\rho/\rho_0)^7 - B$ is a better model of the relation between pressure and density in water, where B is a constant with value about 3×10^8 N m^{-2}.

It is good enough in a first course such as this to stick to the easiest equations of state. Henceforth we shall assume that gases are modelled by *either*

$$\rho = \rho_0$$

when compressibility and temperature effects seem unimportant, *or*

$$p = R\rho T;$$

and that liquids are modelled by

$$\rho = \rho_0.$$

It follows that the rest of this chapter is about 'perfect gases'.

2. Energy and entropy

(a) *Heat energy and the first law of thermodynamics*

Energy comes in several rather different looking forms, but long experience has led to the view that in total there is a conservation of energy. The forms of energy we are concerned with here are

(i) kinetic energy (e.g. $\frac{1}{2}mv^2$);

(ii) potential energy (e.g. mgz);

(iii) work done which does not get converted into either of (i) and (ii) such as work done against friction or work done to compress (and heat) a gas;

(iv) heat.

It is assumed that the first two of these are well understood, from courses in dynamics, and we will concentrate on the other two. There are some preliminary things to be said about heat and temperature, both of which are about hotness (the physiological stimulus); it is useful to compare hotness and 'mechanical push':

	quantity	intensity
hotness	heat	temperature
push	kinetic energy	speed

A glacier has far more 'quantity of push' (kinetic energy) than a bullet, but the bullet has a very high intensity (speed). A hot water bottle has more quantity of heat than a red hot needle, but much less intensity (temperature). In thermodynamics when we deal with energy it is the quantity that is needed, though of course the intensity is also an important observable.

Consider a quantity of gas that is being slowly compressed. During this compression work must be done on the gas, against the pressure in the gas. This work causes no change in obvious kinetic energy (push in slowly from all round, and you don't go far) or in gravitational potential energy. But it is well known that the gas is heated in this compression – the molecules travel faster in their random motion. Thus work has been converted into internal kinetic energy (and possibly also into internal potential energy, as the molecules are now closer together).

But the internal kinetic energy could also be increased by adding heat, by conduction or radiation say, at constant volume. These two methods of adding internal energy will in general act together, an input ΔW from work done on the gas, and an input ΔQ from heat flow into the gas. Notice that heat input is not found to be the differential of a function of state, but is process dependent, and hence the Δ.

It is found experimentally that the two effects act by simple addition: total energy input is

$$\Delta W + \Delta Q.$$

Moreover, it is found that this *is* the differential of a function of state, and we define changes in this function of state E by

$$dE = \Delta W + \Delta Q.$$

This is the first law of thermodynamics: internal energy is a function of state, energy is conserved, and changes in E are found by this formula.

(b) *Entropy and the second law of thermodynamics*

The internal energy, being a function of state, can be put in terms of any two of the other similar variables, since the state of the fluid needs only two variables to define it. Thus

$$E = E(T, \rho)$$

is a possibility, and it is a sensible choice because we expect that internal energy depends on molecular kinetic energy (T) and molecular potential energy (related to ρ).

We ask how the component parts of dE, ΔW and ΔQ depend on other state variables. Start with ΔW. If a sphere of gas of radius r and pressure p is compressed slowly to $r - dr$, the work done on the gas is

$$\Delta W = \text{force} \times \text{distance}$$

$$= \int_{\text{sphere}} p d\mathbf{S} \cdot d\mathbf{r}$$

$$= -p dV$$

where dV is the increase in volume (clearly negative, so that $\Delta W > 0$). It is not hard to show that this formula applies to other shapes than a sphere under compression. The slowness of the compression is in theory very important so that the change is reversible, no energy gets lost into other forms; in practice, all fluid dynamic compressions will be slow enough, except when discontinuities in the flow field are predicted.

The volume V of a particular mass of gas, or the volume v of unit mass of gas, are useful variables related to the density ρ: clearly $v = 1/\rho$. In thermodynamics it is normal to use v, e (and so on) when unit mass of material is being considered; but to use V, E for the total for a system. In fluid dynamics it has been usual to stick to capital letters, and we do so.

Note that T, p, ρ do *not* necessarily change when a larger volume is considered.

We have found then that $\Delta W/p$ *is* the differential of a state variable for such changes

$$\Delta W/p = - dV.$$

In other words, $1/p$ is an integrating factor for ΔW so that $\Delta W/p$ is an exact differential. It would seem likely that the other component of energy, ΔQ, would also have an integrating factor; and this is indeed the case for slow (reversible) changes near thermal equilibrium:

$$\Delta Q/T = dS$$

where S is a new function of state called the entropy. For any given fluid S is just some function of T and ρ (say) which happens to have this fortunate property. It is not directly observable, but is mathematically necessary, and can be given a molecular interpretation in terms of the randomness of the system. Note particularly that for fast changes not near equilibrium we do not need to have $\Delta Q/T = dS$; in shock waves, for example, there is no equality. Moreover, the dissipation of mean flow energy by viscosity, which converts organised motion to random molecular motion, is always irreversible. This is because it is unrealistic to expect random molecular motion (heat) to suddenly organise itself into part of the mean flow. However you will see later that there are occasions when it is useful and sensible to neglect this effect of viscosity, and take as a model that $\Delta Q/T = dS$.

The important formula

$$\Delta Q/T = dS \text{ for reversible changes}$$

can be derived mathematically from the second law of thermodynamics. This law is variously stated:

    (i) 'No process is possible whose sole result is the transfer of heat from a colder to a hotter body' (Clausius). This means, for example, that a refrigerator needs an energy source.

    (ii) 'No process is possible whose sole result is the complete conversion of heat into work' (Kelvin).

Perhaps the best informal statement of this negative law is that 'you can't get something for nothing'. The easiest mathematical form of the law is

$$dS \geqslant \Delta Q/T$$

with equality for reversible changes. Naturally, this implies $dS \geqslant 0$ for an *isolated* system, because there is then no heat inflow.

We have now reached the major results of this chapter

$$\begin{cases} dE = TdS - pdV \\ dS \geqslant 0, \end{cases}$$

and we proceed to examine the consequences for the perfect gas model. These results are true in general, and not just for slow changes near equilibrium, because they are about functions of state.

3. The perfect gas model

(a) *The equation of state*

For reasonable gases at reasonable temperatures it has long been known that Boyle's law is a good approximation:

$$pV = f(T).$$

The easiest definition of the mathematical model known as a 'perfect gas' is to take this formula:

a perfect gas satisfies $pV = f(T)$.

From this and the results of §2 above it may be proved (though this will not be done here) that

$$E = E(T) \quad \text{and} \quad pV = RT.$$

Thus it appears that in a perfect gas there is no contribution to internal energy from potential energy between molecules: the molecules must be thought of as very well separated.

We would like to know how E depends on T for a perfect gas, and also how S depends on T and V (say). For this purpose we need to define 'specific heats'. In some general process, the specific heat c is the ratio of heat added per mass to small temperature change:

$$c = \Delta Q/dT.$$

Two particular specific heats come into the theory conveniently, when the process is

    (i) at constant pressure

$$c_p = \Delta Q/dT, \; p \text{ held constant};$$

    (ii) at constant volume

$$c_v = \Delta Q/dt, \; V \text{ held constant}.$$

(b) *The internal energy*

Let us use the formulae of §2 on these specific heats.

    (i) $c_p = (dE + pdV)/dT$

$$= \left(\frac{\partial E}{\partial T} \right)_p + p \left(\frac{\partial V}{\partial T} \right)_p$$

because here we must be implying

$$E = E(T, p), \quad V = V(T, p)$$

by the denominator dT and the assumption $p = \text{constant}$.

(ii) $c_v = (\partial E/\partial T)_V$

because V is constant and so $dV = 0$, with (this time)

$$E = E(T, V).$$

But we know that for a perfect gas

$$E = E(T).$$

Hence

$$c_v = dE/dT$$

and

$$c_p = dE/dT + p(\partial V/\partial T)_p.$$

But we also have

$$pV = RT$$

for a perfect gas, and so

$$c_p = c_v + R.$$

(iii) Since $c_v = dE/dT$, and $E = E(T)$, we must have $c_v(T)$. Hence

$$E = \int_{T_0}^{T} c_v(\theta)d\theta + E(T_0)$$

and this shows how the internal energy depends on temperature, provided that we know the behaviour of the specific heat c_v.

(c) The entropy for a thermally perfect gas

It is found experimentally that c_v is almost constant at usual temperatures. This may be explained in molecular terms by saying that no new degrees of freedom for the energy of a molecule are excited at these temperatures, so that energy is directly proportional to temperature. It is clear that the model

$$c_v = \text{constant}$$

will be convenient at usual temperatures, so we call a perfect gas with $c_v = \text{constant}$ a 'thermally perfect gas' or a 'calorically perfect gas'. This is an *extra* piece of modelling, which is again good at reasonable temperatures for many gases.

For such a gas we must have

$$E = c_v(T - T_0) + E(T_0)$$

where $c_v = \text{constant}$ applies above the temperature T_0.

Let us calculate the entropy S for such a thermally perfect gas. We have that

$$dE = TdS - pdV$$

and so

$$dS = (dE + pdV)/T$$
$$= c_v dT/T + pdV/T.$$

But $p/T = R/V$ and so

$$dS = c_v dT/T + RdV/V.$$

Hence

$$S = c_v \ln(T/T_0) + R \ln(V/V_0) + S_0,$$

where T_0, V_0, S_0 is some reference state.

We may rewrite this form for S as

$$S - S_0 = \ln(kT^{c_v}V^R)$$

where k is a constant. Now if we rearrange this formula by using

$$\begin{cases} pV = RT \\ R = c_p - c_v, \end{cases}$$

we derive

$$S - S_0 = c_v \ln(KpV^{c_p/c_v})$$

where K is another constant.

It is convenient to define

$$\gamma = c_p/c_v$$

as the (constant in this model) ratio of specific heats of the gas and write the formula for the entropy as

$$S - S_0 = c_v \ln(KpV^\gamma).$$

The value of γ is found to be almost exactly 1.4 for diatomic molecules (such as oxygen and nitrogen, the major constituents of air). This is again related to the molecular structure of air; the specific heat c_v is related to the number n of degrees of freedom for molecular energy excited at that temperature, by

$$c_v = \tfrac{1}{2}nR$$

and so

$$c_p = (\tfrac{1}{2}n + 1)R.$$

For a diatomic molecule at normal temperatures $n = 5$, corresponding to translation in three directions and spin about two axes perpendicular to the line joining the two atoms of the molecule. Hence

$$c_v = 5nR/2, \quad c_p = 7nR/2, \qquad \gamma = 7/5 = 1.4.$$

(d) Adiabatic and isentropic models

In many gas motions there are only small temperature variations, and the time scale over which heat diffuses by molecular motions is considerably longer than the time scale of the fluid motion. Hence it is often a good model to assume there is no heat flow, and to take

$$\Delta Q = 0.$$

This is known as an 'adiabatic' process, and it corresponds to a neglect of the molecular effect of heat conduction.

If you also neglect any irreversibility, i.e. assume that it is small and use a model in which it is absent, you get

$$0 = \Delta Q = T\,dS,$$

and so $dS = 0$. In this case the flow is 'isentropic', or has constant entropy.

You do not need to have $\Delta Q = 0$ and $dS = 0$ together. You will find later that when a gas passes through a 'shock wave', there is no heat addition from outside (and so $\Delta Q = 0$), but the change is not reversible (and so $dS > 0$). In this case the flow is adiabatic, but *not* isentropic.

The model

$$dS = 0 = \Delta Q$$

is, however, useful in large parts of fluid dynamics. If we adopt this model, then it follows from (c) above that

$$pV^\gamma = \text{constant or } p = k\rho^\gamma$$

since $S = S_0$.

This model for gas dynamics is, of course, not always applicable. Heat does flow, and sometimes the heat flow is appreciable, in which case a full equation for E or S must be set up by using the physical laws for heat flow. But when this model is adequate we can see a complete set of equations for gas dynamics: five equations, of which $p = k\rho^\gamma$ is the last, for the five unknowns p, ρ, $\mathbf{v}$ (which has three components).

There are other gas motions in which heat is rapidly communicated throughout the region of gas, or in which the motion is so slow that diffusion of heat can take place fully, and in these motions we may be able to use the simple model

$$T = \text{constant}.$$

This is the 'isothermal' model. It too provides a final equation, $pV =$ constant, to close the set of fluid dynamic equations.

Finally we should note that the liquid model

$$\rho = \text{constant}$$

is closed, with four equations for the four unknowns p, $\mathbf{v}$.

4. The atmosphere

We have completed the little thermodynamics that we need for this course in fluid dynamics. We show it in action by considering the equilibrium of the atmosphere.

Model the atmosphere by a perfect gas; it isn't quite, of course, the water vapour is often too near condensation, for example. Assume that we have static equilibrium – again this is moderately true, good enough as a first try. Thus we have the two equations

$$\begin{cases} p = R\rho T, \\ dp/dz = -\rho g, \end{cases}$$

with z measured vertically upwards from sea level and p, ρ, T depending only on z.

We need some thermodynamic statement to close this system of equations.

(i) $T = $ constant, the isothermal atmosphere. This is on average fairly true for large regions of the atmosphere between 10 km and 20 km up. With this approximation we can substitute

$$p = RT\rho$$

into the static equilibrium equation to get

$$RT d\rho/dz = -\rho g$$

with solution

$$\rho = \rho_0 \exp(-gz/RT),$$

where ρ_0 is chosen to fit the observed density at some point in this layer. The length scale RT/g is called the 'scale height' h and has a value of about 6.65 km for $T = 227$ K (a suitable value at this height).

(ii) In the lower atmosphere there is considerable mixing. This is partly caused by solar heating of the ground, which leads to vertical convection currents. The mixing leaves the atmosphere in a state of approximate 'neutral stability', which means that if a chunk of air is moved up or down *without* mixing (and so adiabatically), it arrives at its new position with just the right density to remain there.

Suppose it is at pressure p and density ρ and moves rapidly up dz. The new pressure is

$$p - \rho g dz$$

because $dp/dz = -\rho g$. Now in this move we have $p = k\rho^{\gamma}$, as there is no mixing, no time for conduction of heat and no irreversibility; so the new density is $\rho + d\rho$, or

$$\rho + (d\rho/dp)dp$$
$$= \rho - \rho g dz/k\gamma\rho^{\gamma-1}.$$

For neutral stability this must be just

$$\rho + (d\rho/dz)dz,$$

the density at this new level, and so

$$d\rho/dz = -(g/k\gamma)\rho^{2-\gamma}.$$

This can be solved to give

$$\rho = \rho_0\{1 - z/H\}^{1/(\gamma-1)}$$

where ρ_0 is the density at ground level and

$$H = k\gamma\rho_0^{\gamma-1}/\{g(\gamma-1)\}$$
$$= RT_0\gamma/\{g(\gamma-1)\},$$

the scale height for this atmosphere. Notice that this model has $\rho = 0$ at $z = H$; hence it cannot apply as far as this. It is easily shown that $H = \gamma h/(\gamma-1) \doteq 29.5$ km, using $T_0 = 288$ K which is a reasonable average value.

(iii) For the real atmosphere, $T(z)$ could be found by firing a rocket, or from satellite observations. From this known (and far from smooth) function, $p(z)$ and $\rho(z)$ can be calculated, by using the given equations:

$$\begin{cases} \rho T = p/R \\ T dp/dz = -pTg = -pg/R. \end{cases}$$

Hence

$$p = p_0 \exp\left\{-\int_0^z [g/RT(\zeta)]d\zeta\right\}.$$

A smoothed 'standard atmosphere' is available in tables. The models of (i) and (ii) above have been used to fit this, using the values·

$$\rho = 0.311 \text{ kg m}^{-3} \text{ at 12 km},$$
$$T = 227 \text{ K at 12 km},$$
$$T = 288 \text{ K at 0 km},$$
$$R = 2.87 \times 10^2 \text{ m}^2 \text{ s}^{-2} \text{ K}^{-1}.$$

The models used are:
  (i) $z \geqslant 10$ km,

$$\rho = 1.890 \exp(-z/6.65),$$

  where z is in km, and ρ is in kg m^{-3},
  (ii) $z \leqslant 10$ km,

$$\rho = 1.182(1 - z/29.5)^{2.5}.$$

The models have matching density at 10 km. The match to the standard atmosphere is good enough, as you can see in fig. VII.2, except at low levels where the average stability is probably not neutral. Notice we have used values at 12 km, which is well into the region where $T \doteq$ constant, but that the two models apply to $z \geqslant 10$ km and $z \leqslant 10$ km. Around 10 km neither will be accurate.

You must not read too much into this modelling exercise. The real atmosphere is very complicated and variable and will never be given by these average values.

Fig. VII.2. Comparison of a standard atmosphere (taken from Batchelor's text) with the model in §4. The standard atmosphere is a dashed line, the model a solid line.

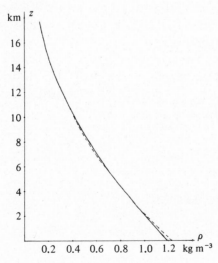

Exercises

1. What kinetic energy is equivalent to the heat energy put into 1 kg of water when raising it at constant pressure from 0°C to 100°C (i.e. 273 K to 373 K)? What rise in temperature would you expect to find at the bottom of a 50 m waterfall?

2. Express E in terms of p, ρ; and S in terms of p, T for a thermally perfect gas. For an isentropic process express T in terms of p.

3. Define F to be $E - TS$ (the Helmholtz free energy) as a new function of state (for any material, not just a perfect gas). Show that

$(\partial F/\partial T)_V = -S$

$(\partial F/\partial V)_T = -p$

and so that

$(\partial S/\partial V)_T = (\partial p/\partial T)_V.$

4. Show from Q3 that

$p^{-1}(\partial p/\partial T)_V = T^{-1}\{1 + p^{-1}(\partial E/\partial V)_T\}.$

This is the pressure coefficient at constant volume, say β. Let α be the expansion coefficient $V^{-1}(\partial V/\partial T)_p$. Show that for a perfect gas which has $pV = f(T)$,

$\alpha = \beta = T^{-1}.$

[Q3 and Q4 prepare for the proof that $E = E(T)$ and $pV = RT.$]

5. A closed tin of height h contains air at atmospheric pressure. It floats vertically in water as shown in fig. VII.3, with a depth k under water. Rust causes a small hole in the base and water leaks in slowly. Show that the tin floats in equilibrium again when the water has risen to fill a height

$hk/(H + k),$

where H is the height of a water barometer.

6. An atmospheric balloon is such that if it were inflated to radius a it would burst. It is inflated to radius $b(<a)$ with a gas lighter than air and is released from ground level in an isothermal atmosphere. Prove that if the balloon rises *slowly* until it bursts, its maximum height is

$3H \ln(a/b)$

Fig. VII.3. Definition sketch for Q5.

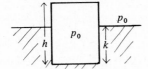

where the difference between the pressures inside and outside the balloon has been neglected and H is the scale height of the atmosphere.

If the balloon ascends rapidly, what is the height reached before bursting?

7. What temperature distribution is needed to produce an atmosphere of constant density? Calculate the temperature at 1 km in this atmosphere when the temperature at ground level is 288 K. Is this atmosphere stable?

8. In a jet engine, air is compressed rapidly to 1/25th of its original volume, and is then used to burn fuel. Find the temperature of the air just before the fuel is burnt
   (i) at sea level;
   (ii) at 12 km.
   Neglect the relative motion of air and aeroplane.

References

This has been a short treatment of a complicated subject, certainly not full enough for a physicist or an engineer. We can only plead that there is not room and not time in an elementary course on fluid dynamics, and offer some further reading.

(a) *A Guide to Thermodynamics*, J. A. Ramsay, Chapman and Hall 1971. This text is an exploration by a physiologist of why thermodynamics is the way it is, and it is an excellent supplement to more formal treatments. Little mathematics is used, and the explanations are very lucid.

(b) *An Introduction to Thermodynamics*, R. S. Silver, C.U.P. 1971. A text based on the ideas of work and energy, with an especial interest in flow in pipes. It is a more mathematical work, really written for engineers, but the first two chapters and parts of the fourth form a good background for elementary fluid dynamics.

(c) *Equilibrium Thermodynamics* (2nd edn), C. J. Adkins, McGraw-Hill 1975. A physicist's text with plenty of background explanation. The first five chapters and parts of the eighth are sufficient for this course.

VIII

The equation of motion

1. The fundamental form

We use the same method to derive the equation for momentum as we used for the continuity (or mass) equation. The rate of increase of momentum inside any volume V, as sketched in fig. VIII.1, is equal to the sum of the

 (i) rate of momentum inflow through S,
 (ii) total forces acting inside V,
 (iii) total forces acting on S.

Now flow rates through S are all controlled by $\mathbf{v} \cdot d\mathbf{S}$, a volume per time, which is proportional to local area and to local outward velocity; since the ith component of momentum per volume is just ρv_i, we must have a first term

(i) $-\displaystyle\int_S \rho v_i \mathbf{v} \cdot d\mathbf{S},$

where the minus converts it to an inflow.

Fig. VIII.1. Definition sketch for deriving the equation of motion.

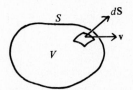

The second term arises from the body force, and is therefore

(ii) $\displaystyle\int_V \rho F_i dV,$

where F_i is the (*i*th component of) the body force per mass.

The interaction across a fluid surface is given by the stress tensor, and since we want the force on V due to fluid outside, and have the normal direction taken as *out* of V, this is

(iii) $\displaystyle\int_S \sigma_{ij} dS_j.$

Thus the equation of motion for the fluid inside S is

$$d/dt \int_V \rho v_i dV = -\int_S \rho v_i \mathbf{v}\cdot d\mathbf{S} + \int_V \rho F_i dV + \int_S \sigma_{ij} dS_j.$$

As before we put all terms as volume integrals, by using the (generalised) divergence theorem, and noting that

$$\rho v_i \mathbf{v}\cdot d\mathbf{S} = \rho v_i v_j dS_j.$$

This leads us to

$$\int_V \left\{ \frac{\partial}{\partial t}(\rho v_i) + \frac{\partial}{\partial x_j}(\rho v_i v_j) - \rho F_i - \frac{\partial}{\partial x_j}\sigma_{ij} \right\} dV = 0,$$

where we have used the fact that V is time independent in the first term. Finally, because V is arbitrary and the integrand is assumed to be smooth, we have

$$\partial/\partial t(\rho v_i) + \partial/\partial x_j(\rho v_i v_j) = \rho F_i + (\partial/\partial x_j)\sigma_{ij}.$$

This is the equation of motion. We may rearrange it by using the mass equation

$$\partial\rho/\partial t + \partial/\partial x_j(\rho v_j) = 0$$

into the form

$$\rho\partial v_i/\partial t + \rho v_j \partial v_i/\partial x_j = \rho F_i + \partial\sigma_{ij}/\partial x_j.$$

Note the occurrence of the term $\rho\mathbf{v}\cdot\nabla\mathbf{v}$, and that the left side is just

$$\rho D\mathbf{v}/Dt,$$

i.e. density times acceleration following the particle.

2. Stress and rate of strain

(a) Isotropic and deviatoric parts of the stress tensor

This equation of motion is of no help at all until we have a form for σ_{ij}. What we have done so far with σ_{ij} is to split it into an isotropic and a deviatoric part

$$\sigma_{ij} = \tfrac{1}{3}\sigma_{kk}\delta_{ij} + (\sigma_{ij} - \tfrac{1}{3}\sigma_{kk}\delta_{ij}).$$

In the case of fluid *statics* we found that the second part had to vanish, and the first part was just

$$- p\delta_{ij},$$

where p is the pressure at any point, the same pressure as comes into equilibrium thermodynamics. It is sensible to keep the same splitting into isotropic and non-isotropic when there is motion, but it is no longer necessary for the isotropic term to be the thermodynamic pressure, as there is not necessarily an exact equilibrium. Let us write

$$\sigma_{ij} = - P\delta_{ij} + s_{ij},$$

where $-P$ is still the mean normal stress $\tfrac{1}{3}\sigma_{kk}$ and so a vital dynamical quantity, whose relation to the thermodynamic pressure p must be investigated. This definition implies that $s_{kk} = 0$, because $\delta_{kk} = 3$, and so

$$\sigma_{kk} = - 3P + s_{kk}.$$

So s_{ij} has zero isotropic part, and we now try to relate the deviatoric term s_{ij} to the local motion.

(b) Physical background

Detailed derivations of the form of s_{ij} belong in more advanced texts. We proceed along plausible lines, without pretending to a full investigation.

We have seen in §1 above that σ_{ij} is about momentum transfers across a surface; the origin of these transfers must be at a molecular level, so we outline a calculation of transfers of x-momentum across a surface $y =$ constant.

Some of the molecules above S in fig. VIII.2 will be moving down and will cross S. They will, on average, have x-momentum proportional to $U + u$; that is, greater than that where they now find themselves. They will lose this extra x-momentum by collisions, and so there is a downward transfer of x-momentum across S. That is, on the larger view, there appears to be a force on the lower region of fluid in the x-direction. There is also

Fig. VIII.2. Molecular transfer of momentum across a surface.

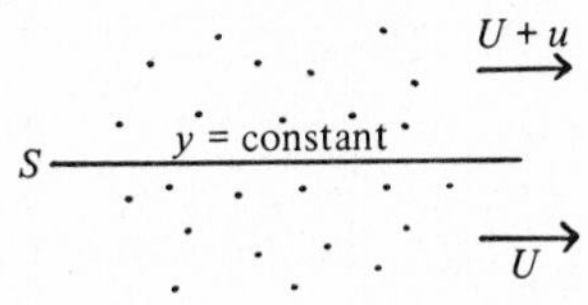

an upward transfer of deficiency of momentum across S, which slows down the molecules above S. The result is a shearing *stress* across S which speeds up fluid below S and slows down fluid above S. The stress must be proportional to u, the velocity difference, and to various parameters of the motion of the molecules. Now in a real flow we do not have a discontinuity of velocity across S, and we must consider what to take for u. Molecules will interchange across a distance comparable to the mean free path l, and in that distance the velocity changes by

$$ldU/dy$$

to a very good approximation, because l is very small compared with the length scale on which U changes. So we expect the shearing stress to be

$$\mu dU/dy,$$

where μ, the viscosity, depends on the parameters of the molecular motion.

(c) *Constitutive relation for a Newtonian fluid*

The above argument, which can be refined for a gas, is a good basis for suggesting that the shear stresses that we need to include in s_{ij} should be proportional to local (both in space and time) values of velocity gradients, at least for fluids of simple molecular constitution and very short time scales of response. Earlier, in Chapter V, we analysed the local motion of a fluid into

(i) local translation;

(ii) local rigid rotation;

(iii) local deformation.

It is obvious that (i) will not have anything to do with local shear stresses; it is plausible that local rigid rotations are not a cause of shear stresses (proofs are available in larger texts); and this leaves the rate of strain tensor e_{ij} as the source of the stresses s_{ij}.

Naturally, we want a tensor relation between e_{ij} and s_{ij}, so that choice of axes is irrelevant. We have already decided to try a linear relation. And for simple fluids we expect to have an isotropic relation, as no preferred directions exist in the molecular motion. This forces the relation to be

$$s_{ij} = A_{ijkl}e_{kl},$$

with

$$A_{ijkl} = \mu\delta_{ik}\delta_{jl} + \mu_1\delta_{il}\delta_{jk} + \mu_2\delta_{ij}\delta_{kl}$$

for isotropy. Hence

$$s_{ij} = \mu e_{ij} + \mu_1 e_{ji} + \mu_2 e_{kk}\delta_{ij}.$$

But we already know that s_{ij} and e_{ij} are symmetric tensors: this gives $\mu = \mu_1$. We have also defined s_{ij} so that $s_{ii} = 0$, and hence $\mu_2 = -2\mu/3$. Finally we have the 'constitutive relation'

$$s_{ij} = 2\mu(e_{ij} - \tfrac{1}{3}e_{kk}\delta_{ij})$$

for a 'Newtonian fluid'.

This is by no means the only possible constitutive relation; the arguments above for linearity and local dependence are good for usual gases, but do not apply to complicated fluids, and the study of non-Newtonian fluids (rheology) is a well-advanced branch of applied mathematics.

(d) Thermodynamic and mechanical pressure

We note that s_{ij} depends only on the deviatoric rate of strain tensor $e_{ij} - \tfrac{1}{3}e_{kk}\delta_{ij}$. This makes us suspect that $\nabla \cdot \mathbf{v}$ or e_{kk} must come in elsewhere. And since $\nabla \cdot \mathbf{v}$ is to do with changes in volume, it might well come into the term $-P\delta_{ij}$, because p and dV have been associated in thermodynamics.

If we argue that

$$P - p$$

must also be linear in e_{ij}, then the only possible relation is

$$P - p = -Ke_{kk},$$

where K is a constant, because e_{kk} is the only linear invariant of e_{ij}. This brief argument can be justified by a full discussion of energy processes in a fluid, but not here.

We have now a full form for σ_{ij}, in terms of two 'constants' μ and K (they will vary with temperature, and perhaps with density):

$$\sigma_{ij} = -(p - Ke_{kk})\delta_{ij} + 2\mu(e_{ij} - \tfrac{1}{3}e_{kk}\delta_{ij}),$$

where p is the thermodynamic pressure and

$$P = p - Ke_{kk}$$

is the mechanical pressure, or mean normal stress.

3. The Navier–Stokes equation

We may now substitute for σ_{ij} in the equation of motion. Remembering that

$$e_{ij} = \tfrac{1}{2}(\partial v_i/\partial x_j + \partial v_j/\partial x_i),$$

$$e_{kk} = \partial v_k/\partial x_k = \nabla \cdot \mathbf{v},$$

we obtain

$$\rho\frac{Dv_i}{Dt} = \rho F_i - \frac{\partial p}{\partial x_i} + \frac{\partial}{\partial x_j}\left\{\mu\frac{\partial v_i}{\partial x_j} + \mu\frac{\partial v_j}{\partial x_i}\right\} + \frac{\partial}{\partial x_i}\left\{(K - \tfrac{2}{3}\mu)\frac{\partial v_k}{\partial x_k}\right\}.$$

This is known as the full Navier–Stokes equation.

In many situations it is fair to neglect the small variations of μ and K with position (due to temperature changes mainly) and rewrite the equation in vector form as

$$\rho Dv/Dt = \rho F - \nabla p + \mu\nabla^2 v + (K + \tfrac{1}{3}\mu)\nabla\nabla\cdot v.$$

The Navier–Stokes equation is commonly written in terms of the mean normal stress P instead of the thermodynamic pressure p, and of course there is no difference when $\nabla\cdot v = 0$; often the difference is small anyway. When we use P, and assume μ to be constant, we have

$$\rho Dv/Dt = \rho F - \nabla P + \mu\nabla^2 v + \tfrac{1}{3}\mu\nabla\nabla\cdot v.$$

You will notice at once that it contains the hydrostatic equation

$$\rho F = \nabla p,$$

just by putting $v = 0$. There are two other obvious reductions of the equation.

(i) When $\nabla\cdot v = 0$, which is appropriate for a liquid and can often be a good model for low speed gas flows, we derive the approximate and widely used form

$$\rho Dv/Dt = \rho F - \nabla p + \mu\nabla^2 v.$$

(ii) Many useful solutions can be found by taking the 'inviscid' model $\mu = K = 0$. The justification for this will emerge during the next two chapters, but it is certainly often true that μ seems to be 'small'. Neglect of the viscous terms leads to

$$\rho Dv/Dt = \rho F - \nabla p,$$

which is known as Euler's equation, also widely used.

We shall spend some time solving, and attempting to solve, the equations in (i) and (ii). We shall not even attempt to solve the full equation of motion – though some of our solutions for (i) and (ii) will in fact be solutions of the full equation.

The constant μ has dimensions $ML^{-1}T^{-1}$ and its values for some common fluids are, at 288 K:

Air	Water	Mercury	Olive oil	Glycerine
1.8×10^{-5}	1.1×10^{-3}	1.6×10^{-3}	0.10	2.33

all measured in kg m^{-1} s^{-1}. These figures are not very meaningful, except for a comparison between fluids. As we discussed earlier, we should consider non-dimensional groups if we want to assess the significance of the viscosity μ. For example, if a stream of speed U passes a body of size a, then

$$\rho U a / \mu$$

is dimensionless, and must be some sort of overall test of the size of μ. This is a 'Reynolds number'. If we take air flowing at 1 m s^{-1} past a cylinder of diameter 10^{-2} m, then the Reynolds number R has value

$$6.7 \times 10^2$$

and so the viscosity appears to be a small parameter which might be neglected. In fact we shall see that experience is needed to tell when the viscosity may be neglected, even when the apparent Reynolds number is quite large.

The constant K has the same dimensions as μ, and has not been measured accurately for many fluids. This is because it is connected with $\nabla \cdot \mathbf{v}$, and this is always small. Measurements in sound waves (which depend on compression of air, i.e. non-zero $\nabla \cdot \mathbf{v}$) give K a value near that for μ.

The Navier–Stokes equation in form (i) above is often divided by the density ρ to give

$$D\mathbf{v}/Dt = \mathbf{F} - \rho^{-1}\nabla p + \nu\nabla^2\mathbf{v}$$

by defining the 'kinematic viscosity' ν as μ/ρ. This minor change makes the equation into a more convenient shape, especially when ρ is constant. The Reynolds number in terms of ν is typically Ua/ν, and values of ν are (at 288 K):

Air	Water	Mercury	Olive oil	Glycerine
1.5×10^{-5}	1.1×10^{-6}	1.2×10^{-7}	1.1×10^{-4}	1.8×10^{-3}

all in m^2 s^{-1}. In this view of viscosity, as an ability to change velocity (rather than momentum), air is much more viscous than water, and mercury really has very little viscosity compared with other fluids.

4. Discussion of the Navier–Stokes equation

(a) Boundary conditions

Differential equations need boundary conditions. The Navier–Stokes equation can come in so many physical situations that all we will do here is give some of the common boundary conditions.

(i) At a rigid boundary the fluid velocity and the boundary velocity match exactly. That is, the normal components are the same, because fluid does not penetrate into the solid or leave a gap (usually). But also the *tangential* velocity component in fluid and solid is identical: this is called the 'no-slip condition'. It can be visualised in terms of molecules of fluid and of solid being somewhat entangled at the boundary; but this is rather too simple a view to be quite real, and such a stopping of the fluid on a molecular scale need not show up in the continuum model, whose scales are much larger. However, all predictions using the no-slip condition have turned out to be accurate, and it is universally accepted as the appropriate boundary condition.

(ii) Rigid boundaries are only one model of real boundaries. Flexible boundaries occur widely; for example a ship's plates deform in a heavy sea, and our arteries expand and contract as our hearts beat. For flexible boundaries there must be equality of velocities and also the fluid stress on the surface must match the stress in the material of the boundary. A particular simple example of this is in the expansion or contraction of an effectively empty bubble in a liquid: there are no stresses inside the bubble, and so the fluid at the boundary can have no stresses.

(iii) Asymptotic boundary conditions occur. There may be no motion at large distances, or a uniform stream there. Equally we may ask for 'no singularity' as $r \to 0$, as we did for the Rankine vortex in Chapter V.

(iv) Pressure boundary conditions are also common. At the sea surface we would wish to take the pressure as everywhere constant – this is a reasonable model as atmospheric pressures change little compared with the pressures involved in the waves on the surface. And flow along pipes is usually discussed in terms of the pressure difference between the two ends.

(b) Gravity and dynamic pressure

The major body force in most flows is gravity; this is sometimes the main cause of a flow (as in the flow of rivers, or the rise of cumulus clouds), but usually most of the gravitational body force is balanced by a hydrostatic pressure field. It is often useful to subtract out these two influences before starting to discuss the motion.

Start with a fluid of *constant* density ρ. Take the pressure field when there is *no* motion to be p_1, and the pressure when there is motion to be

$p_1 + p_2$. Then we have

$$\rho\mathbf{g} = \nabla p_1,$$

assuming that the body force is gravity; and also

$$\rho D\mathbf{v}/Dt = \rho\mathbf{g} - \nabla(p_1 + p_2) + \mu\nabla^2\mathbf{v}.$$

The two equations together give

$$\rho D\mathbf{v}/Dt = -\nabla p_2 + \mu\nabla^2\mathbf{v};$$

only the 'dynamic pressure' p_2 appears in this equation. We can in many cases solve this equation and then just add on the hydrostatic pressure if we need the total pressure. For example, the forces on a well-submerged submarine (so that it causes no surface waves, obviously controlled by gravity) consist of lift and drag and sideways forces independent of depth, together with a hydrostatic pressure which does depend on depth; the response of the submarine to its control surfaces is independent of depth.

When the fluid has density variations, as in the atmosphere, we must consider more carefully what density field ρ_1 to use in the hydrostatic equation

$$\rho_1\mathbf{g} = \nabla p_1.$$

In the lower atmosphere it is sensible to use the density appropriate to the neutrally stable atmosphere discussed in Chapter VII; any other density field will tend to cause motions upwards or downwards. Then if we write $\rho = \rho_1 + \rho_2$ when the velocity is non-zero, we can easily derive the equation

$$(\rho_1 + \rho_2)D\mathbf{v}/Dt = \rho_2\mathbf{g} - \nabla p_2 + \text{viscous terms.}$$

In this case gravity cannot be entirely cancelled out with a hydrostatic pressure field; density variations cause buoyancy forces in the fluid.

At this stage of fluid dynamics we shall say no more about buoyancy effects. Our later examples will assume constant density, or else a flow so confined that vertical motions are of little importance. But it must be noted that in reality buoyancy is often a most important effect.

(c) *Motion relative to the Earth*

Most flows of interest take place on the Earth, and we normally use axes fixed on the Earth. Thus we are not using inertial axes, and so Newton's law of motion will not be

$$\rho D\mathbf{v}/Dt = \text{forces}$$

when measurements are made in a rotating or accelerating system. The major effect is the Earth's rotation, at angular velocity $\mathbf{\Omega}$. This effectively introduces extra vorticity into any flow, which can be important or dominant when the scale of the flow is large.

It is shown in texts on dynamics that D/Dt must be replaced by

$$D/Dt + \mathbf{\Omega} \times$$

when measurements are made in a rotating system. The Navier–Stokes equation becomes

$$\rho D\mathbf{v}/Dt + 2\rho\mathbf{\Omega} \times \mathbf{v} + \rho\mathbf{\Omega} \times (\mathbf{\Omega} \times \mathbf{r}) = \rho\mathbf{F} - \nabla p + \text{viscous terms},$$

where the term in $\mathbf{\Omega} \times (\mathbf{\Omega} \times \mathbf{r})$ is a 'centrifugal' term, and we need to replace

$$\mathbf{v} = D\mathbf{r}/Dt$$

by

$$\mathbf{v} + \mathbf{\Omega} \times \mathbf{r}$$

in the derivation.

The centrifugal term is usually taken into the hydrostatic equation, because it is non-zero even when $\mathbf{v}$ is zero. So on subtracting out the hydrostatic equation as in (*b*) above (and neglecting buoyancy) we are left with

$$\rho D\mathbf{v}/Dt + 2\rho\mathbf{\Omega} \times \mathbf{v} = -\nabla p_2 + \text{viscous terms}.$$

Suppose we have a weather system with length scale 1000 km and velocity of general size 10 m s^{-1}. Since $\Omega = 7.27 \times 10^{-5}$ s^{-1}, the sizes of the terms

$$\mathbf{v}\cdot\nabla\mathbf{v} \text{ and } 2\mathbf{\Omega} \times \mathbf{v}$$

are about (in m s^{-2})

$$10^2/(1000 \times 10^3) \text{ and } 1.5 \times 10^{-4} \times 10,$$

i.e.

$$10^{-4} \text{ and } 1.5 \times 10^{-3}.$$

Thus the 'Coriolis force' $2\mathbf{\Omega} \times \mathbf{v}$ is the more important term in determining weather patterns. If you repeat the calculations for smaller length scales, like 1 m, you find that the Coriolis term is insignificant; the flow of water out of a bath is *most* unlikely to be affected by the rotation of the Earth, and so will be the same in northern and southern hemispheres even though $\mathbf{\Omega} \times \mathbf{v}$ changes direction.

If the weather pattern above has almost steady circular streamlines, as many depressions have to a fair approximation, we may take the motion

Fig. VIII.3. A simple model for a depression on a rotating Earth.

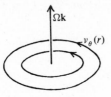

to be approximated by

$$2\rho\boldsymbol{\Omega} \times \mathbf{v} = -\nabla p_2.$$

Here we neglect viscous effects as well as buoyancy and the mass-acceleration terms. Putting

$$\mathbf{v} = v_\theta(r)\hat{\boldsymbol{\theta}}$$

$$\boldsymbol{\Omega} = \Omega\mathbf{k}$$

in a cylindrical system on a flat Earth (sketched in fig. VIII.3) leads to

$$p_2 = -\int 2\rho\Omega v_\theta(r)dr.$$

This very simple model of a weather pattern on a rotating Earth explains how you can have the pressure gradient outwards from the centre of a depression, with the wind not moving from high pressure to low but going along circles of constant pressure: the Coriolis force balances the pressure gradient.

(d) Thermodynamic equations

The full problem of fluid dynamics must include some equations of state and some thermodynamic statement. We have discussed equations of state in Chapter VII above, and we have started the thermodynamics there too. It is worth giving a fuller thermodynamic statement here, if only to show why we almost always reject it in favour of the simple models described above.

In a real fluid heat flows from regions of higher temperature to regions of lower temperature. It is found that a thermal conductivity k – one would expect $k(\rho, T)$ on thermodynamic grounds – relates heat flow rate and temperature gradient. Moreover in a viscous fluid the internal viscous forces (molecular collisions) convert organised motion to random molecular motion, i.e. heat. This generation of heat in the fluid is proportional to squares of rates of strain and to the viscosity μ:

$$\Phi = 2\mu(e_{ij}e_{ij} - \tfrac{1}{3}e_{kk}e_{ll})$$

is the rate of addition of heat per volume. This formula may be derived, but will not be here, by considering the rate of working of the various terms in the Navier–Stokes equation.

These considerations lead to the two (alternative) thermodynamic equations

(i) $\rho DE/Dt = -p\nabla \cdot \mathbf{v} + \Phi + \nabla \cdot (k\nabla T) + K(\nabla \cdot \mathbf{v})^2,$

(ii) $\rho T DS/Dt = K(\nabla \cdot \mathbf{v})^2 + \Phi + \nabla \cdot (k\nabla T).$

It is clear that to solve one of these in conjunction with the Navier–Stokes equation, the mass equation and an equation of state is really too much to be contemplated.

Exercises

1. Consider a gas at rest and use methods like those in §2(b) to show that heat flow in a gas should be of the form $-k\nabla T$ for some constant k.

2. Show that heat energy changes in a fixed volume dV are given, for temperature change dT, by

   $\rho c_v dT dV.$

   Use the methods of §1 to set up an equation for temperature in a fluid at rest. Assume that there is a source of heat

   $Q(\mathbf{x}, t)$

   per volume per time at each point of the fluid.

3. Show that Euler's equation in a rotating frame of reference and when ρ is constant may be written as

   $\partial \mathbf{v}/\partial t + (\boldsymbol{\omega} + 2\boldsymbol{\Omega}) \times \mathbf{v} = \mathbf{F} - \nabla\{\tfrac{1}{2}\mathbf{v}^2 + p/\rho - \tfrac{1}{2}\Omega^2 r^2\}$

   where r is distance from the axis of rotation.

4. Find a suitable weather map and determine how closely the horizontal components of ∇p and $2\boldsymbol{\Omega} \times \mathbf{v}$ match. What would cause differences?

5. For an emptying bath compare the 'vorticity' $\boldsymbol{\Omega}$ and naturally occurring vorticities $\boldsymbol{\omega}$ due to the exit of the bather.

6. Derive equations §4(d) (i) and (ii) from the law of heat flow in Q1 and the laws of thermodynamics.

7. The stream function

   $\psi = Ur \sin \theta - r^{-1}Ua^2 \sin \theta$

   gives a model flow past the cylinder $r = a$. The rate of strain tensor has components

   $e_{rr} = \partial v_r/\partial r$

   $e_{\theta\theta} = r^{-1}(v_r + \partial v_\theta/\partial \theta)$

$e_{r\theta} = \frac{1}{2}r\partial/\partial r(v_\theta/r) + \frac{1}{2}r^{-1}\partial v_r/\partial\theta.$

Calculate the components of the deviatoric stress tensor.

8. Calculate the viscous force on the cylinder for Exercise 7. Is it realistic?

References

(a) *An Introduction to Fluid Dynamics*, G. K. Batchelor, C.U.P. 1967. This text gives full derivations and discussions of all the equations in this chapter.

(b) *Mechanics of Continuous Media*, S. C. Hunter, Ellis Horwood (John Wiley) 1976. This text sets the Navier–Stokes equation into the context of other constitutive relations for continuous media.

IX

Solutions of the Navier–Stokes equations

---

1. Flows with only one coordinate

The Navier–Stokes equations of motion are difficult equations, so we only deal with the very easiest cases. No useful general solution is known, and existence and uniqueness theorems are only available in very special cases.

Firstly, we restrict the physics somewhat. We consider fluids of constant uniform density, so that $\nabla \cdot \mathbf{v} = 0$. This promptly removes a large range of interesting problems where compressibility or buoyancy are important. Then we demand that there shall be no changes in the viscosity: really we cannot do this as viscous dissipation of energy heats the fluid and so changes the viscosity.

And we further restrict the mathematics by setting up simple boundary conditions, so that we have at most two variables to deal with. Even then we cannot do many problems, but among those we can do are some which give us important ideas on which we can build up other large areas of theory.

So we will be trying to solve

$$\begin{cases} \nabla \cdot \mathbf{v} = 0, \\ \rho D\mathbf{v}/Dt = -\nabla p + \mu \nabla^2 \mathbf{v}, \end{cases}$$

where p is now the dynamic pressure and ρ and μ are constants. This is a set of four equations for the four unknown functions p and $\mathbf{v}$; this is unlike

Fig. IX.1. Definition sketch for flow in a channel.

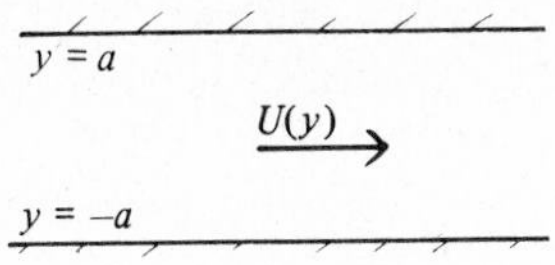

ordinary dynamics in that the force $-\nabla p$ must be found as part of the solution, and is not given as something known. Later we will see that we could

(i) eliminate p by taking $\nabla \times$, and solving the resulting equation;

(ii) express p in terms of $\mathbf{v}$ by taking $\nabla \cdot$ and then solving for p;

but for the moment we attack the problem directly in simple cases.

We start with some flows in which $\mathbf{v}$ is along one coordinate direction and depends only on another coordinate.

(a) *Flow in an infinite channel*

This is a two dimensional flow

$$\mathbf{v} = U(y)\mathbf{i}$$

between the fixed walls $y = \pm a$, as indicated in fig. IX.1. The equation

$$\nabla \cdot \mathbf{v} = 0$$

is satisfied automatically, and the three Navier–Stokes equations are

$$\begin{cases} 0 = -\partial p/\partial x + \mu d^2 U/dy^2, \\ 0 = -\partial p/\partial y, \\ 0 = -\partial p/\partial z. \end{cases}$$

In this case $\partial \mathbf{v}/\partial t$ and $\mathbf{v} \cdot \nabla \mathbf{v}$ are both zero, and $\mu \nabla^2 \mathbf{v}$ is a single term.

The second and third of these equations show that $p(x)$ is appropriate; but the term $\mu d^2 U/dy^2$ is a function of y only. Hence both dp/dx and $\mu d^2 U/dy^2$ must be constants: so

$$dp/dx = -G$$

where G is the pressure gradient – the pressure is higher on the left to push the fluid in the direction shown.

So we have found the form of the pressure:

$$p = p_0 - Gx;$$

and the equation for $\mathbf{v}$ is therefore

$$\mu d^2 U/dy^2 = -G.$$

The boundary conditions must be
    (i) $U = 0$ on the walls $y = \pm a$, the no-slip condition,
    (ii) values of the pressure at the two ends of the channel, say $p = p_0$

    at $x = 0$ and $p = p_1$ at $x = l$.

Thus the final solution is

$$\begin{cases} p = p_0 - (p_0 - p_1)x/l, \\ U = \tfrac{1}{2}(p_0 - p_1)(a^2 - y^2)/\mu l. \end{cases}$$

The velocity profile is parabolic, and dU/dy is non-zero at the walls, having the value $\pm (p_0 - p_1)a/\mu l$. Now the shear stress on the wall $y = -a$ is

$$\sigma_{12} = s_{12} = \mu(dU/dy)_{y=-a} = \mu(p_0 - p_1)a/\mu l,$$

while on the wall $y = a$ it is $(-\sigma_{12})_{y=a}$ because the normal to this wall is in the $-y$ direction. Hence each wall has a force per area

$$(p_0 - p_1)a/l$$

acting on it in the direction of the flow.

The forces on the fluid in the channel, in the direction of flow, are:
    (i) $-2l(p_0 - p_1)a/l$, half from each wall, with the factor l for the length of each wall;
    (ii) $p_0 \times 2a$ at the end $x = 0$;
    (iii) $-p_1 \times 2a$ at the end $x = l$.

These forces have zero sum, as they must for a non-accelerated flow. So in this case the wall stresses $(p_0 - p_1)a/l$ could have been deduced directly from the pressure gradient and the width without solving the equation of motion.

This flow has some relation to reality when the speeds are not too high, but the requirement of independence of z is rather unreasonable, and the next example avoids this restriction. In both examples the flow has infinite extent in the x-direction: in real pipes and channels there is an 'inlet region' where there *is* variation with x, and these solutions do not apply.

(b) *Flow in a pipe*
In a pipe of circular cross-section we may assume a flow

$$\mathbf{v} = U(r)\mathbf{k},$$

Fig. IX.2. Definition sketch for flow in a pipe.

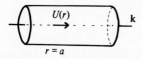

sketched in fig. IX.2, with boundary value

$$U(a) = 0$$

and with pressure gradient

$$G = (p_0 - p_1)/l.$$

For this flow we shall need to calculate the element s_{rz} of the stress tensor in cylindrical polar components, in order to calculate the force on the pipe. We shall make no attempt to produce general formulae, which may be found in larger texts, but shall proceed in the manner of Chapter I §6.

The rate of strain tensor e_{ij} was derived from the tensor $\partial v_i/\partial x_j$, which in vector notation is the transpose of ∇v. Now ∇v in this example is

$$(\hat{\mathbf{r}}\partial/\partial r + r^{-1}\hat{\boldsymbol{\theta}}\partial/\partial\theta + \mathbf{k}\partial/\partial z)U(r)\mathbf{k}.$$

Since $\mathbf{k}$ is a constant vector, this reduces to

$$\nabla v = dU/dr\ \hat{\mathbf{r}}\mathbf{k}.$$

But e_{ij} is the symmetric part of $\partial v_i/\partial x_j$, and so the rate of strain tensor is just

$$\tfrac{1}{2}(\hat{\mathbf{r}}\mathbf{k} + \mathbf{k}\hat{\mathbf{r}})dU/dr,$$

because reversing two suffices in $\partial v_i/\partial x_j$ corresponds to reversing the order of unit vectors (or taking the transpose of a matrix).

So for this flow

$$\begin{cases} e_{rz} = \tfrac{1}{2}dU/dr, \\ s_{rz} = \mu dU/dr. \end{cases}$$

The Navier–Stokes equation requires $\nabla^2 v$. We could calculate it as

$$\nabla\cdot(\nabla v),$$

by using the methods just employed for ∇v, but there is no need, as it is

$$\nabla^2\{U(r)\mathbf{k}\}$$

and $\mathbf{k}$ is a constant vector. That is, all we need is

$$\{\nabla^2 U(r)\}\mathbf{k} = (d^2U/dr^2 + r^{-1}dU/dr)\mathbf{k}.$$

Consequently the Navier–Stokes equations are

$$\begin{cases} 0 = -\partial p/\partial z + \mu\{d^2U/dr^2 + r^{-1}dU/dr\}, \\ 0 = -\partial p/\partial r, \\ 0 = -r^{-1}\partial p/\partial\theta. \end{cases}$$

These equations are solved exactly as in (a) above to give

$$\begin{cases} p = p - Gz, \\ U = \tfrac{1}{4}G(a^2 - r^2)/\mu. \end{cases}$$

This is the well-known Poiseuille formula for steady flow of viscous liquid down a tube. It is accurately true for maximum velocities U_{max} such that the Reynolds number

$$R = 2aU_{max}/v < 2000.$$

For higher speeds than this the flow becomes increasingly unstable – with extreme experimental care it is possible to get steady flow of this type up to $R = 100\,000$. It was from experiments of this type that Reynolds first realised that this non-dimensional group controlled the type of flow that was seen in any given experiment with geometrically similar apparatus.

The shear stress on the wall in Poiseuille flow is

$$- s_{zr} = - \mu(dU/dr)_{r=a}.$$

From the solution this has value $\frac{1}{2}Ga$ and so the force on unit length of the pipe is $2\pi a \times \frac{1}{2}Ga$. As before it may be demonstrated that the pressure stresses at the ends balance the shear stresses on the walls.

The maximum speed in Poiseuille flow is

$$\tfrac{1}{4}Ga^2/\mu,$$

but it is more realistic to talk in terms of the mean flow rate

$$U_m = (\pi a^2)^{-1} \int_{r=0}^{a} U(r) \times 2\pi r dr = Ga^2/8\mu ;$$

mean flow rates are easy to measure because all you need is a watch and a graduated jar. In terms of U_m the velocity profile is

$$U(r) = 2U_m(1 - r^2/a^2).$$

(c) *Flow between rotating cylinders*

Consider the two-dimensional steady flow

$$\mathbf{v} = V(r)\hat{\boldsymbol{\theta}}$$

between concentric cylinders, sketched in fig. IX.3. The flow can only be

Fig. IX.3. Simple flow between rotating cylinders.

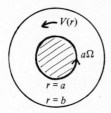

steady if at least one of the cylinders is forced to rotate, so take as example

$$\begin{cases} V = a\Omega & \text{on} \quad r = a, \\ V = 0 & \text{on} \quad r = b. \end{cases}$$

This flow can easily be set up between long cylinders, and for moderate Reynolds numbers it is of the type given above.

Notice that we now need no pressure boundary conditions – there is no flow through the apparatus.

In this case the element $s_{r\theta}$ of the stress tensor is needed to calculate the moment of the stresses on either cylinder. Proceeding as in (b) above, we have

$$\nabla \mathbf{v} = (\hat{\mathbf{r}}\partial/\partial r + r^{-1}\hat{\boldsymbol{\theta}}\,\partial/\partial\theta)V(r)\hat{\boldsymbol{\theta}},$$

where we must now use $\partial\hat{\boldsymbol{\theta}}/\partial\theta = -\hat{\mathbf{r}}$. Simple calculations give

$$\nabla \mathbf{v} = \hat{\mathbf{r}}\hat{\boldsymbol{\theta}}dV/dr - \hat{\boldsymbol{\theta}}\hat{\mathbf{r}}V(r)/r.$$

Hence

$$\begin{cases} e_{r\theta} = \tfrac{1}{2}(dV/dr - V/r), \\ s_{r\theta} = \mu(dV/dr - V/r). \end{cases}$$

For this flow we also need

$$\mathbf{v}\cdot\nabla\mathbf{v} \quad \text{and} \quad \nabla^2\mathbf{v}.$$

The first of these is

$$(r^{-1}V\partial/\partial\theta)\{V(r)\hat{\boldsymbol{\theta}}\}$$
$$= -(V^2/r)\hat{\mathbf{r}}.$$

The second may be calculated as $\nabla\cdot\nabla\mathbf{v}$, where $\nabla\mathbf{v}$ is given above. Alternatively it may easily be found from the formula (in Chapter I)

$$\nabla^2\mathbf{v} = \nabla\nabla\cdot\mathbf{v} - \nabla\times(\nabla\times\mathbf{v}).$$

Since $\nabla\cdot\mathbf{v} = 0$, only $\nabla\times(\nabla\times\mathbf{v})$ must be found, which is quite easy since

$$\nabla\times\mathbf{v} = (r^{-1}d/dr)(rV)\mathbf{k}.$$

Using either method, the Navier–Stokes equations are

$$\begin{cases} -\rho V^2/r = -\partial p/\partial r, \\ \quad 0 = -r^{-1}\partial p/\partial\theta + \mu(d^2V/dr^2 + r^{-1}dV/dr - V/r^2), \\ \quad 0 = -\partial p/\partial z. \end{cases}$$

From the second equation $\partial p/\partial\theta$ can only depend on r, and so that p can be continuous as you go round from $\theta = 0$ to $\theta = 2\pi$, it must be that $\partial p/\partial\theta = 0$ and so p is a function of r only. This reduces the Navier–Stokes equation to

$$\begin{cases} dp/dr = \rho V^2/r, \\ d^2V/dr^2 + r^{-1}dV/dr - V/r^2 = 0. \end{cases}$$

The solution of these is easily found to be

$$\begin{cases} V = Ar + B/r, \\ p = p_0 + \rho \int_a^r \{V(s)\}^2/s \, ds. \end{cases}$$

When we fit the boundary values at $r = a$ and $r = b$, we find

$$V(r) = \frac{-\Omega a^2 r}{b^2 - a^2} + \frac{\Omega a^2 b^2}{(b^2 - a^2)r}.$$

An expression for $p(r)$ can now be calculated, but it gives little more information.

This solution has a number of interesting points.

(i) It is a combination of rigid rotation $v_\theta = Ar$ and line vortex flow $v_\theta = B/r$. Either can be achieved separately by choosing somewhat different boundary values $-b \to \infty$ gives vortex flow, and $V(b) = \Omega b$ instead of 0 gives rigid body motion.

(ii) The shear stresses must sum to give a clockwise couple on the inner cylinder, and an equal and opposite couple on the outer cylinder, since the flow does not accelerate. The magnitude of the couple is easily calculated as

$$\int_{\theta=0}^{2\pi} s_{\theta r} a \, d\theta$$

per unit length of cylinder, where $s_{\theta r}$ is taken at $r = a$. Now

$$s_{\theta r} = \mu(dV/dr - V/r)$$
$$= -\mu \times 2B/r^2.$$

Hence the couple has magnitude
$$4\pi\mu a^2 b^2/(b^2 - a^2).$$

(d) Instability of flow between rotating cylinders

Experiments show that as the speed of rotation of the inner cylinder increases, there comes a value of $\Omega b^2/\nu$ (which acts as a Reynolds number here in giving a dimensionless measure of viscosity) at which the simple flow

$$\mathbf{v} = V(r)\hat{\boldsymbol{\theta}}$$

Fig. IX.4. Taylor vortices between rotating cylinders.

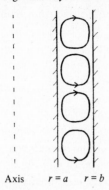

Axis $r = a$ $r = b$

ceases to exist, and is replaced by another steady flow in which the space between the cylinders is filled with vortices, as indicated in fig. IX.4. This instability of the simple circular flow can be quite easily explained, even though the calculation of the exact speed of rotation which gives instability is rather hard.

In the simple circular flow the pressure gradient just balances the centrifugal force (i.e. it provides the correct inward acceleration)

$$dp/dr = \rho V^2/r.$$

Now suppose that some slight disturbance is given to the flow (a lorry passing outside, or a sound wave, might do). Assume that a ring of fluid at radius r moves out a little to r'. If the transfer takes place reasonably quickly, so that viscous diffusion has little time to act, the angular momentum of this ring will be preserved, giving it the new speed

$$Vr/r'$$

at the new radius r'. So the new centrifugal force (central acceleration) is

$$\rho V^2 r^2/r'^3.$$

If the pressure gradient is larger than this at r', then the ring is forced back towards its original position, and there is stability. But if the pressure gradient at r' is smaller, then the ring moves on out, and the original situation was unstable. So the flow is *unstable* if

$$\rho V^2 r^2/r'^3 > \rho V'^2/r'.$$

This requirement is that

$$V^2 r^2, \quad \text{or} \quad |Vr|,$$

decreases as r increases, which is to say

$$d|Vr|/dr < 0.$$

This is Rayleigh's criterion for circular flows: there is stability to this kind of disturbance if the circulation Vr increases outwards, instability is possible if Vr decreases.

In the flow we have analysed above, we have

$$Vr = \Omega a^2 (b^2 - r^2)/(b^2 - a^2)$$

so that

$$d|Vr|/dr = -2\Omega a^2 r/(b^2 - a^2)$$

which is certainly negative. We therefore expect instability if the flow is fairly fast, because then viscosity will have little time in which to act; but stability if the flow is rather slow, as then viscosity has time to diffuse away angular momentum differences. Exact analysis confirms this general description and gives a value

$$\Omega b^2/v \doteqdot 1100$$

for the stability boundary. This value is well confirmed experimentally.

What has happened in this flow is that a new mechanism of (Taylor) vortices has come into action, which is more efficient at transporting the angular momentum from one cylinder to the other. If Ω is increased still further, then this flow with Taylor vortices itself becomes unstable, yielding place to a flow in which waves travel round the cylinder. At yet higher speeds the flow finally becomes apparently random, and is described as turbulent. Turbulence is a yet more efficient way of transferring angular momentum from inner to outer cylinder.

If you solve the similar circular flow to the one in (*c*) which has the *outer* cylinder rotating and the inner one fixed, then you find

$$d|Vr|/dr = 2b^2 \Omega r/(b^2 - a^2)$$

which is certainly positive. Rayleigh's criterion suggests that this flow is stable. What happens as Ω increases in this flow is that turbulent patches suddenly start to appear at a critical value of $\Omega b^2/v$, not preceded by any regular arrangement of vortices. Then at a slightly higher speed the motion becomes fully turbulent.

These two flows, with one or other cylinder rotating, provide good examples of the two ways in which stability of simple flows breaks down as the speed increases. One way is via a succession of more complicated but still regular flows, and finally to turbulence. The other way is a direct jump to patches of turbulence and then to full turbulence.

2. Some flows with two variables

The flows described in §2 above involved only one coordinate, and this gave ordinary differential equations to solve. We go on now to

Fig. IX.5. Flow starting from rest, above a plane which moves in its own plane.

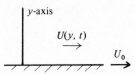

discuss three flows in which there is only one space coordinate, but there is also time variation.

(a) A plane started from rest

In this flow an infinite plane starts to move at speed U_0 in its own plane at $t = 0$; the fluid is initially at rest. We assume a solution of the form sketched in fig. IX.5,

$$\mathbf{v} = U(y, t)\mathbf{i}$$

with boundary values

$$\begin{cases} U(0, t) = U_0 & \text{for} \quad t > 0, \\ U(y, 0) = 0 & \text{for} \quad y > 0, \\ U(y, t) \to 0 & \text{as} \quad y \to \infty, \quad \text{all} \quad t. \end{cases}$$

The equation of continuity, $\nabla \cdot \mathbf{v} = 0$, is satisfied identically in this example as in the previous ones; the Navier–Stokes equations are

$$\begin{cases} \rho \partial U / \partial t = -\partial p / \partial x + \mu \partial^2 U / \partial y^2, \\ 0 = -\partial p / \partial y, \\ 0 = -\partial p / \partial z. \end{cases}$$

As before we find that p depends only on x and dp/dx is a constant. In this example there is no reason to require an imposed pressure field in the direction of motion – the flow is generated by the motion of the plane. So we take

$$p = \text{constant}.$$

This leaves us with

$$\partial U / \partial t = v \partial^2 U / \partial y^2,$$

a standard partial differential equation which describes processes of diffusion. In this case momentum is being diffused across the fluid by the (molecular) action of viscosity.

An equation such as this has three usual methods of (analytic) solution. We may separate variables and seek an eigenfunction expansion; in this case this method seems inappropriate, as there is no countable set of eigenvalues. We may use a Laplace transform; but this method has possibly not been met at this level of mathematics. We may seek a 'similarity solution'; this method is often useful in fluid dynamics, and this example is a good one for showing how it works.

The problem depends on two parameters

$$U_0 \quad \text{and} \quad \nu$$

and two coordinates

$$y \quad \text{and} \quad t.$$

It is moreover linear in U_0 and U: double U_0 and you merely have to double U to get the new solution. So we expect that $U(y, t)/U_0$ will be independent of U_0, and it must also be dimensionless. Now the quantities ν, y, t can be formed into (essentially) only one dimensionless group, which we take to be

$$\eta = y/(4\nu t)^{1/2}.$$

Naturally we could have chosen 2η or η^2 or $\sin \eta$ as a dimensionless group, but this one turns out to be most convenient. So we are expecting to find a solution

$$\begin{cases} U(y, t) = U_0 f(\eta), \\ \eta = y/(4\nu t)^{1/2}. \end{cases}$$

When we substitute this form into the partial differential equation for U we find that

$$f''(\eta) = -2\eta f'(\eta),$$

which is a simple *ordinary* differential equation for $f'(\eta)$. The boundary conditions must also be transformed into terms of f and η. Now

$$\begin{cases} y \to \infty & \text{for fixed} \quad t > 0 \quad \text{is} \quad \eta \to \infty \ \dots \text{(i)} \\ t \to \infty & \text{for fixed} \quad y > 0 \quad \text{is} \quad \eta \to 0 \ \ \dots \text{(ii)} \\ t \to 0 & \text{for fixed} \quad y > 0 \quad \text{is} \quad \eta \to \infty \ \dots \text{(iii)} \\ y \to 0 & \text{for fixed} \quad t > 0 \quad \text{is} \quad \eta \to 0 \ \ \dots \text{(iv)}. \end{cases}$$

What we need is

$$\begin{cases} U(y, t) \to 0 & \text{in case (i)}, \\ U(y, t) \to 0 & \text{in case (iii)}, \\ U(y, t) \to U_0 & \text{in case (iv)}. \end{cases}$$

That is

$$\begin{cases} f(\eta) \to 0 & \text{as} \quad \eta \to \infty, \\ f(\eta) \to 1 & \text{as} \quad \eta \to 0. \end{cases}$$

The problem has thus been reduced to an ordinary differential equation for $f(\eta)$, with suitable boundary conditions. This is a check on our assumed form of solution: it has produced no inconsistency.

The solution for $f'(\eta)$ is

$$f'(\eta) = Ae^{-\eta^2}$$

for some constant A; this is done by separation of variables. The solution for $f(\eta)$ is now found by integration:

$$f(\eta) = - \int_{\eta}^{\infty} Ae^{-s^2} ds,$$

where the limits have been chosen to give $f \to 0$ as $\eta \to \infty$. Finally we use $f(0) = 1$ to evaluate A, and find

$$\begin{cases} f(\eta) = 2\pi^{-1/2} \int_{\eta}^{\infty} e^{-s^2} ds, \\ U(y, t) = 2U_0 \pi^{-1/2} \int_{y/(4vt)^{1/2}}^{\infty} e^{-s^2} ds. \end{cases}$$

This formula for $U(y, t)$ is not very simple. Fortunately integrals like

$$\int_{\eta}^{\infty} e^{-s^2} ds$$

are readily available in books of tables, for example

$$\Phi(x) = (2\pi)^{-1/2} \int_{-\infty}^{x} e^{-t^2/2} dt$$

is a standard function in the theory of the normal distribution of statistics, representing the probability of achieving a value less than x from a normal distribution of zero mean and unit variance. So there is no difficulty in plotting $f(\eta)$ from standard tables:

$$f(\eta) = 2\{1 - \Phi(\eta\sqrt{2})\}.$$

But $f(\eta)$ can also be described in terms of another standard function (less commonly found in books of tables)

$$f(\eta) = 1 - \operatorname{erf} \eta$$

where the 'error function' erf is defined by

$$\operatorname{erf} \eta = 2\pi^{-1/2} \int_{0}^{\eta} e^{-s^2} ds.$$

The graph of $f(\eta)$ in fig. IX.6 shows the form of the velocity profile for

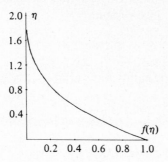

Fig. IX.6. Velocity profile in dimensionless form for the flow due to a moving plane.

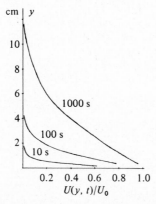

Fig. IX.7. The velocity profile at various times for water in §2(a).

any value of t: as t increases, the scale on the vertical axis just represents larger and larger y values, because $\eta = y/(4vt)^{1/2}$. This is best shown by an example. Consider water, for which $v = 1.1 \times 10^{-6} \text{ m}^2 \text{ s}^{-1}$; the edge of the region affected by the motion of the plane may be taken to be $\eta = 2.0$, as beyond there $U/U < 0.01$. For various times this edge at $\eta = 2$ corresponds to very different y values, as shown in the table below, and in fig. IX.7.

Time t (s)	10	100	1000
Distance at which $\eta = 2$, (m)	13.3×10^{-3}	4.2×10^{-2}	13.3×10^{-2}

We see that even after 1000 s the effect of the motion of the plane is only noticeable for some 10 cm. For the momentum to diffuse out to 1 m takes 5.7×10^4 s, or over 15 hours.

This rather special flow shows what turns out to be a general result.

Important variations in a flow can often be found to occur in a thin 'boundary layer', and the diffusion of this boundary layer into the rest of the fluid is usually a very slow process.

(b) Diffusion of a vortex sheet

The second flow that we consider in this section is really the same one again. Suppose that at $t = 0$ we have a vortex sheet along the plane $y = 0$, so that for $y > 0$ the velocity is U_0 and for $y < 0$ the velocity is $-U_0$. How does this vortex sheet decay as time increases (that is, assuming that it is stable)?

The governing equations are just the same as in (a) above, reducing as before to

$$\partial U / \partial t = v \partial^2 U / \partial y^2.$$

But the boundary conditions are now

$$\begin{cases} U(y, t) \to U_0 & \text{for fixed} \quad t > 0 \quad \text{and} \quad y \to \infty \\ U(y, t) \to -U_0 & \text{for fixed} \quad t > 0 \quad \text{and} \quad y \to -\infty. \end{cases}$$

In terms of the variable η, and using

$$U(y, t) = U_0 f(\eta)$$

as before, the equation and boundary conditions are

$$\begin{cases} f''(\eta) = -2\eta f'(\eta) \\ f(\eta) \to \pm 1 & \text{as} \quad \eta \to \pm \infty. \end{cases}$$

We get the same type of solution as before, but with slight adjustments to fit the new boundary values:

$$f(\eta) = 2\pi^{-1/2} \int_0^\eta e^{-s^2} ds = \text{erf} \, \eta.$$

The velocity profile is shown in fig. IX.8. The vortex sheet slowly spreads

Fig. IX.8. Velocity profile for a diffusing vortex sheet.

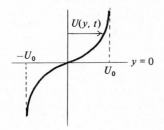

Fig. IX.9. The distribution of vorticity in a diffusing vortex sheet.

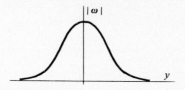

out, having a width (between $\eta = -2$ and $\eta = +2$) of

$$8(vt)^{1/2}$$

at any time t.

Since this flow was introduced in terms of a vortex sheet, it is worth discussing the later distribution of vorticity. For this flow we have

$$\mathbf{v} = U(y, t)\mathbf{i}$$

and so

$$\boldsymbol{\omega} = \nabla \times \mathbf{v} = -\partial U/\partial y \ \mathbf{k}.$$

Now

$$\partial U/\partial y = U_0(4vt)^{-1/2}f'(\eta)$$
$$= U_0(\pi vt)^{-1/2}\exp(-y^2/4vt).$$

This is a normal distribution, sketched in fig. IX.9, with 'width' increasing at the rate $(vt)^{1/2}$, and at $t = 0$ there is a singularity, with all the vorticity concentrated at $y = 0$.

This discussion of a vortex sheet and its development may seem rather far removed from reality, but large concentrations of vorticity can occur in real flows, and it is useful to solve a model problem to see how they would diffuse into the surrounding fluid. However this *is* only a model problem, and all it should be taken to show is the likely dependence of the thickness on $(vt)^{1/2}$ in more realistic flows.

(c) Decay of a line vortex

A more realistic version of the flow discussed in (a) above can be achieved by making a cylinder rotate in a fluid at rest. The flow will be round in circles, as sketched in fig. IX.10,

$$\mathbf{v} = U(r, t)\hat{\boldsymbol{\theta}}$$

and the Navier–Stokes equations reduce to

$$\begin{cases} dp/dr = \rho U^2/r \\ \partial U/\partial t = v\{\partial^2 U/\partial r^2 + r^{-1}\partial U/\partial r - U/r^2\}, \end{cases}$$

Fig. IX.10. Simple flow outside a rotating cylinder.

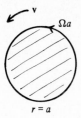

with

$$
\begin{cases}
U \to 0 \text{ as } r \to \infty \text{ for fixed } t > 0, \\
U = \Omega a \text{ on } r = a \text{ for } t > 0, \\
U = 0 \text{ for } r > a \text{ at } t = 0.
\end{cases}
$$

This problem is more realistic, because a cylinder takes up less room than an infinite plane; but it is mathematically harder because we now have one more dimensional parameter, the radius. There are now *two* dimensionless groupings,

$$r^2/vt \quad \text{and} \quad r/a;$$

and so we cannot expect a solution as simple as that in (*a*). The problem *can* be solved, but the mathematics required is perhaps more than can reasonably be expected at this stage; so instead of solving the problem here, we attempt to simplify it.

Earlier, in Chapter V, we discussed the Rankine vortex, which had vorticity zero for $r > a$ and constant vorticity for $r < a$. And we saw that in the limit

$$a \to 0, \quad \Omega a^2 \quad \text{fixed}$$

the Rankine vortex reduced to the line vortex singularity. We shall do the same thing here to eliminate one of the dimensionless groups; the problem we shall solve is that of the spreading out of a line vortex of circulation K, whose vorticity at $t = 0$ forms a singularity at $r = 0$.

We now expect to get a solution $U(r, t)$ that is proportional to K, because the problem is linear, and depending only on the dimensionless group

$$r^2/vt.$$

However since the dimensions of K are $L^2 T^{-1}$, we must (to keep the dimensions right) have

$$U = Kr^{-1}f(r^2/4vt).$$

When we substitute this form into the partial differential equation for

$U(r, t)$, it simplifies to

$$f''(\zeta) + f'(\zeta) = 0,$$

taking $\zeta = r^2/4vt$, and so

$$f(\zeta) = A + Be^{-\zeta}.$$

The conditions that must be imposed on this solution are

$$\begin{cases} f(\zeta) \to 1 & \text{as} \quad \zeta \to \infty, \\ f(\zeta) \to 0 & \text{as} \quad \zeta \to 0, \end{cases}$$

because these correspond to

$$\begin{cases} U(r, t) \sim K/r \text{ as } r \to \infty \text{ for fixed } t, \\ U(r, t) \to 0 \text{ as } r \to 0 \text{ for fixed } t. \end{cases}$$

The solution for this model of the spreading out of a line vortex is therefore

$$U(r, t) = Kr^{-1}[1 - \exp(-r^2/4vt)]$$

where K is the circulation round the vortex at large distances. The form of f is shown in fig. IX.11.

We note that this solution is very like the Rankine vortex in two important ways:

(i) for large r we have $U(r, t) \sim K/r$, which has zero vorticity; this, in effect, applies for $r^2/4vt > 5$, as then the exponential term is less than 10^{-2};

(ii) for small r we have

$$U(r, t) \doteqdot Kr/4vt;$$

(by expanding the exponential), which is rigid rotation – with

Fig. IX.11. Velocity profile for a viscous vortex (solid line) and a Rankine vortex model (dashed line). The Rankine vortex has inner vorticity $\Omega = K/2vt$ and radius $a = (4vt)^{1/2}$.

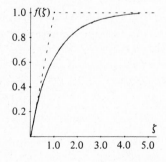

decreasing angular velocity $K/4vt$; this applies when $r^2/4vt$ is very small, say $r^2/4vt < 1/5$.

Between these two regions of large and small $r^2/4vt$, this viscous solution provides a smooth solution. The graph in fig. IX.11 shows these asymptotic regions and the smooth transition layer very clearly.

The solution derived above for the diffusion of a line vortex under the action of viscosity is clearly much more satisfactory than the previous vortex solutions – it has no singularity at $r = 0$ and no discontinuity of any kind for any r. But it is still only a model of any real vortex flow. The most important assumptions that have been made in deriving this solution have been that there is no outer boundary to the flow and that it is two-dimensional. An outer boundary at rest would impose zero velocity there, and viscous action near the boundary would cause a general slowing of the outer parts of the flow as time went on. In a real vortex flow in a stirred mug of coffee, the outer boundary reduces the flow to rest after a time; but in this flow the effect of the bottom of the mug is also important – this effect will be discussed again later. There will of course be other minor effects in the coffee mug vortex due to heat flow and buoyancy, and due to surface tension.

3. A boundary layer flow

Take next a somewhat unrealistic problem to show some very real and important mathematics. We consider flow in an underground channel between two slightly porous walls (say of chalk). Suppose water percolates in through one wall and out through the other at the same speed u (taken to be a constant), while the stream flows along the channel at speed $U(y)$; the situation is sketched in fig. IX.12. The velocity is taken to be

$$\mathbf{v} = U(y)\mathbf{i} + u\mathbf{j}$$

and the boundary conditions on $U(y)$ are that there is no tangential velocity at either wall,

$$U(0) = U(a) = 0.$$

Fig. IX.12. Definition sketch for a combination of percolation $u\mathbf{j}$ and channel flow $U(y)\mathbf{i}$.

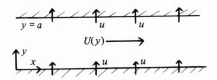

There must be a pressure gradient G to force the fluid along against the viscous resistance, and so the problem has the parameters

$$u, a, G, \rho, v.$$

There are two Reynolds numbers available here, one involving G and v, the other involving u and v; they are

$$Ga^3/\rho v^2 \quad \text{and} \quad ua/v.$$

The former is likely to be rather large: take as example $G = 10$ m of fall per km, $a = 1$ m and get $Ga^3/\rho v^2 = 10^8$. The latter will be assumed to be large also: $u = 10^{-4}$ m s^{-1} and $a = 1$ m give $ua/v = 100$. The three relevant equations here are that of continuity, which is automatically satisfied, and two momentum equations which are

$$\begin{cases} x\text{-direction, } udU/dy = -\rho^{-1}\partial p/\partial x + vd^2U/dy^2, \\ y\text{-direction, } 0 = -\partial p/\partial y. \end{cases}$$

As before, we find that p is independent of y, and dp/dx is independent of x – we have given it value $-G$ above. Thus we are left with the equation

$$vd^2U/dy^2 - udU/dy = -G/\rho.$$

This has solution, with the above boundary values,

$$U(y) = \frac{Ga}{\rho u}\left\{\frac{y}{a} - \frac{1 - \exp(uy/v)}{1 - \exp(ua/v)}\right\}.$$

Now since ua/v is large, the second term in the bracket only becomes noticeable when uy/v is almost equal to ua/v; for example with $ua/v = 100$ and $y = 0.9a$, this term has value 4.5×10^{-5}. Hence over *most* of the channel we have

$$U(y) = Gy/\rho u$$

Fig. IX.13. Inviscid solution $V(\eta) \doteqdot \eta$ up to $\eta = 0.95$ followed by a boundary layer from $\eta = 0.95$ to $\eta = 1$, for the flow of §3. $V(\eta) = \rho U u/Ga$ and $\eta = y/a$.

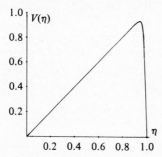

Fig. IX.14. The boundary layer at a larger scale.

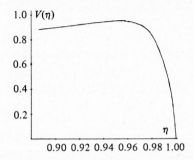

to a very good approximation. The velocity profile is shown, in dimensionless form, in figs. IX.13 and 14. It is only very near $y = a$ that the velocity $U(y)$ suddenly reduces to its value 0 at the wall. If ua/v were larger, say 1000, y/a would have to be even nearer to 1 before the second term became noticeable.

In this problem we can also attempt to solve the related non-viscous, 'inviscid', problem. Take $v = 0$ in the governing equation above to get

$$ud U/dy = G/\rho$$

and so

$$U(y) = Gy/\rho u + A.$$

We can no longer satisfy both boundary conditions; it seems reasonable to have $A \geqslant 0$ to avoid negative velocities, and to have A as small as possible – with $A = 0$ we can at least satisfy one boundary value.

The difference between the viscous solution and the inviscid solution is negligible over most of the channel. It is only in the thin 'boundary layer' near $y = a$ that viscosity is important, and in this boundary layer the velocity profile $U(y)$ is brought at an exponential rate down to zero.

The example is, of course, very special and also very simple. But it shows the major features of many real situations:

(i) an inviscid 'outer' solution which cannot match all the boundary conditions;

(ii) a thin boundary layer where the velocity is forced by viscosity to fit the boundary value;

(iii) the thickness of the boundary layer depending in some fashion on a Reynolds number which is much greater than 1;

(iv) the exponential approach of $U(y)$ to the boundary value in the boundary layer.

This problem is also typical of many in fluid dynamics (and elsewhere) in that it has the general form

$$\varepsilon d^2 U/dy^2 + adU/dy + bU = c,$$

where ε is a small number (when the equation is put in dimensionless form). If we put $\varepsilon = 0$ to get an approximate solution, then both boundary conditions on U cannot be satisfied, and $U(\varepsilon, y)$ is *not* well approximated by $U(0, y)$ over the whole range of y. This type of problem is known as a 'singular perturbation problem', and techniques for dealing with such problems have been developed relatively recently.

4. Flow at high Reynolds number

(a) *Convection and diffusion of vorticity*

The examples in §2 all show that vorticity spreads out from a region where it is initially concentrated, reaching a distance of $\alpha(vt)^{1/2}$ in time t, where α is a number around 3. In these cases the vorticity is uniform in the direction of flow and the diffusion is across the flow, and the flow is one which evolves in time.

The example in §3 shows a steady flow in which the vorticity is mainly concentrated into a thin layer, in which it remains because the flow does not change either with time or downstream.

In more general flows there are two competing tendencies: the vorticity diffuses out from regions of large vorticity caused by the no-slip condition, and it is convected along by the velocity field in accordance with Kelvin's theorem (which is still approximately true in many flows even though non-zero viscosity prevents it from being exactly true). We may expect, from the examples, that in time t the vorticity diffuses a distance of order $(vt)^{1/2}$; and that in this time it is convected a distance of order Ut downstream, where U is the general size of the velocity field.

Let us make the discussion more specific by choosing an example. Consider a stream of speed U flowing past a smoothly shaped ('streamlined') body whose dimensions are around L in size (about $2L$ and $\frac{1}{2}L$ in fig. IX. 15). The stream takes a time of order

$$L/U$$

to pass the body, and in this time the vorticity (which must exist near the

Fig. IX.15. Length and velocity scales in a high Reynolds number flow.

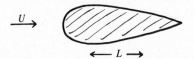

body as there is no slip actually at the surface and a stream of speed U quite near the surface) will diffuse across the stream a distance of order

$$(vL/U)^{1/2}.$$

When this diffusion distance is very much less than the size of the body, we have a boundary layer of vorticity near the body. This condition is

$$(vL/U)^{1/2} \ll L$$

or

$$UL/v \gg 1.$$

(b) The wake of a streamlined body when $R \gg 1$

The argument so far suggests that if the Reynolds number $UL/v = R$ is large, then there will be a thin boundary layer around the body in which vorticity is concentrated. This vorticity will be convected away from the body as a wake, whose thickness at distance x downstream may again be estimated by balancing the effects of convection and diffusion (boundary layers and wake are shown in fig. IX.16). Distance x takes time x/U, during which cross-stream diffusion reaches a distance $(vx/U)^{1/2}$. This is small compared to x when the Reynolds number is large, but is no longer necessarily small compared to L. The wake is thin in the sense that it spreads to fill the narrow parabolic region

$$y = \pm (vx/U)^{1/2}$$

or

$$y^2 = (v/U)x.$$

In non-dimensional form this is $(y/L)^2 = (v/UL)(x/L)$, or $y' = R^{-1/2}x'^{1/2}$, using dimensionless variables $y' = y/L$ and $x' = x/L$. The velocity profiles at sections AA', BB' and CC' will be approximately as shown in fig. IX.17. Note the increased velocity just outside the boundary layers at BB', to get all the fluid round the body; and the decreased velocity in the wake at CC'. Note also that in these sketches UL/v is *not* very large, or else the boundary layer would be too thin to mark in: the thickness at

Fig. IX.16. Boundary layers and wake in a flow at moderately large Reynolds number.

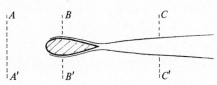

Fig. IX.17. Velocity profiles for the flow of fig. IX.16.

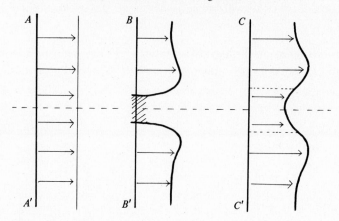

the rear end of the boundary layer is only $(\nu L/U)^{1/2}$, and if we take a glider wing with $U = 15 \text{ m s}^{-1}$ and $L = 1$ m, we get this thickness to be only 1 mm (the Reynolds number is about 10^6).

When the Reynolds number is around 1 in size, the two effects of diffusion and convection are about equal, and there is no sense in talking of boundary layers. The region in which vorticity from the boundary is important is now of the same sort of size as the body. And when the Reynolds number is small, then it looks as though convection is a small effect and diffusion dominates: boundary vorticity reaches to large distances and the 'wake' is so wide as to be almost undetectable. There are, of course, important flows at small Reynolds numbers – lubrication and seepage flows, and the flow of blood in capillaries are obvious examples. But there are many more flows at larger scale and speed for which the Reynolds number is large and the ideas of thin boundary layers and thin wakes are important.

(c) Secondary flow and separation

There is another general pattern to be noted in the straight flows in §§2, 3: the pressure is constant across the boundary layers. Naturally in the flow in circles in §2(c) there must be a pressure gradient to provide the inward acceleration, but changes of pressure across a thin layer at the surface of a rotating cylinder can be shown to be small. In fact it can be proved that this is a general result, that pressure changes are small across a boundary layer, though the calculations required are too much for this text. This property has two important consequences which we discuss in terms of examples.

Consider a vortex in the semi-infinite region $z > 0$, with its centre along the z-axis and the plane $z = 0$ as a rigid boundary. The vortex velocity field is

$$v_\theta = Kr^{-1}\{1 - \exp(-r^2/4vt)\}$$

as we have seen, and there must be a boundary layer near $z = 0$ to reduce this velocity to zero on the boundary $z = 0$, which is assumed to be fixed. If the vortex was set up at $t = 0$ by stirring the fluid, and t is not too large, this boundary layer will be thin, and so the pressure *on* the boundary will equal the pressure in the flow outside the boundary layer, given by

$$dp/dr = \rho v_\theta^2/r.$$

Now this pressure field near the boundary cannot be balanced by a circular motion, as the velocities are too low near the boundary. Hence there must be a motion inwards, with a viscous stress balancing much of the pressure field. This 'secondary motion' can be observed in a stirred glass of fluid, where small particles are convected towards the centre along the bottom. It can also be seen on a larger scale in the motion of air round a depression, where the surface winds have an inward component towards the low pressure region; though in this case there are so many effects present that it is hard to be sure that this is always a secondary flow in a boundary layer.

Consider next the flow of a stream past a cylinder. We have had a model for this previously

$$\psi = Ur\sin\theta - Ua^2\sin\theta/r,$$

which gives a tangential velocity

$$2U\sin\theta$$

on the surface of the cylinder. The fluid accelerates round the leading half of the cylinder from 0 to $2U$, then retards round the trailing half from $2U$ to 0. These accelerations must be due to a pressure field round

Fig. IX.18. Initial flow past a cylinder: the pressures and velocities outside the boundary layer.

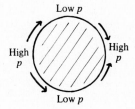

the cylinder, and in particular the pressure must *increase* from $\theta = \pm\frac{1}{2}\pi$ towards the rear stagnation point at $\theta = 0$, as is shown in fig. IX.18. If the flow is at high Reynolds number there will be a thin boundary layer, and inside this the pressure field will be similar. *Inside* the boundary layer speeds are low, and the pressure field cannot be balanced by the acceleration, as happens *outside* the boundary layer. Hence we expect a secondary flow from $\theta = 0$ towards $\theta = \pm\frac{1}{2}\pi$, as in fig. IX.19.

This secondary flow can be clearly seen in films of the early stages of flow round circular cylinders at high Reynolds numbers, and its effect is to cause the boundary layer from the leading half cylinder to *separate* from the surface as a vortex sheet; this is sketched in fig. IX.20. The exact point of separation is very hard to calculate, but is at a value of $|\theta| < \frac{1}{2}\pi$. The

Fig. IX.19. Initial flow past a cylinder: the pressures induce a secondary flow near the surface.

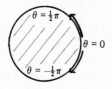

Fig. IX.20. Flow past a cylinder: the secondary flow causes the boundary layer to separate as a vortex sheet.

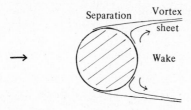

Fig. IX.21. The general character of the flow past a cylinder at high Reynolds number.

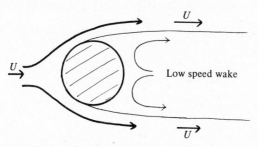

resulting vortex sheet separates an external region where the speed is approximately the free stream speed U from a wake where the velocities are *much* lower, and the wake has a generally circulating flow with non-zero vorticity, as sketched in fig. IX.21. The detail of the wake flow depends greatly on the Reynolds number, and requires numerical calculation or experiment.

As a summary, we may say that in high Reynolds number flow we may expect:

   (i) thin boundary layers;

  (ii) thin wakes from streamlined bodies;

 (iii) secondary flows;

 (iv) boundary layer separation into vortex sheets, and wide wakes, when the pressure increases in the boundary layer, as it does over the rear of a blunt body.

But in low Reynolds number flow we expect widely diffused vorticity and wide smooth wakes.

Exercises

1. Plane Couette flow. The wall $y = 0$ is fixed, and the rigid wall $y = a$ moves at steady speed V in its own plane. Solve the Navier–Stokes equations for the case $\rho =$ constant to show that a possible flow is

   $\mathbf{v} = Vy/a\mathbf{i}.$

   Calculate the stress on each wall.

2. Annular Poiseuille flow. Fluid of constant density is forced between the rigid cylinders $r = a$ and $r = b$ by a pressure gradient G. Solve the Navier–Stokes equations to find the velocity $U(r)\mathbf{k}$. Verify that the stresses on the walls balance the pressure difference over a length l.

3. Flow down a slope. A liquid of constant density flows down a plane which slopes at angle α to the horizontal, as indicated in fig. IX.22. The free surface has no shear stress on it apart from a pressure p_0, and is at a uniform distance from the plane. For this flow you need to keep in the gravitational field, as it is now dynamically active. Set up and solve equa-

Fig. IX.22. Definition diagram for Q3.

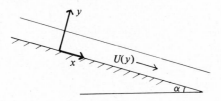

tions for $U(y)$, and verify that the forces on a length l of the fluid layer are in equilibrium.

4. Repeat Q3 for flow down a full pipe of circular section at angle α to the horizontal.

5. Show that the steady circular flow inside a rotating cylinder is rigid body motion, and that the steady flow outside is that associated with a line vortex. Determine the couples associated with these two motions and explain your results.

6. The plane $y = 0$ oscillates so that its velocity is in the plane $y = 0$ and of magnitude $V \cos \omega t$. Show that the velocity of fluid above the plane is

$$U(y, t) = \mathscr{R}e\{V \exp[i\omega t - y(i\omega/v)^{1/2}]\},$$

where $i^2 = -1$ and $\mathscr{R}e$ means 'real part of'.

7. Show that the stress on the plane in §2(a) is

$$\mu U_0/(\pi v t)^{1/2}$$

against the direction of motion.

   This stress is very large for small values of t, so it is worth trying to solve a related problem for which the stress is kept constant and the velocity increases from zero at $t = 0$. We are now imposing

$$\mu(\partial U/\partial y)_{y=0} = -\mu S$$

where S is a constant with dimension (time)$^{-1}$. The problem has parameters

S and v

and coordinates

y and t.

Try to find a solution in the form

$$U(y, t) = ySF(\eta)$$
$$\eta = y/(4vt)^{1/2}.$$

You may not find any simple functional form for $F(\eta)$, but you should at least find a simple form for $F'(\eta)$, and you should have some conditions on F and F' to fix the constants of integration.

8. The gap between the cylinders $r = a$ and $r = b$ is full of liquid. The inner

Fig. IX.23. Definition diagram for Q8.

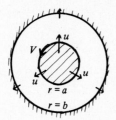

cylinder rotates at rate V/a and the outer is fixed. Liquid seeps through the cylinders to give a radial velocity

ua/r

throughout the flow, where u is a constant. The situation is sketched in fig. IX. 23. Show that the equation of continuity is satisfied by

$\mathbf{v} = U(r)\hat{\boldsymbol{\theta}} + (ua/r)\hat{\mathbf{r}},$

and show that the pressure is given by

$dp/dr = \rho U^2/r + \rho u^2 a^2/r^3.$

Find the equation for $U(r)$ and show it has a solution of the form

$U(r) = Ar^{R+1} + B/r$

where R is the Reynolds number of the seepage flow. Determine A and B from the boundary values, and interpret your results when R is large.

9. Fluid flows against a wall on which a plate AB has been welded. Discuss the relative sizes of the pressures at A, B, C, D, as marked in fig. IX.24. Hence discuss the likely secondary flow associated with the boundary layer on the plate AB, and show why separation of the boundary layer from the plate is to be expected. [Plate 7 in Batchelor's text shows this separation when a plate is present.]

10. Test the line vortex of §2(c) for stability by using Rayleigh's criterion in §1(d).

11. Fluid is at rest in a long channel with rigid walls $y = \pm a$ when a pressure gradient $-G$ is suddenly imposed at $t = 0$. Show that the velocity $U(y, t)\mathbf{i}$ satisfies the equation

$\partial U/\partial t = v\partial^2 U/\partial y^2 + G/\rho$

for $t > 0$, and state the boundary and initial conditions for this flow.

As $t \to \infty$ we expect to get the flow appropriate for a pressure gradient in a channel

$U_1(y) = G(a^2 - y^2)/2\mu,$

Fig. IX.24. Sketch of the initial flow for Q9.

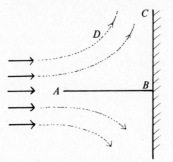

so seek a solution in the form

$$U(y, t) = U_1(y) + V(y, t):$$

what equation and boundary values does V satisfy?

Show that $V(y, t)$ may be found by separation of variables in terms of a Fourier cosine series and exponentials in time. How long does it take for the flow U_1 to be established? Explain this answer physically. [This exercise may be left until after Chapter XI.]

References

(*a*) Film loops 38.5052 and 28.5053 show the instability of circular flow between rotating cylinders, and the eventual onset of turbulence at very high Reynolds numbers. (Film loops FM-31 and 32.)

(*b*) The film loops 28.5074 and 38.5048 show the secondary flow associated with a sink–vortex combination. (FM-70 and 26.)

(*c*) The film loop 28.5042 shows the boundary layer of vorticity at the side of a water channel. (FM-14A. FM-6 and FM-88 are also useful on boundary layers.)

(*d*) The film loops 28.5040 and 38.5038 and plates 5, 6, 7, 8, 9, 10 of *Introduction to Fluid Dynamics*, G. K. Batchelor, C.U.P. 1967, show boundary layer separation and the secondary flow that precedes it, and also wakes from various bodies. (FM-12 and 10. FM-4 is useful on separated flows.)

(*e*) More general solutions involving viscosity can be found in Batchelor's text, chapters 4 and 5. His treatment takes some 200 pages, and shows how little is being attempted here.

(*f*) The mathematical treatment of singular perturbation problems can be read in chapter 9 of *Mathematics Applied to Deterministic Problems in the Natural Sciences*, C. C. Lin and L. A. Segel, Macmillan 1974.

X
Inviscid flow

1. Euler's equation

The examples of the last chapter suggest that viscosity is important in a flow in the following cases.

(i) When the overall Reynolds number is *low*, then viscous diffusion acts over most of the flow.

(ii) When the overall Reynolds number is *high*, then viscosity is important in thin boundary layers, vortex sheets and wakes.

(iii) When the flow is *enclosed*, as in a pipe or a cylinder, then the available diffusion time is large and viscosity is important in the whole flow, after some initial region or time. This may also be expected to be the case in a closed eddy, such as those observed at some speeds behind a cylinder in a stream, or in a Hill's vortex.

Thus we may hope that in high Reynolds number flow there will be large regions of the flow in which the viscosity is unimportant. A good example of such a flow is that of air past a streamlined strut, as shown in fig. X.1. At reasonably high speeds the boundary layer and wake will be so thin that, outside these thin regions, the flow will be given to a very good approximation by neglecting viscous terms in the equation of motion, and solving for the flow past the strut.

Fig. X.1. Boundary layer and wake at high Reynolds number.

It should be noted here that viscosity and heat conduction are both molecular effects of the same type – in one case momentum is transferred by random molecular motion, in the other case energy is transferred. So that when viscosity is neglected it also usually correct to neglect heat conduction. However there are some fluids (e.g. mercury) for which the diffusion coefficient in the heat equation is much larger than that in the momentum equation, and so heat boundary layers will be much thicker than momentum ones.

If we neglect viscous effects, the Navier–Stokes equation reduces in complexity to

$$\partial \mathbf{v}/\partial t + \mathbf{v}\cdot\nabla\mathbf{v} = \mathbf{F} - \rho^{-1}\nabla p.$$

This is known as Euler's equation, and it is to be solved with the mass conservation equation

$$\partial\rho/\partial t + \nabla\cdot(\rho\mathbf{v}) = 0$$

and some thermodynamic equations, which will usually be either

$$\rho = \text{constant}$$

for a liquid, or

$$\begin{cases} DS/Dt = 0 \\ p = R\rho T \end{cases}$$

for a gas, in which heat conduction and irreversibility are also neglected.

This reduced equation of motion was for many years taken to be the proper equation of motion for a fluid for the *whole* flow. The existence of viscosity was known, but it was taken to be a small effect, which would only modify the resulting flow slightly. It is only since it has been recognised that this small effect is dominant in regions near boundaries, that it has been possible to explain even in outline the flow of a stream past a cylinder. Without some appreciation of boundary layers, secondary flow, and separation, it is dangerous to look at solutions of Euler's equation: they may give quite wrong results. However, with careful use, Euler's equation can be a helpful simplification of the Navier–Stokes equation, giving sufficiently accurate solutions over large regions of the flow.

2. The vorticity equation

(a) Derivation and two-dimensional form of vorticity equation

It is not obvious that a flow without viscosity will be a flow without vorticity, and it need not even be true if the flow is set up properly.

So it is sensible to ask how vorticity changes, in a region where Euler's equation may be used. For the sake of simplicity we shall also use the model of constant density, and assume that the body force is gravitational or some other which is derived from a potential, so that $\mathbf{F} = -\nabla\Phi$.

The equations we are using are therefore

$$\begin{cases} \partial \mathbf{v}/\partial t + \mathbf{v}\cdot\nabla\mathbf{v} = -\nabla\Phi - \rho^{-1}\nabla p, \\ \rho = \text{constant.} \end{cases}$$

Let us take $\nabla \times$ Euler; it is easy to show that

$$\nabla \times \partial\mathbf{v}/\partial t = \partial\boldsymbol{\omega}/\partial t,$$

where $\boldsymbol{\omega} = \nabla \times \mathbf{v}$ is the vorticity, and it may be shown with more difficulty (see Chapter I, Q6), that

$$\nabla \times (\mathbf{v}\cdot\nabla\mathbf{v}) = \mathbf{v}\cdot\nabla\boldsymbol{\omega} - \boldsymbol{\omega}\cdot\nabla\mathbf{v},$$

where we have used $\nabla\cdot\mathbf{v} = 0$ and $\nabla\cdot\boldsymbol{\omega} = 0$. Since $\nabla \times \nabla f = 0$ for any scalar function f (see Chapter I, Q5), we have shown that the equation for vorticity in this simplified model is

$$\partial\boldsymbol{\omega}/\partial t + \mathbf{v}\cdot\nabla\boldsymbol{\omega} = \boldsymbol{\omega}\cdot\nabla\mathbf{v},$$

or

$$D\boldsymbol{\omega}/Dt = \boldsymbol{\omega}\cdot\nabla\mathbf{v}.$$

The interpretation of this vorticity equation is that the rate of change of vorticity as you follow the fluid is given by the term $\boldsymbol{\omega}\cdot\nabla\mathbf{v}$. Before trying to understand this term in a general situation, let us try it out in the easier cases we have considered previously, of two-dimensional or axisymmetric flows. In such flows there are stream functions, and we have already seen in Chapter V how vorticity behaves in general terms.

In two dimensions we have

$$\mathbf{v} = v_1(x, y)\mathbf{i} + v_2(x, y)\mathbf{j}$$

and so $\boldsymbol{\omega} = \omega(x, y)\mathbf{k}$. Hence the term $\boldsymbol{\omega}\cdot\nabla\mathbf{v}$ in the vorticity equation is

$$\omega(x, y)\partial/\partial z\{v_1(x, y)\mathbf{i} + v_2(x, y)\mathbf{j}\},$$

which is clearly zero, as the term $\{\dots\}$ is independent of z. This gives the interesting result that in two-dimensional flow

$$D\boldsymbol{\omega}/Dt = 0,$$

or vorticity is conserved. Remember that this is to be applied away from regions where viscosity is important, and that we have assumed that the density is constant. Note the alternative method of deriving this result in Chapter V, Q8.

(b) Example

Let us take an example to show this equation in action. The seepage boundary layer flow of Chapter IX §3 has vorticity

$$\omega = -(G/\rho u)\mathbf{k}$$

outside the boundary layer. Let us take this flow on into a region where the rock is impermeable; at the start of this region we shall still have constant vorticity $-G/\rho u$ outside the boundary layers, because this vorticity has moved in with the fluid, so this is the condition at the start of the new region. Consequently outside the boundary layer

$$\mathbf{v} = U(y)\mathbf{i} = (Gy/\rho u)\mathbf{i}$$

at this point, because the boundary layers cannot have changed suddenly and so we still need $U(0) = 0$.

Suppose that the channel contracts smoothly and slowly to width $\frac{1}{2}a$ over a distance L (as in fig. X.2) which is not so long that the boundary layers on the walls diffuse out into the flow very much. That is,

$$(\nu L/U_{\text{average}})^{1/2}$$

(an estimate of the diffusion distance) must be much less than $\frac{1}{4}a$: this is equivalent to

$$16L/a \ll R,$$

where R is the Reynolds number based on the width a and the average velocity. This condition can easily be satisfied.

In the bulk of the channel then, we have that the vorticity is preserved as it moves along with the fluid. So that at the end of the contraction, where the velocity will again be parallel to the parallel walls, the vorticity is still $-(G/\rho u)\mathbf{k}$. If the velocity is now

$$\mathbf{v} = V(y)\mathbf{i},$$

this means that

$$-dV/dy = -G/\rho u,$$

so that

$$V(y) = V_0 + Gy/\rho u,$$

Fig. X.2. A flow through a contracting section in a channel.

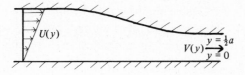

where V_0 is a constant. Because the vorticity is constant (moving with the fluid) and because the velocity at both ends of the contraction is parallel to **i** and varying only with y, we have been able to find its form at the downstream end rather easily. The constant V_0 can also be found by using the conservation of mass. The mass flow into the contraction is

$$\int_0^a \rho U(y)dy$$

which, neglecting the boundary layer, has value

$$\tfrac{1}{2}Ga^2/u.$$

The mass flow out of the contraction is, in the same way, $\tfrac{1}{2}\rho V_0 a + Ga^2/8u$. Since they must be equal,

$$V_0 = 3Ga/4\rho u.$$

Thus there is a slip velocity V_0 at the wall $y = 0$, which has to be taken up in a boundary layer there; and at the other wall the slip velocity is now

$$5Ga/4\rho u.$$

For this rather simple flow we have in effect solved the Euler equation by using the vorticity equation, which was derived from it. The only further information we can get from the Euler equation is about the pressure, and we will find this later. Notice that in this example we have *not* solved for **v** throughout the flow: that would be hard to do, and in this case is unnecessary.

(c) *The stream function equation in steady two-dimensional flow*

In a two-dimensional steady flow then, we have that ω is constant for any fluid particle. But we also know that there is a stream function ψ which is constant on a streamline, and that the particles move along the streamlines (because the flow is steady). So in this very simple kind of flow, both ψ and ω are constant along the streamlines.

Now if $\psi(x, y)$ is constant on the curve $y = h(x)$, then

$$\psi\{x, h(x)\} = \text{constant}$$

and so, differentiating,

$$\frac{\partial \psi}{\partial x} + \frac{\partial \psi}{\partial y}\frac{dh}{dx} = 0.$$

This simple partial differential equation has the solution

$$\psi(x, y) = F\{y - h(x)\},$$

for some function F. But in a similar fashion

$$\omega(x, y) = G\{y - h(x)\}$$

for some function G, because ω is also constant on the same curve, the streamline. It follows directly that

ω is a function of ψ

in two-dimensional steady flows for which Euler's equation holds and ρ is constant. Finally, since

$$\omega = -\nabla^2\psi$$

in such flows, we have the equation

$$\nabla^2\psi = f(\psi)$$

for some function f.

This equation for ψ is of course just equivalent to the equation for ω in such flows,

$$\mathbf{v}\cdot\nabla\omega = 0,$$

or to Euler's equation itself. And though it looks simple, it is non-linear except in the special cases

$$\nabla^2\psi = \alpha\psi + \beta$$

for some constants α and β. So we can only rarely solve it.

The flow in (b) above through the contraction is an example of this work. Upstream we have

$$\psi(y) = \tfrac{1}{2}Gy^2/(\rho u)$$

and

$$\nabla^2\psi = d^2\psi/dy^2 = G/(\rho u).$$

This is the case

$$\alpha = 0, \beta = G/(\rho u)$$

in the linear relation

$$\nabla^2\psi = \alpha\psi + \beta.$$

A more complicated example is provided by another flow through the same contraction. Take the upstream velocity profile to be

$$\mathbf{v} = U(y)\mathbf{i}$$

where

$$U(y) = U_0 \sin(\pi y/a),$$

Fig. X.3. Sinusoidal velocity in a channel.

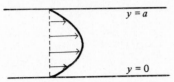

as sketched in fig. X.3. This is rather similar to the Poiseuille flow of Chapter IX. In this part of the flow, and hence downstream as well for distances such that the diffusion of vorticity from the walls is small,

$$\nabla^2 \psi = f(\psi)$$

where (upstream)

$$\psi = (U_0 a/\pi)\{1 - \cos(\pi y/a)\}$$

and

$$\nabla^2 \psi = (U_0 \pi/a)\cos(\pi y/a).$$

So we deduce the form of the function f:

$$f(\psi) = -(\pi/a)^2 \psi + \pi U_0/a.$$

Downstream of the contraction, we have to solve

$$\nabla^2 \psi = -(\pi/a)^2 \psi + \pi U_0/a,$$

with $\psi = 0$ on $y = 0$, and $\psi = 2U_0 a/\pi$ on $y = \frac{1}{2}a$ (the upper wall in fig. X.4) to get the two walls to be streamlines of the flow. In this region we again have a velocity profile of the form

$$\mathbf{v} = V(y)\mathbf{i}$$

and so we must solve

$$d^2\psi/dy^2 + (\pi/a)^2\psi = \pi U_0/a$$

with the given boundary values. The solution is

$$\psi = (aU_0/\pi)\{1 + \sin(\pi y/a) - \cos(\pi y/a)\},$$

and so

$$V(y) = U_0\{\cos(\pi y/a) + \sin(\pi y/a)\}.$$

Fig. X.4. The channel contracts from width a to width $\frac{1}{2}a$.

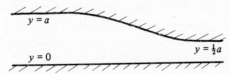

This velocity profile has a slip velocity U_0 at each wall, which must be taken up by boundary layers.

(d) The vorticity equation in axisymmetric flow

In an axisymmetric flow the term $\boldsymbol{\omega} \cdot \nabla \mathbf{v}$ is not so easy. Let the velocity field be

$$\mathbf{v} = U(r, z)\hat{\mathbf{r}} + W(r, z)\mathbf{k}.$$

Then

$$\boldsymbol{\omega} = (\partial U/\partial z - \partial W/\partial r)\hat{\boldsymbol{\theta}} = \omega\hat{\boldsymbol{\theta}},$$

and

$$\boldsymbol{\omega} \cdot \nabla \mathbf{v} = (\partial U/\partial z - \partial W/\partial r)r^{-1}\partial/\partial\theta(U\hat{\mathbf{r}} + W\mathbf{k}).$$

The only non-zero term comes from $\partial\hat{\mathbf{r}}/\partial\theta$, and may be written as

$$(\omega U/r)\hat{\boldsymbol{\theta}}.$$

Hence the vorticity equation in this case has the form

$$D(\omega\hat{\boldsymbol{\theta}})/Dt = (U/r)\omega\hat{\boldsymbol{\theta}}.$$

We may now interpret this equation: as a piece of fluid moves, its vorticity increases if U and ω have the same sign. That is, a velocity *out* from the axis increases positive vorticity and decreases negative vorticity. More-over, the rates of change are proportional to both U/r and ω. Consider a vortex line round the z-axis; the rate at which its length increases is $2\pi U$, and so it increases at rate U/r per unit length. The rate of increase of vorticity is therefore equal to the rate of stretching of the vortex line (per length) multiplied by the vorticity. In short

$$D(\omega\hat{\boldsymbol{\theta}})/Dt = (\omega\hat{\boldsymbol{\theta}}/l)Dl/Dt,$$

where l is the length of a vortex circle (a closed vortex line round the z-axis) at which the vorticity is $\omega\hat{\boldsymbol{\theta}}$. This may be rewritten as

$$D/Dt(\omega\hat{\boldsymbol{\theta}}/l) = 0,$$

or just as $\omega/l = $ constant following the fluid in an axisymmetric flow. And since l is proportional to r (the distance from the axis), we finish with

$$\omega/r = \text{constant}.$$

This is a partial justification for the form of the vorticity chosen in a Hill's vortex in Chapter V. At least we have satisfied the vorticity form of Euler's equation by taking $\omega/r = $ constant in the region of non-zero vorticity. In reality there would be viscous effects throughout this closed region, and so vorticity would be spread out by diffusion. But for any short time the diffusive effects would be small and so, as a vortex ring moved round in the Hill's vortex, its vorticity would change according to $\omega/r = $ constant.

(e) The vorticity equation in general flows

In general flows the interpretation of the right hand side of

$$D\omega/Dt = \boldsymbol{\omega}\cdot\nabla\mathbf{v}$$

is similar to that just given for the easier axisymmetric flows. Take coordinate s along a vortex line, so that

$$\boldsymbol{\omega}\cdot\nabla = \omega\partial/\partial s.$$

Then if we write $\mathbf{v} = v_1\hat{\mathbf{s}} + v_2\hat{\mathbf{n}}$, where $\hat{\mathbf{s}}$ is along the line and $\hat{\mathbf{n}}$ is perpendicular to it, we see that there is a term in

$$\omega\hat{\mathbf{s}}\partial v_1/\partial s$$

as well as the terms in $\omega\partial/\partial s(v_2\hat{\mathbf{n}})$ and $\omega v_1\partial\hat{\mathbf{s}}/\partial s$. Now $\partial v_1/\partial s$ is about changes in v_1 as you go along the vortex line, and these changes stretch the line and give (if positive) an increase of vorticity. The other terms in $\partial/\partial s(v_2\hat{\mathbf{n}})$ and $\omega v_1\partial\hat{\mathbf{s}}/\partial s$ in fact cause (this is too hard to demonstrate here) the vorticity to change just enough to keep up with the rotation of the vortex line: we have seen previously that vortex lines move with the fluid, and these terms are the ones that keep the two together.

It will be noticed that the vorticity equation is purely in terms of $\mathbf{v}$ and its derivatives. This, therefore, is an equation that one might hope to solve for $\mathbf{v}$; whereas Euler's equation also contains p (and perhaps $\mathbf{F}$ or Φ), and can only give p and $\mathbf{v}$ in terms of each other. This suggests that we should solve the vorticity equation for $\mathbf{v}$ and then derive p from Euler's equation, using the continuity equation to eliminate unsuitable flows on the way. Such a programme of action is reasonable, but hard, because the vorticity equation contains non-linearities in $\mathbf{v}\cdot\nabla\omega$ and $\omega\cdot\nabla\mathbf{v}$, and has second derivatives of $\mathbf{v}$ in the $\mathbf{v}\cdot\nabla\omega$ term; moreover there are three components to solve for. Even the 'easy' cases, when $\mathbf{v}$ is derived from a stream function, are hard, and we leave them for further study elsewhere.

3. Kelvin's theorem

Our earlier discussions of the motion of vortex lines, in Chapter V, were in terms of Kelvin's theorem – that circulation is constant following the fluid. No proof was given at that stage, so we shall now set out to prove Kelvin's theorem from Euler's equation for a fluid of constant density.

Fig. X.5. Definition sketch for Kelvin's theorem.

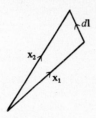

We have to show that

$$D/Dt \oint \mathbf{v} \cdot d\mathbf{l} = 0.$$

We approach the problem by calculating

$$D/Dt(\mathbf{v} \cdot d\mathbf{l})$$

for a small section of the closed circuit (shown in fig. X.5), and then integrating round the circuit. Consider fluid at $\mathbf{x}_1$: in a short time δt it moves to

$$\mathbf{x}_1 + \mathbf{v}(\mathbf{x}_1)\delta t$$

And fluid at $\mathbf{x}_2$ moves to

$$\mathbf{x}_2 + \mathbf{v}(\mathbf{x}_2)\delta t.$$

Thus the element $d\mathbf{l} = \mathbf{x}_2 - \mathbf{x}_1$ is changed to

$$\mathbf{x}_2 + \mathbf{v}(\mathbf{x}_2)\delta t - \mathbf{x}_1 - \mathbf{v}(\mathbf{x}_1)\delta t = d\mathbf{l} + \{\mathbf{v}(\mathbf{x}_2) - \mathbf{v}(\mathbf{x}_1)\}\delta t.$$

Now if we put $\mathbf{x}_2 = \mathbf{x}_1 + d\mathbf{l}$ and use Taylor's theorem, we get (to first order)

$$\mathbf{v}(\mathbf{x}_2) = \mathbf{v}(\mathbf{x}_1) + d\mathbf{l} \cdot \nabla \mathbf{v}.$$

Hence $d\mathbf{l}$ is changed to $d\mathbf{l} + d\mathbf{l} \cdot \nabla\mathbf{v}\delta t$ by the fluid motion in time δt. Which is just to say that

$$D/Dt(d\mathbf{l}) = d\mathbf{l} \cdot \nabla\mathbf{v}.$$

It is easily shown that

$$D/Dt(\mathbf{v} \cdot d\mathbf{l}) = d\mathbf{l} \cdot D\mathbf{v}/Dt + \mathbf{v} \cdot D(d\mathbf{l})/Dt,$$

but the last term needs careful treatment; it is

$$v_i D(dl_i)/Dt = v_i dl_j \partial v_i / \partial x_j$$

from above, which may be rearranged as

$$dl_j \partial / \partial x_j (\tfrac{1}{2} v_i v_i).$$

So we may now collect up terms and write $D/Dt(\mathbf{v} \cdot d\mathbf{l})$ as

$$d\mathbf{l} \cdot \{D\mathbf{v}/Dt + \nabla(\tfrac{1}{2}v^2)\}.$$

Finally we use Euler's equation to replace $D\mathbf{v}/Dt$ by $-\nabla p/\rho - \nabla\Phi$, and adding up round the circuit leaves (remember that ρ is a constant)

$$D/Dt \oint \mathbf{v} \cdot d\mathbf{l} = - \oint \nabla\{p/\rho + \Phi - \tfrac{1}{2}v^2\} \cdot d\mathbf{l}.$$

This last integral is zero, because it integrates exactly to

$$[p/\rho + \Phi - \tfrac{1}{2}v^2],$$

and the difference of this function round a closed circuit is zero. Hence in this approximation we have proved

$$D/Dt \oint \mathbf{v} \cdot d\mathbf{l} = 0,$$

which is Kelvin's theorem.

It is worth asking, and not very hard to find out, what this equation becomes if you don't make so many approximations.

(i) If ρ is not constant but we still have constant entropy S (or some other reason) so that

$$\rho = f(p),$$

then the term

$$\rho^{-1}\nabla p$$

in Euler's equation can be replaced by

$$\nabla \int \{f(p)\}^{-1} dp.$$

This is still a term that can be integrated round the circuit to give zero and so Kelvin's theorem is not modified.

(ii) If ρ and p are not so simply related as in (i), e.g. if S is not constant because of heat flow or irreversibility, the term in

$$\oint \rho^{-1}\nabla p \cdot d\mathbf{l}$$

is not zero. Stokes' theorem gives it the form

$$- \int \rho^{-2}(\nabla\rho \times \nabla p) \cdot d\mathbf{S},$$

which shows that there is a change in circulation when the surfaces

$p = $ constant are not the same as the surfaces $\rho = $ constant. This can certainly happen in the lower atmosphere, though Kelvin's theorem is still a good first approximation.

(iii) If you add a term $\nu\nabla^2 \mathbf{v}$ to Euler's equation to get the Navier–Stokes equation for constant density ρ, then you must add a term

$$\nu \oint \nabla^2 \mathbf{v} \cdot d\mathbf{l}$$

in Kelvin's theorem. Writing this in the form

$$\nu \int \nabla^2 \boldsymbol{\omega} \cdot d\mathbf{S}$$

with the help of Stokes' theorem shows that in regions where vorticity is being noticeably diffused, Kelvin's theorem will be inaccurate. It will also be inaccurate if the time available for diffusion is large enough for this term to become important.

The conclusion then is that Kelvin's theorem will be a good approximation in high Reynolds number flows over reasonable lengths of time provided that the circuit round which you measure the circulation stays clear of regions where viscosity or heat flow are important.

You should return to the examples in Chapter V §5(a) and (b) to see how the approximations needed for Kelvin's theorem are not satisfied in these real situations.

4. Bernoulli's equation

(a) *Pressure in a steady flow*

It was said above that Euler's equation should be thought of as an equation for the pressure once the velocity was known. This can be shown quite simply in the easiest case, and we do so now.

Consider steady flow of a fluid of constant density. Euler's equation may be rearranged by using the identity

$$\mathbf{v} \cdot \nabla \mathbf{v} = \nabla(\tfrac{1}{2}v^2) - \mathbf{v} \times \boldsymbol{\omega}$$

(see Chapter I); we get

$$\mathbf{v} \times \boldsymbol{\omega} = \nabla\{\tfrac{1}{2}v^2 + \Phi + p/\rho\}.$$

This rather unpromising equation in fact reveals the pressure, as follows. Take a vector $\hat{\mathbf{s}}$ along a streamline, so that $\hat{\mathbf{s}}$ is parallel to $\mathbf{v}$; then

$$\hat{\mathbf{s}} \cdot (\mathbf{v} \times \boldsymbol{\omega}) = 0$$

because there are two parallel vectors in the scalar triple product. Consequently

$$\hat{s}\cdot\nabla\{\tfrac{1}{2}\mathbf{v}^2 + \Phi + p/\rho\} = 0$$

which means that the derivative along the streamline of $\tfrac{1}{2}\mathbf{v}^2 + \Phi + p/\rho$ is zero. We have proved that

$$\tfrac{1}{2}\mathbf{v}^2 + \Phi + p/\rho = \text{constant}$$

along any streamline in a steady flow of inviscid fluid of constant density. This is Bernoulli's equation in its most basic form. Provided that the constant on a streamline is known from the boundary conditions of a problem, the pressure at any other point on that streamline is given in terms of the velocity $\mathbf{v}$ and force potential Φ at that other point.

(b) Example

As an example of this theorem, let us consider the flow of water from a barrel through a small pipe in the side. The flow is almost steady at any time, and if the pipe is not too narrow or too long, the effects of viscosity will be confined to boundary layers. Streamlines will start at the upper surface, and converge through the pipe, roughly as shown in fig. X.6; we will consider the streamline PQ.

At P, the downward speed is V, the gravitational potential (above that at Q) is gh, and the pressure is atmospheric pressure p_0. At Q, the speed out of the pipe is v, the gravitational potential is zero, and the pressure is again p_0 (as we have seen that pressure is continuous across boundary layers). Hence Bernoulli's equation is

$$\tfrac{1}{2}V^2 + gh + p_0/\rho = \tfrac{1}{2}v^2 + p_0/\rho.$$

But we can derive another relation between V and v from conservation of mass. The downward flow at the surface is given in terms of the cross-sectional area $A(h)$ of the barrel by

$$\rho V A(h)$$

Fig. X.6. A flow out of a barrel.

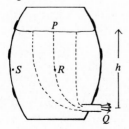

and this must equal the flow out of the pipe, so that

$$VA(h) = va,$$

where a is the area of cross-section of the pipe. Taking these two equations together gives an equation for $h(t)$, because $V = -\,dh/dt$. This equation for h is

$$V^2(A^2/a^2 - 1) = 2gh$$

or

$$dh/dt = -\,\{2gh(A^2/a^2 - 1)^{-1}\}^{1/2}.$$

The solution is easy when the barrel is of constant cross-sectional area A, and is made even easier when $A \gg a$ – which it will need to be to justify the model of steady flow. In this simplified case we may take

$$dh/dt = -\,(2gha^2/A^2)^{1/2}$$

and

$$h = \{-\,(\tfrac{1}{2}ga^2/A^2)^{1/2}t + h_0^{1/2}\}^2$$

taking $h = h_0$ at $t = 0$. The time to empty the barrel on this simple model is

$$(A/a)(2h_0/g)^{1/2}.$$

Before we leave this example of Bernoulli's equation, in which the pressure seems to have played little part, let us ask what the pressure is at the point R on the streamline PQ. With R as shown in fig. X.6, the local velocity will still be close to V and so relatively small. So at R we have, from Bernoulli's equation for PR,

$$\tfrac{1}{2}V^2 + gh_R + p_R/\rho = \tfrac{1}{2}V^2 + gh + p_0/\rho.$$

That is, the pressure at R is just the hydrostatic pressure

$$p_0 + \rho g(h - h_R).$$

The upper part of the flow is controlled by the weight of the fluid, which is almost exactly in balance with the pressure. Further down, between R and the pipe entry, the pressure falls to atmospheric pressure p_0 while the fluid accelerates to speed v.

Finally, consider the point S on the inside of the barrel. It is on a streamline from the fluid surface to the pipe, since a solid wall must always be a streamline. So p_S will also be approximately the hydrostatic pressure appropriate to that depth. This cannot be the entire truth about pressure forces on the sides of the barrel, as there must be a net sideways force on the fluid to cause a jet to emerge with sideways momentum from the pipe. The pressures on the side wall near the exit pipe must be reduced below those on the parts away from the pipe, because of the locally higher speeds.

It must be emphasized that this is an approximate piece of work. In reality, the shape of the exit pipe matters, and there are viscous forces to be overcome. But this example does show the power and use of Bernoulli's equation, and gives fairly successful answers.

(c) *Further relations on the pressure*

Bernoulli's equation was derived from

$$\mathbf{v} \times \boldsymbol{\omega} = \nabla(\tfrac{1}{2}\mathbf{v}^2 + \Phi + p/\rho).$$

We may make further deductions from this equation.

(i) $\tfrac{1}{2}\mathbf{v}^2 + \Phi + p/\rho$ is constant along a vortex line. This is not often used, as the vortex lines are less directly observable, and not much can be made out of this statement.

(ii) $\tfrac{1}{2}\mathbf{v}^2 + \Phi + p/\rho$ is constant *throughout* the fluid when $\mathbf{v} \times \boldsymbol{\omega} = 0$. The main case here is when $\boldsymbol{\omega} = 0$, as flows with non-zero $\boldsymbol{\omega}$ for which $\mathbf{v} \times \boldsymbol{\omega} = 0$ are rather specialised. We return to study the case $\boldsymbol{\omega} = 0$ in the next chapter.

(iii) Regions of *uniform* vorticity can occur in two-dimensional flows as we have seen in examples previously. In such cases we have

$$\begin{cases} \mathbf{v} = \nabla\psi \times \mathbf{k}, \\ \boldsymbol{\omega} = \omega\mathbf{k}, \\ \mathbf{v} \times \boldsymbol{\omega} = -\omega\nabla\psi. \end{cases}$$

Hence the equation above reduces in the case $\omega = $ constant to

$$\nabla(\omega\psi + \tfrac{1}{2}\mathbf{v}^2 + \Phi + p/\rho) = 0,$$

so that

$$\omega\psi + \tfrac{1}{2}\mathbf{v}^2 + \Phi + p/\rho$$

is constant everywhere.

Bernoulli's equation can also be regarded in another way, which does not directly bring in the vorticity. The basic form of Euler's equation that we are considering here is (neglecting gravity)

$$\mathbf{v}\cdot\nabla\mathbf{v} = -\rho^{-1}\nabla p$$

where ρ is a constant. Let us take coordinates (in two dimensions)

$$\begin{cases} s \text{ as distance along the streamlines,} \\ n \text{ as distance perpendicular to the streamlines.} \end{cases}$$

This will of course be (in general) a curved system of coordinates.

In this system

$$\mathbf{v} = v\hat{\mathbf{s}},$$

and so

$$\mathbf{v}\cdot\nabla\mathbf{v} = v\partial/\partial s(v\hat{\mathbf{s}}).$$

Now

$$\partial\hat{\mathbf{s}}/\partial s$$

is not zero, because the unit vector $\hat{\mathbf{s}}$ changes direction as you go along a streamline, and it is a standard result in vector geometry that

$$\partial\hat{\mathbf{s}}/\partial s = -R^{-1}\hat{\mathbf{n}},$$

where R is the local radius of curvature of the streamline. So

$$\mathbf{v}\cdot\nabla\mathbf{v} = (v\partial v/\partial s)\hat{\mathbf{s}} - R^{-1}v^2\hat{\mathbf{n}}.$$

On taking components of the equation of motion, we get

$$\begin{cases} \rho v\partial v/\partial s = -\partial p/\partial s, \\ -\rho v^2/R = -\partial p/\partial n. \end{cases}$$

The first of these integrates to give

$$p + \tfrac{1}{2}\rho v^2 = \text{constant on a streamline}$$

and the second shows how the pressure varies from one streamline to the next.

The second result here can sometimes be useful in interpreting a given flow when there is vorticity present; if there is no vorticity, then Bernoulli's equation already gives all the information, as

$$p + \tfrac{1}{2}\rho v^2$$

is constant everywhere. Take as example the flow, initially uniform across the channel, through a contraction, sketched in fig. X.7. The upper stream-line $ABCDE$ starts straight before A, is concave inwards at B, has zero curvature at C, is convex inwards at D, and becomes straight again after E. The lower streamline $PQRST$ is always straight. The pressure before AP

Fig. X.7. Definition sketch for §4(c).

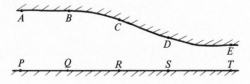

is uniform across the channel, and it is again uniform after ET but with a lower value because the speed is higher.

The streamlines at CR are locally of zero curvature approximately, and so

$$p_C = p_R.$$

But there is curvature at both BQ and DS, so that

$$p_B > p_Q,$$
$$p_D < p_S,$$

since the curvature has opposite signs. It follows that

$$v_Q^2 - v_P^2 > v_B^2 - v_A^2$$

because there has been a greater pressure drop from P to Q than from A to B. Thus the flow initially distorts with the fluid near the lower wall getting ahead of fluid near the upper wall. And after this we have

$$\mathbf{v}_C^2 \doteqdot \mathbf{v}_R^2$$

as the pressures are almost equal, though here $\mathbf{v}_C$ is not parallel to $\mathbf{v}_R$ so that the fluid near R is still moving *along* the channel faster than fluid near C. However the fluid near the upper wall accelerates faster from C to E because $p_D < p_S$, and so finally arrives at E with the same speed as the lower fluid. If two particles start near A and P with the same velocity, the one near P is ahead when they pass out of the contraction.

(d) Bernoulli's equation and energy

It looks as though Bernoulli's equation is about energy per mass of a fluid particle as it moves along a streamline in steady flow, because it contains an obvious kinetic energy term $\frac{1}{2}v^2$, and an equally obvious potential energy term Φ from the external force field. This must mean that the remaining term p/ρ is also an energy of some sort–it can only be an internal energy, and as most of the internal energy is tied up in molecular motions, it must be the amount of available internal energy which can be exchanged into kinetic or potential energy. In a compressible fluid this available internal energy will be related to the thermodynamic processes going on; these may usually be modelled either by the isothermal relation $p = \text{constant} \times \rho$ or the isentropic relation $p = \text{constant} \times \rho^{\gamma}$.

In either case we have

$$\rho = \rho(p)$$

and so the term

$$\rho^{-1}\nabla p$$

in Euler's equation is expressible as

$$\nabla\left\{\int[\rho(p)]^{-1}dp\right\}$$

or

$$\nabla\left\{\int\rho^{-1}(dp/d\rho)d\rho\right\}.$$

For flows of this type, Bernoulli's equation is derived from

$$\mathbf{v}\times\omega=\nabla\left(\tfrac{1}{2}\mathbf{v}^2+\Phi+\int\rho^{-1}dp\right),$$

and the basic form of it is

$$\tfrac{1}{2}\mathbf{v}^2+\Phi+\int\rho^{-1}dp=\text{constant}$$

along a streamline. In the isentropic model, which is the more useful one as slow isothermal flows will be influenced by viscosity over wide areas,

$$\int\rho^{-1}dp=\frac{\gamma}{\gamma-1}\frac{p}{\rho}.$$

This is larger than p/ρ, since γ is usually 1.4, and so the pressure term is more important in Bernoulli's equation for a compressible gas.

Since Bernoulli's equation is about energy, it should be derivable from the thermodynamic equations that we have seen previously. This can indeed be done, though we must add terms representing kinetic and potential energy into the first law statement about internal energy

$$dE=\Delta Q+\Delta W.$$

We shall not pursue this line further, as the derivation from the equation of motion is sufficient.

5. Examples using Bernoulli's equation

(a) The Rankine vortex

Let us take a look at the pressure field in the Rankine vortex, which has velocity distribution (derived in Chapter V §3)

$$\begin{cases}\mathbf{v}=\tfrac{1}{2}r\Omega\hat{\theta},\ r<a,\\ \mathbf{v}=\tfrac{1}{2}\Omega a^2/r\ \hat{\theta},\ r>a.\end{cases}$$

This is a steady flow and there are no boundaries at which viscous or heat conduction effects are likely to be important; the interface $r = a$ is a discontinuity of velocity gradient, and so viscous forces will in reality act there, but we shall neglect this region, and assume that Euler's equation applies everywhere.

Assume also that gravity acts in the z-direction, with potential energy

$$\Phi = gz.$$

Then for the outer region the Bernoulli equation applies everywhere, since $\omega = 0$:

$$\tfrac{1}{8}\Omega^2 a^4/r^2 + gz + p/\rho = \text{constant}.$$

If we call the constant p_0/ρ, then the pressure for $r > a$ is given by

$$p = p_0 - \rho\Omega^2 a^4/8r^2 - \rho gz.$$

For the region $r < a$, the vorticity is constant,

$$\omega = \nabla \times \mathbf{v} = \Omega\mathbf{k}$$

and the stream function may be taken as

$$\psi = -\tfrac{1}{4}r^2\Omega,$$

so that the version of Bernoulli's equation that allows for constant vorticity gives

$$-\tfrac{1}{4}\Omega^2 r^2 + \tfrac{1}{8}\Omega^2 r^2 + gz + p/\rho = \text{constant}$$

for the whole region $r < a$. We choose the constant to make the pressure continuous at $r = a$ – a discontinuity of pressure across a surface would give an infinite acceleration to an element of the surface. This gives a value of

$$-\tfrac{1}{4}\Omega^2 a^2 + p_0/\rho$$

for the constant, and so for $r < a$ we derive

$$p = p_0 - \tfrac{1}{4}\rho\Omega^2 a^2 + \tfrac{1}{8}\rho\Omega^2 r^2 - \rho gz.$$

It is important to notice that in this inner region

$$\tfrac{1}{2}\mathbf{v}^2 + \Phi + p/\rho$$

is still constant on each streamline $r = \text{constant}$, but the value of the Bernoulli constant now changes from streamline to streamline in such a way that the final equation for the pressure looks rather different. This is caused by the presence of vorticity in this region.

If such a Rankine vortex is set up in a large quantity of water with a free surface, then the pressure at the free surface must be everywhere

Fig. X.8. The water surface above a Rankine vortex.

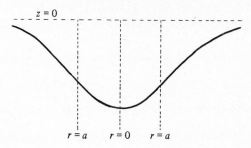

atmospheric, say p_0. This must also be the pressure inside the water at the surface, and so the shape of the surface will be given by

$$p_0 = p_0 - \rho\Omega^2 a^4/8r^2 - \rho gz, \quad r > a,$$
$$p_0 = p_0 - \tfrac{1}{4}\rho\Omega^2 a^2 + \tfrac{1}{8}\rho\Omega^2 r^2 - \rho gz, \quad r < a.$$

That is

$$\begin{cases} z = \Omega^2 a^4/8gr^2, & r > a, \\ z = -\Omega^2 a^2/4g + \Omega^2 r^2/8g, & r < a. \end{cases}$$

The surface at large distances is $z = 0$, dipping smoothly down, first according to

$$z \propto r^{-2},$$

and then near the centre of the vortex in a parabola, as shown in fig. X.8. This is close to what is observed. However this model does have a discontinuity in d^2z/dr^2 at $r = a$.

It is worth remarking that the viscous vortex of Chapter IX §2 has a pressure field which would not have this discontinuity; but this pressure field has no simple form, so we shall not attempt to see how the discontinuity is smoothed away.

(b) Force on a cylinder in a stream

Bernoulli's equation is often used to give the pressure force on the surface of a body which is in some flow. This seems a little unfair, since at the surface there will be significant viscous effects and Bernoulli's equation assumes an inviscid fluid. However, we have seen that pressure is continuous through a thin boundary layer, so that for flow at high Reynolds number we may indeed use Bernoulli's equation just outside the boundary layer, and then say that this gives the correct surface pressure. This of course depends on the boundary layer remaining on the surface,

Fig. X.9. The force due to the pressure in a model flow past a cylinder.

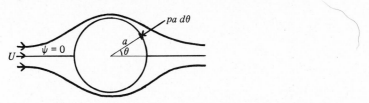

and so such a method should not be used for unstreamlined bodies unless you are sure that the flow remains attached.

The simple model of flow past a cylinder that we had in Chapter IV, and which is sketched in fig. X.9, is

$$\psi = Ur \sin \theta - (Ua^2 \sin \theta)/r.$$

This is a reasonable model for the initial flow past a cylinder, before separation has had time to take place. Hence it is fairly sensible to calculate the force on the cylinder (parallel to the stream, by symmetry) from this stream function and Bernoulli's equation.

The surface is the streamline $\psi = 0$, and on it we have that

$$p + \tfrac{1}{2}\rho v^2$$

is a constant. Now $\mathbf{v}$ on the surface is $-2U \sin \theta \, \hat{\boldsymbol{\theta}}$, and the constant may be found from values at infinity, where $p = p_0$ (say) and $\mathbf{v} = U\mathbf{i}$. Thus we have

$$p + 2\rho U^2 \sin^2 \theta = p_0 + \tfrac{1}{2}\rho U^2$$

on the surface $r = a$.

The force parallel to the stream is (per unit length of cylinder)

$$-\int_{-\pi}^{\pi} pa \cos \theta d\theta,$$

because the force on each element of area $ad\theta$ is $pad\theta$ acting towards the centre. When we use the formula for p, we find that there is *no* force. This result should not be surprising because of the symmetry of the flow pattern upstream and downstream. In reality of course there is a force; this apparent contradiction just reflects the poorness of this model – during the initial period when the streamlines are of about the right shape, the flow is not steady; later the flow is more steady, but the boundary layers have separated to give a wide wake.

A similar calculation can be done when there is a circulation κ round the cylinder, and

$$\psi = Ur \sin \theta - (Ua^2 \sin \theta)/r - (\kappa/2\pi)\ln(r/a).$$

Fig. X.10. Definition sketch for the force components on a cylinder with circulation round it.

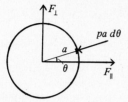

Provided that

$$|\kappa| \leqslant 4\pi U a$$

there is a stagnation point on the cylinder and a streamline from infinity forms the surface of the cylinder. Then on this streamline we may use Bernoulli's equation (under the same assumptions as before) to get

$$p + \tfrac{1}{2}\rho(2U \sin\theta - \kappa/2\pi a)^2 = p_0 + \tfrac{1}{2}\rho U^2.$$

We must now calculate both components of the force on the cylinder, say $F_\parallel$ and $F_\perp$ as indicated on fig. X.10. As before

$$F_\parallel = -\int_{-\pi}^{\pi} pa \cos\theta d\theta = 0$$

but

$$F_\perp = -\int_{-\pi}^{\pi} pa \sin\theta d\theta$$

$$= -\rho U \kappa/\pi \int_{-\pi}^{\pi} \sin^2\theta d\theta$$

(other terms vanish)

$$= -\rho U \kappa.$$

This transverse (lift) force is easily understood. The circulation gives a

Fig. X.11. Distortion of the streamlines when there is a circulation, and the resulting force.

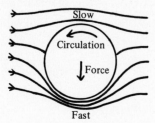

higher speed (shown by closer streamlines in fig. X.11) on one side of the cylinder. This higher speed is associated with lower pressure since

$$p + \tfrac{1}{2}\rho v^2 = \text{constant},$$

and hence there is a force from the high pressure (slow) side to the low pressure (fast) side. This force is sometimes called the Magnus force, and it is readily observed in reality to have approximately this value. Note that its direction is given by

$$U\mathbf{i} \times \boldsymbol{\omega}$$

where $\boldsymbol{\omega}$ is a vorticity that would give the right sense of circulation round the cylinder: a term $\rho\mathbf{v} \times \boldsymbol{\omega}$ can be seen as a force per volume.

This transverse force on a cylinder can be observed at certain flow velocities of a stream past a cylinder. For Reynolds numbers around 100, vorticity is shed from the region near the cylinder into the wake, first vorticity of one sign, then an equal amount of the opposite sign, and so on; as vorticity is shed, a circulation in the opposite sense is left round the cylinder (by Kelvin's theorem, as a large circuit round the cylinder and wake has initially – and so always – no circulation round it). These circulations of alternating sign give alternating forces on the cylinder, which may lead to sound (Aeolian tones) or vibration. Some modern factory chimneys are fitted with spiral flanges to break up this regular eddy shedding and prevent such vibrations.

Another example of this transverse force occurs with aircraft wings, for which the flow is sketched in fig. X.12. These are designed to have higher velocities over the top surfaces than over the bottom, and hence there is an upward force on the wing. The calculation of such forces for simple wing shapes is left until later, when better techniques are available.

(c) *Measuring devices*

There are many practical devices which can be described in terms of Bernoulli's equation. A 'pitot tube' is a robust and simple device for measuring (e.g. aeroplane) speeds by way of the pressures in a flow. A

Fig. X.12. Flow past a wing.

Fig. X.13. Diagram of a pitot tube.

probe with two holes in it as shown in fig. X.13 is put into the flow and the pressure difference $p_2 - p_1$ is measured by some gauge that requires no flow through the holes. Now at A the flow is brought to rest (relative to the pitot tube), and so the Bernoulli constant on the streamline through A is just p_2/ρ. This streamline also passes through B, where the velocity just outside the boundary layer is close to U. Hence, using Bernoulli's equation we have

$$p_2/\rho = \tfrac{1}{2}U^2 + p_1/\rho$$

and so

$$U = \{2(p_2 - p_1)/\rho\}^{1/2}.$$

The simplicity of this result is never, of course, achieved exactly in reality. For example, the position and size of the holes at A and B are important, as is the accuracy with which the whole instrument is aligned with the flow.

   Another device is used for measuring flow in pipes. Put a contraction in area into the system, and measure the pressures before the contraction and in it, as in fig. X.14. For a streamline passing near the pressure tappings, Bernoulli's equation gives

$$\tfrac{1}{2}V_1^2 + p_1/\rho = \tfrac{1}{2}V_2^2 + p_2/\rho.$$

Now if we assume that the flow is uniform across the pipe at each tapping (and this is not too far from being true), then conservation of mass shows that

$$\rho A_1 V_1 = \rho A_2 V_2.$$

Eliminate V_2 from these two equations, and you get

$$V_1 = \left\{ \frac{2(p_1 - p_2)A_2^2}{\rho(A_1^2 - A_2^2)} \right\}^{1/2}.$$

This formula is somewhat inaccurate as the conditions of Bernoulli's

Fig. X.14. Diagram of a Venturi meter.

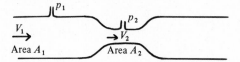

equation are not really met in this flow, and the velocities are not really uniform across the pipe. But a device based on such pressure measurements is indeed used for velocity measurement, and if it is calibrated against known velocities, then it will be satisfactory. The device is known as a Venturi meter.

(*d*) *Simple experiments*
The general form of Bernoulli's equation is
'high speed = low pressure'.

and this can be well demonstrated by many simple experiments.
(i) Hang up two table tennis balls on threads so that they are about 1 cm apart, roughly as sketched in fig. X.15, and blow gently between them. They will move together. The jet of air going between them is at high speed and so low pressure p_1, while the non-adjacent sides of the balls have the higher, atmospheric, pressure p_0 acting. Hence there is a force pushing the balls together.
(ii) Bend a sheet of paper as shown in fig. X.16 and hold it by its edges on a table, and then blow down the gap between sheet and table. The sheet curves down towards the table because the pressure underneath is low where the speed is high, and the pressure on top is still atmospheric pressure.

Fig. X.15. Flow between suspended table tennis balls makes them move together.

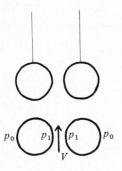

Fig. X.16. Blow through a tunnel formed by a sheet of paper.

Fig. X.17. Balancing a ball on a jet of air.

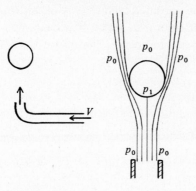

(iii) A light plastic ball can be balanced quite stably on top of a jet of air (see fig. X.17). There are two things to explain here, that the ball stays up and that it is stable.

The airflow round the ball separates and leaves a stagnant wake above the ball, where the pressure is almost the atmospheric pressure p_0. The pressure is also p_0 just outside the initial jet, and by continuity of pressure across boundary layers it will also be about p_0 *inside* the initial jet. So the Bernoulli constant we use on streamlines is

$$p_0 + \tfrac{1}{2}\rho V^2.$$

Further up near the bottom of the ball, the velocities are much smaller, and so the pressures are greater than p_0:

$$p_1 = p_0 + \tfrac{1}{2}\rho V^2.$$

This higher pressure provides the force needed to hold the ball up.

Fig. X.18. The ball of fig. X.17 is displaced slightly.

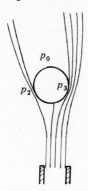

If the ball moves to one side in the jet, then the two sides have different speeds near them, the higher speed being at the centre of the jet, as in fig. X.18. Hence the lower pressure is p_3 at the centre of the jet, and there is a force proportional to $p_2 - p_3$ restoring the ball to the jet centre. That is, the centre is a stable position.

Note in this example the vortex sheets at the edge of the jet and at the boundary of the wake, across which the Bernoulli constant changes discontinuously, but across which the pressure is almost continuous.

The flow on the right side will also attach itself to the ball (Coanda effect). This gives curved streamlines and so a reduced surface pressure; an asymmetric wake also contributes to the force pushing the ball back into the jet.

(iv) The previous experiment provides no surprise: you blow upwards at a ball and it feels an upward force on it. In the next experiment however, you blow *down* to exert an upward force! The apparatus consists of a cotton reel, a post card and a drawing pin. The pin is used to keep the centre of the card at the centre of the hole in the reel. If you then blow hard through the hole, as indicated in fig. X.19, the card stays up by itself, despite its weight pulling it down and the downward jet of air impinging on it.

In this case also we look for a place to establish the Bernoulli constant for some suitable streamline. Where the jet of air leaves the outer rim of the reel, the pressure is atmospheric pressure p_0, and as before the Bernoulli constant here must be

$$p_0 + \tfrac{1}{2}\rho U^2.$$

Where the air leaves the central hole, the speed is

$$W = Ub/a,$$

because the same volume must flow through the circle of radius

Fig. X.19. Blowing through a cotton reel onto a card.

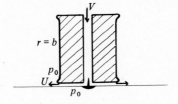

Fig. X.20. Sketches for the experiment in §5(*d*)(iv).

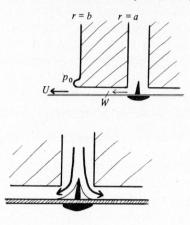

b as flows through the one of radius *a*. Hence the pressure near the central hole is p_1 where

$$p_1 + \tfrac{1}{2}\rho W^2 = p_0 + \tfrac{1}{2}\rho U^2.$$

Thus

$$p_1 = p_0 - \tfrac{1}{2}\rho U^2(b^2/a^2 - 1),$$

and this reduction of pressure below atmospheric enables the card to stay up. See fig. X.20 for details of the flow.

Note that the air flow will probably separate from the spike of the drawing pin, and this will help to give a smooth transition to the flow between the base of the reel and the card.

The same surprising 'defiance of gravity' can be shown with a funnel and a table tennis ball, as in fig. X.21. Blowing hard down the funnel will hold the ball up. The explanation is much the same as the one for cotton reel and card.

Fig. X.21. A table tennis ball held up by the flow through a funnel.

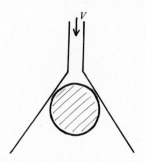

6. A model for the force on a sphere in a stream

(a) The model flow

As we have seen, the flow of a stream past a sphere will separate to leave a wake region, roughly as in fig. X.22. The flow outside the region markedly affected by viscosity (the vortex sheets, boundary layers and wake) looks rather like the flow past the Rankine body that we discussed in Chapter IV §7(b). This flow will at least approximate to the flow past a sphere, so we shall do some force calculations for the Rankine body flow rather than for the separated flow round a sphere, which is very hard to calculate in detail.

At high speeds U we may assume that there is a thin boundary layer on the body, at least for all places reasonably near the nose. We calculate on a streamline just outside this boundary layer, and assume continuity of pressure across it. What we shall calculate is the pressure force on the body in the direction of the stream – since we do not calculate the boundary layer in detail, we cannot calculate the viscous force.

(b) Velocities in the model

We shall use cylindrical polar coordinates r, z to start with, and describe the flow in terms of the stream speed U and the distance a from the origin to the nose of the body. The coordinates are defined in fig. X.23. Then the stream function is

$$\Psi = \tfrac{1}{2}Ur^2 + Ua^2\{1 - z(z^2 + r^2)^{-1/2}\},$$

Fig. X.22. The general character of flow past a sphere.

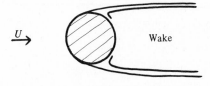

Fig. X.23. The dividing streamline for flow past a source.

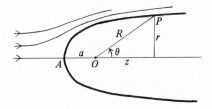

and the surface of the body is $\Psi = 2Ua^2$ or

$$r^2 = 2a^2\{1 + z(z^2 + r^2)^{-1/2}\}.$$

These formulae were derived in Chapter IV, except that there m was used as a parameter instead of a, where $a^2 = m/4\pi U$. It was also shown that

$$r \to 2a \quad \text{as} \quad z \to \infty:$$

the final radius of the body is $2a$.

To apply Bernoulli's theorem, we need to know the value of $\mathbf{v}^2$ on the body, and this is a straightforward calculation (which needs some care). In cylindrical coordinates

$$\mathbf{v} = r^{-1}(-\partial\Psi/\partial z, \partial\Psi/\partial r, 0)$$

and so

$$\mathbf{v}^2 = r^{-2}\{(\partial\Psi/\partial z)^2 + (\partial\Psi/\partial r)^2\}.$$

Now

$$\begin{cases} \partial\Psi/\partial z = -Ua^2r^2(z^2 + r^2)^{-3/2} \\ \partial\Psi/\partial r = Ur + Ua^2zr(z^2 + r^2)^{-3/2} \end{cases}$$

and $\mathbf{v}^2$ may be shown to have the value

$$U^2\{1 + 2a^2z(z^2 + r^2)^{-3/2} + a^4(z^2 + r^2)^{-2}\}.$$

At this stage it is easiest to go into spherical coordinates briefly: we put

$$z = R\cos\theta, r = R\sin\theta$$

(see fig. X.23) and derive

$$\mathbf{v}^2 = U^2(1 + 2a^2\cos\theta/R^2 + a^4/R^4)$$

with the body surface being

$$R^2 = 2a^2(1 + \cos\theta)/\sin^2\theta = 2a^2/(1 - \cos\theta).$$

Hence on the surface $\mathbf{v}^2$ has the value

$$\mathbf{v}^2(P) = U^2\{1 + \cos\theta(1 - \cos\theta) + \tfrac{1}{4}(1 - \cos\theta)^2\}.$$

(c) Force on the Rankine body

Apply Bernoulli's equation on the streamline from upstream infinity that passes close to A (the nose) and P:

$$p_0 + \tfrac{1}{2}\rho U^2 = p + \tfrac{1}{2}\rho\mathbf{v}^2(P).$$

The pressure at P is therefore

$$\begin{aligned} p &= p_0 - \tfrac{1}{2}\rho U^2\{\cos\theta(1 - \cos\theta) + \tfrac{1}{4}(1 - \cos\theta)^2\} \\ &= p_0 - \tfrac{1}{8}\rho U^2(1 - \cos\theta)(3\cos\theta + 1). \end{aligned}$$

Fig. X.24. Pressure changes round the Rankine body.

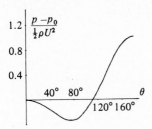

This pressure, whose graph is given in fig. X.24, is *less* than p_0 when $3 \cos \theta + 1 > 0$, i.e. for $\theta < 109.47°$; that is, far from the nose there is a suction as compared with p_0, while near the nose there is a pressure greater than p_0.

It remains to be seen which of these has the greater effect. To find out, we must integrate a force element parallel to the stream over the surface of the body. We use as surface element (see fig. X.25) a strip round the axis which has width dl and radius r: its area is $2\pi r dl$, and the force on it parallel to the axis is

$$p \sin \alpha \times 2\pi r dl = p 2\pi r dr.$$

The integration will thus be easiest in terms of r rather than θ. Now on the surface

$$r^2 = 2a^2(1 + \cos \theta)$$

and so we put

$$\cos \theta = \tfrac{1}{2} r^2/a^2 - 1.$$

Fig. X.25. The force on a surface element of the body.

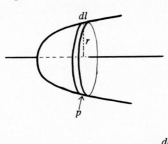

Hence the total force component parallel to the stream is

$$F_z = \int_{r=0}^{2a} 2\pi pr \, dr,$$

where

$$p = p_0 - \tfrac{1}{8}\rho U^2(2 - \tfrac{1}{2}r^2/a^2)(3r^2/2a^2 - 2).$$

The integral is easily evaluated, and we find

$$F_z = 4\pi p_0 a^2.$$

There is *no* contribution from the stream U; the suction and the pressure exactly balance. The non-zero force is just due to the pressure p_0 acting over the area $4\pi a^2$ of the cross-section of the body, which has radius $2a$ as $z \to \infty$.

(d) Model for the force on a sphere

This result seems surprising at first, much more so than the similar result for a cylinder or sphere in a stream. In this case there is no upstream–downstream symmetry, and yet the forces manage to balance exactly. This can in fact be proved for any such semi-infinite body generated by sources in a stream; the argument is in Batchelor's text, pp. 461–2, and will not be given here.

We should notice that the pressure on the surface is high at the nose $(\theta = 180°)$ and then decreases until $\theta \doteqdot 70°$ $(\cos \theta = \tfrac{1}{3})$. Thereafter the pressure increases, as is shown in fig. X.26, and by our previous arguments we might expect the boundary layer to separate since the secondary flow will be running against the main flow. However, boundary layers can remain attached to surfaces when the adverse pressure gradient is not large, and the pressure gradient is not $dp/d\theta$ but dp/dl, which is here quite small.

Let us finish this rather long example on Bernoulli's equation by discussing the force on a sphere in a stream in the light of what we now know.

Fig. X.26. Relative pressures round the Rankine body.

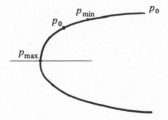

Fig. X.27. Pressures in the flow past a sphere.

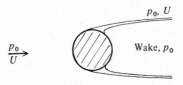

The flow round a sphere separates and leaves a wake (as in fig. X.27) in which the pressure is about p_0, as it is continuous across the shear layers which bound the wake, outside which we have speeds of about U and so pressures about p_0. Model the flow round the leading half of the sphere by part of the flow round the Rankine body – this cannot be too good, as the nose is *not* hemispherical, but it should give at least a plausible answer. We choose to take the flow as far as the point where $p = p_0$, to give continuity of pressure in our model. The force due to the flow U up to this point (where $\cos \theta = -\frac{1}{3}$) is

$$F_z = \int_{r=0}^{2a/\sqrt{3}} 2\pi(p - p_0)r \, dr$$
$$= 2\rho U^2 A/9$$

where A is the area of cross-section at this point. Hence we expect the force on the sphere to be *about*

$$2\rho U^2 A/9$$

where A is now the area of cross-section of the sphere. The forces due to the pressure p_0 must, of course, cancel out.

This estimate for the drag on a sphere is in fact surprisingly good. For Reynolds numbers Ud/v (where d is the sphere's diameter) between 10^3 and 10^5 the correct value is between 0.23 and 0.20 times $\rho U^2 A$, and our estimate is 0.22. Beyond a Reynolds number of 10^5 the separation point moves in such a way that the modelling is no longer appropriate. Details may be found in Batchelor's text on pp. 341–2.

Exercises

1. The stream function

   $$\psi = Ur \sin \theta - \tfrac{1}{2}\tau r^2 \sin^2 \theta - Ua^2 \sin \theta/r - \tfrac{1}{4}\tau a^4 \cos 2\theta/r^2$$

   represents a flow round the cylinder $r = a$ (where $\psi = -\tfrac{1}{4}\tau a^2$). Show that at large distances upstream the vorticity is uniform, and verify that the vorticity (in accordance with the inviscid vorticity equation) is everywhere constant.

2. A swirling flow along a pipe is assumed to have velocity

$$\mathbf{v} = U(r)\hat{\boldsymbol{\theta}} + V(r)\mathbf{k}$$

in cylindrical polars. Does this satisfy the vorticity equation and the continuity equation?

3. Take '$\nabla \cdot$ Euler' for the case when $\rho =$ constant and the body force is derived from a potential Φ with

$$\nabla^2\Phi = 0.$$

This gives an equation for the pressure p in terms of the velocity $\mathbf{v}$; show that it can be written as

$$\nabla^2 p = -\rho(\partial^2/\partial x_i \partial x_j)(v_i v_j).$$

Look up solutions of Poisson's equation

$$\nabla^2\phi = f(\mathbf{r})$$

to see that this equation can be solved to give p when $\mathbf{v}$ is known.

4. Show that, for a two-dimensional flow with a stream function ψ,

$$\nabla^2\psi = f(\psi)$$

if and only if

$$\mathbf{v} \cdot \nabla\omega = 0.$$

5. For a non-swirling axisymmetric steady flow with $\rho =$ constant we have

$$\omega/r = \text{constant following a particle}$$

and

there is a Ψ for the flow.

Show that in this flow

$$D^2\Psi = r^2 f(\Psi).$$

6. A steady swirling flow along a pipe enters a contraction; describe the effect on the vorticity components.

7. What is the inviscid vorticity equation when ρ is not a constant? Deduce that when $\rho = f(p)$ there is no extra generation of vorticity. Derive the vorticity equation when ρ is a constant and viscosity is included.

Fig. X.28. Diagram for Q8.

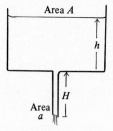

8. Water flows out of the reservoir shown in fig. X.28 down a pipe of cross-sectional area a. What is the speed of the issuing jet of water? Estimate the time to empty the reservoir. Describe what happens if there is a small hole in the pipe half way down.

9. For the contraction flow of §2(b) use an appropriate form of Bernoulli's equation to calculate the pressure.

10. In two-dimensional flow of constant vorticity there is a form of Bernoulli's equation involving ω and ψ. Is there an analogue in axisymmetric flow, using ω and Ψ?

11. Calculate the force due to the pressure on a sphere in a stream, using the model

$$\Psi = \tfrac{1}{2}Ur^2 \sin^2\theta - Ua^3 \sin^2\theta/2r.$$

12. Calculate the force due to the pressure on the cylinder in Q1 above.

13. A strong wind blows past a building. In the middle of the exposed face there is a pair of swing doors fitted with spring to return them to the closed position. Explain why (in terms of air pressures) the doors are blown open slightly by the wind.

14. Outline a calculation similar to that in §6 for flow of a uniform stream past a cylinder.

15. Water flows along a horizontal pipe, through a contraction and out into the atmosphere soon after. The situation is sketched in fig. X.29. The volume flow rate is Q. Estimate the height h for which water can be sucked into the main flow from a lower reservoir open to the atmosphere.

Fig. X.29. Diagram for Q15.

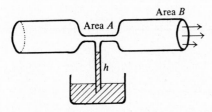

Fig. X.30. Sketch of the double eddy in Q16.

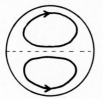

16. This exercise requires some elementary knowledge of Bessel functions, and can wait until this work has been done in Chapter XIII.
Show that one solution of

$$\nabla^2 \psi = c\psi$$

in plane polar coordinates is

$$\psi = CJ_1(kr) \sin \theta.$$

This may be used as the flow inside a 'double eddy' (sketched in fig. X.30) contained inside $r = a$ if k is suitably chosen. Match on a suitable outer solution to find a solution for such a flow in a uniform stream.

References

(a) The first reference is to the real world. Try to explain flow phenomena and forces in terms of Bernoulli's equation and boundary layer separation. You will find that there are *many* cases where this gives a reasonable first answer.
(b) A full discussion of many of the points in this chapter is given in *Introduction to Fluid Dynamics*, G. K. Batchelor, C.U.P. 1967. A detailed study of vorticity dynamics is in his Chapter 5 pp. 266–82. Bernoulli's theorem is dealt with in Chapter 3 pp. 151–64.
(c) Bernoulli's equation is vital in engineering, where it is taken as the basic relation between pressure and velocity, and where the quantity

$$p + \tfrac{1}{2}\rho v^2 + \rho gz$$

is called the 'total head'. Dip into some texts on fluid mechanics for engineering students to see a practical view of this material.
(d) The film *Pressure Fields and Fluid Acceleration* is useful, as are the loops FM-36 on flows through changes in area, FM-33 and 34 on stagnation pressure and Coanda effect, FM-37 and 38 on streamwise and normal pressure gradients.

XI

Potential theory

1. The velocity potential and Laplace's equation

(a) Occurrence of irrotational flows

If a closed curve in an inviscid fluid for which $\rho = f(p)$ has at some time no circulation round it, then by Kelvin's theorem there is never any circulation round the curve. Of course there are no such 'ideal' fluids around, but we have seen that at high Reynolds number and away from boundaries and other awkward regions, a fluid will behave in a near enough ideal fashion. Hence we expect that there can be large regions of flow which have no circulation round any circuit, and hence no vorticity. Thus it is well worth discussing irrotational flows, which have

$$\nabla \times \mathbf{v} = 0.$$

Naturally we must not try to use irrotational flow theory in those regions where we have already seen vorticity to be inevitably developed from the no-slip condition and the diffusive action of viscosity, such as in boundary layers, wakes, eddies and enclosed regions. But away from these regions we can use irrotational flow theory provided that the flow is started, or arrives, with no vorticity. For example, the following flows are closely irrotational.

(i) Flow of air round a streamlined aeroplane wing or body. The aircraft flies into air that is effectively at rest, and the boundary layers and wake are thin enough to be neglected at a first approximation. Such flows will be discussed in detail in Chapter XVII.

(ii) Waves on the surface of reasonably deep water. The boundary condition at the surface does not bring in a noticeable boundary layer because the air is so much less dense than the water. Water waves are dealt with in Chapter XIII.

(iii) Sound waves in air (or water) are of such short time scale that diffusive effects have no time to act, and hence irrotational flow is an appropriate model in many cases. Sound waves are considered in Chapter XII.

(iv) In certain cases irrotational flow theory is useful for some regions of the flow of a uniform stream past a blunt body. In the last chapter, the calculation of the force on a sphere in a stream used an irrotational flow as a model up as far as the separation of the boundary layers. And there will be further examples in this chapter.

(b) Definition of the velocity potential

It is a theorem, familiar in dynamics, that if **v** is a smooth vector field with $\nabla \times \mathbf{v} = 0$, then **v** can be derived from a potential function

$$\mathbf{v} = \nabla \phi.$$

Notice that in fluid dynamics it is common (but not universal) to take a + sign in this equation between **v** and ϕ, even though we have previously used

$$\mathbf{F} = -\nabla \Phi$$

when considering simple body forces. It is proper to call ϕ 'the velocity potential', but this is usually shortened to 'the potential'.

There is a slight awkwardness in the definition of ϕ, particularly in two dimensions, which is best shown by an example. Take the (line vortex) velocity

$$\mathbf{v} = A\hat{\boldsymbol{\theta}}/r$$

Fig. XI.1. Circulating flow between two cylinders.

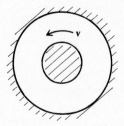

in the region between the cylinders $r = a$ and $r = b$, as shown in fig. XI.1. This has $\nabla \times \mathbf{v} = 0$ in the fluid, and it is not hard to see that

$$\phi = A\theta + B$$

has $\nabla\phi = A\hat{\boldsymbol{\theta}}/r$, for any constant B. Now *if* we restrict θ by $-\pi < \theta \leqslant \pi$, ϕ is a single-valued discontinuous function. If we do not, ϕ is many-valued at each point, though in some sense it is continuous. The problem has come about because the space is doubly connected – there are circuits in the fluid which cannot be shrunk to zero while remaining in the fluid. These 'irreducible' circuits can occur whenever the space is not simply connected; for example the mug shown in fig. XI.2 has irreducible circuits around its handle. But they occur most frequently in fluid dynamics in two dimensions, as any body in the flow gives rise to them. Now it is not so convenient to have a many-valued function ϕ, and therefore we choose to define ϕ to be single-valued by making the space simply connected in a rather arbitrary fashion. For the flow between the two cylinders we will indeed restrict θ by $-\pi < \theta \leqslant \pi$, which is equivalent to putting the cut shown in fig. XI.3 along the negative x-axis. For the mug handle we put a similar restriction on circuits round the handle by putting a cut in the plane of the handle. Having done this, we may define ϕ uniquely by

$$\phi(\mathbf{r}) = \int_{a}^{r} \mathbf{v}(\mathbf{s}) \cdot d\mathbf{s},$$

Fig. XI.2. An irreducible circuit in three dimensions.

Fig. XI.3. A cut between two cylinders to prevent irreducible circuits.

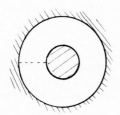

where we have chosen to have $\phi(\mathbf{a}) = 0$, and where any allowable route (i.e. not crossing any cut) is taken from $\mathbf{a}$ to $\mathbf{r}$. It may be proved, using Stokes' theorem, that this definition is independent of the path from $\mathbf{a}$ to $\mathbf{r}$ – this is left as an exercise.

(c) Laplace's equation

If the fluid is also incompressible, we have $\nabla \cdot \mathbf{v} = 0$ as well, and so

$$\nabla^2 \phi = 0.$$

We concentrate for the rest of this chapter on this rather simple equation, usually called Laplace's equation.

What we have done is to use the now known properties of the Navier–Stokes and Euler's equation at high Reynolds number to replace these complicated non-linear vector equations by the simpler, linear, scalar equation $\nabla^2 \phi = 0$. This reduction to the case $\nabla \times \mathbf{v} = 0$, $\nabla \cdot \mathbf{v} = 0$ is of course not universally applicable; but where it is, the mathematics is greatly simplified. Not only can we derive some interesting solutions for flow fields, but we can also prove general theorems which are rarely available for the full equations.

The typical procedure will now be to solve $\nabla^2 \phi = 0$ in some region, and then to use the Euler equation to derive the pressure p. In principle, we should also use the Navier–Stokes equations to fit a boundary layer wherever the potential theory solution does not fit the no-slip condition, but this is beyond the scope of the present text. Naturally enough, when we solve for ϕ we must have some boundary conditions. These will, as previously discussed, usually be on the velocity or the pressure, and not on ϕ itself. In fact, the commonest condition is on the velocity normal to a solid surface, $\partial \phi / \partial n$ on S. We cannot expect to have any matching of tangential velocity on the surface, as there will usually be a thin boundary layer where the potential flow is joined on to the no-slip condition. Another common type of boundary condition will be 'at infinity', giving the way in which $\mathbf{v}$ (and so ϕ) behaves at large distances from the region where the flow is mainly taking place. For example, in the line vortex flow outside a cylinder we have

$$\mathbf{v} = A\hat{\boldsymbol{\theta}}/r,$$

$$\phi = A\theta,$$

on choosing $\phi = 0$ when $\theta = 0$. In this case we see that

$$\text{'}|\mathbf{v}| \to 0 \text{ like } r^{-1}, \text{ as } r \to \infty\text{'},$$

usually written as

$$|\mathbf{v}| = O(1/r);$$

and

ϕ is bounded as $r \to \infty$

or

$\phi = O(1)$.

2. General properties of Laplace's equation

(a) *Uniqueness of the solution in a finite region*

We start with general properties of potential flows, i.e. of solutions of $\nabla^2 \phi = 0$. One of the most important general properties is the uniqueness of the solution of $\nabla^2 \phi = 0$ under reasonable boundary conditions. This uniqueness does not occur for the Navier–Stokes equations; for example the flow between rotating cylinders can have many forms as we have seen in Chapter IX, because the simple flow is unstable at high speeds.

Let us take first the case of the flow inside a surface S, on which the normal velocity $\mathbf{v} \cdot \hat{\mathbf{n}}$ is given. Suppose there are two smooth solutions to the system

$$\begin{cases} \nabla^2 \phi = 0 \text{ inside } S, \\ \partial \phi / \partial n = \text{given function } f(\mathbf{r}) \text{ on } S. \end{cases}$$

Let them be ϕ_1 and ϕ_2, and consider the difference

$$\phi_3 = \phi_1 - \phi_2.$$

Then

$$\begin{cases} \nabla^2 \phi_3 = \nabla^2(\phi_1 - \phi_2) = 0 \text{ inside } S, \\ \partial \phi_3 / \partial n = \partial / \partial n(\phi_1 - \phi_2) = 0 \text{ on } S. \end{cases}$$

Now consider the kinetic energy T associated with ϕ_3. This is

$$T = \tfrac{1}{2}\rho \int_V (\nabla \phi_3)^2 dV,$$

where V is the region inside S. Now it is easily shown that

$$(\nabla \phi_3)^2 = \nabla \cdot (\phi_3 \nabla \phi_3) - \phi_3 \nabla^2 \phi_3,$$
$$= \nabla \cdot (\phi_3 \nabla \phi_3)$$

because $\nabla^2 \phi_3 = 0$. Hence

$$T = \tfrac{1}{2}\rho \int_V \nabla \cdot (\phi_3 \nabla \phi_3) dV$$

$$= \tfrac{1}{2}\rho \int_S \phi_3 \partial \phi_3 / \partial n \, dS$$

from the divergence theorem. But we already know that $\partial \phi_3 / \partial n = 0$ on

S, and so

$$T = 0.$$

Since T is an integral of a non-negative quantity, it can only be zero if the quantity is everywhere zero. That is,

$$\nabla\phi_3 = 0 \text{ inside } S.$$

The conclusion is therefore that ϕ_1 and ϕ_2 have the *same* velocity field throughout the region, and so there is essentially only one solution to the problem. Of course ϕ_1 and ϕ_2 can differ by a constant, but this is of no consequence: the observable is the velocity $\nabla\phi$ and not the potential ϕ.

This simple proof of uniqueness relies on the smoothness of the solutions ϕ_1 and ϕ_2, as it uses the divergence theorem. The proof cannot apply to the solutions

$$\phi = A\theta$$

for the region

$$a < r < b,$$
$$-\pi < \theta \leqslant \pi,$$

with zero normal velocity at $r = a$ and $r = b$. These solutions (one for each different value of A) are not continuous across the 'cut' $\theta = \pi$ (see fig. XI.4). Clearly, then, there can be no uniqueness theorem for such a doubly connected region, because any circulation in the region can give a solution.

This proof of uniqueness assumed that there was at least one solution, i.e. it assumed an existence theorem for the system. We shall not consider existence theorems in this text – we shall either find a solution (and hence one certainly exists) or else assume that it is 'physically reasonable' that there should be a solution.

(b) Uniqueness for an infinite region

There is a similar uniqueness theorem for an unbounded region of fluid outside a surface S on which the normal velocity is given. Take the

Fig. XI.4. The region for §2(a).

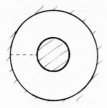

Fig. XI.5. Definition sketch for §2(b).

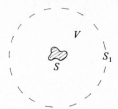

case of three dimensions, smooth solutions and

$$\phi = O(1/r)$$

as $r \to \infty$. The method is the same, but to keep a finite region of fluid V to apply the divergence theorem in, we introduce a surface S_1 at $r = R$, where r is measured from some point in or near S, and where R is assumed to be very large. Fig. XI.5 shows the surfaces S and S_1.

Assume again two solutions ϕ_1 and ϕ_2 to

$$\begin{cases} \nabla^2 \phi = 0 \text{ outside } S, \\ \phi = O(r^{-1}) \text{ as } r \to \infty, \\ \partial\phi/\partial n = f(\mathbf{r}) \text{ on } S. \end{cases}$$

Then $\phi_3 = \phi_1 - \phi_2$ satisfies

$$\begin{cases} \nabla^2 \phi_3 = 0 \text{ outside } S \\ \phi_3 = O(r^{-1}) \text{ as } r \to \infty \\ \partial\phi_3/\partial n = 0 \text{ on } S. \end{cases}$$

Notice that the difference between two functions which are both $O(r^{-1})$ is itself $O(r^{-1})$: for example

$$3/r \text{ and } 2/r + 1/r^2$$

are both $O(r^{-1})$ as $r \to \infty$, and their difference

$$1/r - 1/r^2$$

is also $O(r^{-1})$. This is because

$$f(\mathbf{r}) = O(r^{-1}) \text{ as } r \to \infty$$

means that $f(\mathbf{r})$ is no larger than k/r as r becomes very large, for some constant k.

We consider again the kinetic energy

$$T = \tfrac{1}{2}\rho \int_V (\nabla\phi_3)^2 dV$$

in the region V between S and S_1. As before we may convert this to

$$\tfrac{1}{2}\rho \int_V \nabla\cdot(\phi_3\nabla\phi_3)dV,$$

and the divergence theorem now gives

$$T = -\tfrac{1}{2}\rho \int_S \phi_3\partial\phi_3/\partial n\,dS + \tfrac{1}{2}\rho \int_{S_1} \phi_3\partial\phi_3/\partial r\,dS_1,$$

where the negative sign arises from the different directions of the normals out of V at S and S_1. But the first integral is zero because $\partial\phi_3/\partial n$ is zero on S. And the second integral is estimated as follows:

$$\begin{cases} dS_1 = O(R^2) \text{ as a sphere has surface } 4\pi R^2, \\ \phi_3 = O(R^{-1}) \text{ at distance } R, \text{ for large } R, \\ \partial\phi_3/\partial r = O(R^{-2}) \text{ because } \phi_3 = O(R^{-1}). \end{cases}$$

Hence the second integral is $O(R^{-1})$, and so as $R \to \infty$, it tends to zero.

Thus the total kinetic energy in the infinite region outside S is zero, and so (as before) $\nabla\phi_3$ must vanish everywhere outside S, and the two solutions ϕ_1 and ϕ_2 have the same velocity field.

Notice that this theorem has been proved for three dimensions and smooth solutions. For an infinite two-dimensional region there are difficulties associated with circulations and also with infinite total kinetic energies. For a non-simply connected region in three dimensions (such as that outside a torus), the circulation round irreducible circuits must be given as an extra condition.

(c) *Kelvin's minimum energy theorem*

Associated with the uniqueness theorems is a minimum energy theorem due to Kelvin, which is included here as an example of the simple general results that follow from the equation $\nabla^2\phi = 0$.

Consider incompressible fluid between S_1 and S_2, shown in fig. XI.6 (in three dimensions, to avoid circulations). Let

$$\mathbf{v} = \nabla\phi$$

Fig. XI.6. Sketch for §2(c).

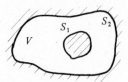

be the solution for given normal velocities on S_1 and S_2. Let $\mathbf{v}'$ be any other velocity which satisfies

$$\nabla \cdot \mathbf{v}' = 0 \text{ in } V$$

and matches the given normal velocities on S_1 and S_2. Let T and T' be the associated kinetic energies. Then

$$T' - T = \tfrac{1}{2}\rho \int_V (\mathbf{v}'^2 - \mathbf{v}^2) dV.$$

We rearrange this by writing

$$\mathbf{v}'^2 - \mathbf{v}^2 = (\mathbf{v}' - \mathbf{v})^2 + 2(\mathbf{v}' - \mathbf{v}) \cdot \mathbf{v}.$$

Now $\int_V (\mathbf{v}' - \mathbf{v}) \cdot \mathbf{v}\, dV$ can also be rewritten, because $\mathbf{v} = \nabla\phi$ and $\nabla \cdot \mathbf{v} = \nabla \cdot \mathbf{v}' = 0$, as $\int_V \nabla \cdot \{\phi(\mathbf{v}' - \mathbf{v})\} dV$. Finally, using the divergence theorem on this last integral, we get integrals, over the boundaries, of

$$\phi(\mathbf{v}' - \mathbf{v}) \cdot d\mathbf{S}$$

which are zero because $\mathbf{v}$ and $\mathbf{v}'$ have the same normal components on S_1 and S_2.

Collecting these results, we arrive at

$$T' - T = \tfrac{1}{2}\rho \int_V (\mathbf{v}' - \mathbf{v})^2 \, dV,$$

which is certainly positive if $\mathbf{v}'$ differs from $\mathbf{v}$. Hence T' exceeds T, and the kinetic energy of the potential theory solution is the least for all possible motions inside V.

This interesting 'economical' property of the irrotational solution is in fact a property of many naturally occurring differential equations, and could properly be studied in a course on differential equations rather than in a course on fluid dynamics.

(d) Further properties of solutions of Laplace's equation

The uniqueness theorem justifies the method of images that we have met above in Chapter IV, but only if we conduct the argument in terms of ϕ. For, since there can be only one solution to a problem in potential theory, it cannot matter how we derive it. If the method of images gives us a function which:

(i) satisfies $\nabla^2\phi = 0$ in the fluid, except at given singularities;

(ii) satisfies appropriate boundary values on finite surfaces in the fluid;

(iii) has the right behaviour at infinity (if the fluid is unbounded);

then it must be the correct solution. In two dimensions it is easy to extend this result to the stream function, since

$$\nabla \times \mathbf{v} = \omega \mathbf{k} = 0$$

implies

$$\nabla^2 \psi = 0,$$

and a uniqueness theorem for ψ is easily constructed, in the same way as for ϕ.

The time t does not enter into Laplace's equation for the potential

$$\nabla^2 \phi = 0.$$

This means that if any variation with t arises in the boundary conditions, then the same variation with t occurs throughout the whole flow. For example

$$\phi = ua^2 r^{-1} f(t), r > a$$

satisfies $\nabla^2 \phi = 0$ outside $r = a$ and

$$\partial\phi/\partial r = -uf(t) \text{ on } r = a.$$

A change in $f(t)$ at the boundary is instantly communicated to the whole of the fluid. This is, of course, unreasonable. It is, as we shall see below in Chapter XII, connected with the modelling of the actual fluid by an incompressible fluid.

3. Simple irrotational flows

We have seen in Chapter V that most of the simple stream functions of Chapter IV corresponded to $\nabla \times \mathbf{v} = 0$, i.e. were irrotational. It is not hard to calculate the corresponding potentials. We do first the two-dimensional examples, in which we use

$$\mathbf{v} = \nabla\phi = (\partial\phi/\partial x, \partial\phi/\partial y)$$

in cartesian components, or

$$\mathbf{v} = \nabla\phi = (\partial\phi/\partial r, r^{-1}\partial\phi/\partial\theta)$$

in plane polar components.

(i) The uniform stream along the x-axis has

$$\partial\phi/\partial x = U, \partial\phi/\partial y = 0$$

and so

$$\phi = Ux$$
$$= Ur\cos\theta,$$

on choosing a constant of integration appropriately. Remember that we had

$\psi = Uy = Ur\sin\theta$.

(ii) The simple source of (volume) flux m has

$\partial\phi/\partial r = m/(2\pi r)$,

$\partial\phi/\partial\theta = 0$

to give radially outward velocity and total flux m through a circle. Hence

$\phi = (m/2\pi)\ln(r/a)$

for some constant a. For the same flow

$\psi = (m/2\pi)\theta$.

(iii) The dipole along the x-axis is derived as before by differentiating the source solution with respect to x and changing sign. Hence it has potential

$\phi = -Ar^{-1}\cos\theta$

for strength A. For general vector strength $\boldsymbol{\mu}$, a dipole has potential

$\phi = -\boldsymbol{\mu}\cdot\nabla\ln r$

$\quad = -\boldsymbol{\mu}\cdot\mathbf{r}/r^2$.

Remember that $\psi = Ar^{-1}\sin\theta$ for the dipole along the x-axis.

(iv) The line vortex has velocity components

$\partial\phi/\partial r = 0$,

$r^{-1}\partial\phi/\partial\theta = C/r$,

where the circulation of the vortex is $2\pi C$, and this gives

$\phi = C\theta$

where we had previously

$\psi = -C\ln(r/a)$.

The two-dimensional flows above have somewhat similar forms for both ϕ and ψ, and we shall see why in Chapter XVI. In three-dimensional axisymmetric flow there is less obvious relationship between ϕ and Ψ, because the relation between $\mathbf{v}$ and Ψ is not particularly simple.

(v) The uniform stream along the z-axis still has

$\phi = Uz = Ur\cos\theta$

where spherical polar coordinates with $\theta = 0$ along the stream direction are used in the last formula.

(vi) The simple source of volume flux m has

$$\partial\phi/\partial r = m/(4\pi r^2)$$

and so

$$\phi = -m/(4\pi r)$$

(again in spherical polars).

(vii) The dipole of strength $\boldsymbol{\mu}$ now has potential

$$\phi = \boldsymbol{\mu}\cdot\nabla(1/r) = -\boldsymbol{\mu}\cdot\mathbf{r}/r^3.$$

In spherical polars this is

$$-\mu\cos\theta/r^2$$

for a dipole of strength μ along the line $\theta = 0$.

4. Solutions by separation of variables

(a) An example of potential flow

The modelling in Chapters IV and V was essentially descriptive: we see that the flows look roughly like some simple form, and so we seek a mathematical function of the same general behaviour, and hope that the two are a good enough fit. In this chapter, the modelling is analytic: a detailed analysis has revealed that $\nabla^2\phi = 0$ is often a good model for the dynamics of a fluid, and we proceed to solve this equation to predict the properties of the flow. What we have seen so far is that some of the guesses (descriptive models) in the earlier work also fit the dynamic equation $\nabla^2\phi = 0$, and so their reasonable fit to reality is now explained.

Our next example is not one we have attempted before. We shall consider first a two-dimensional flow against a wall, sketched in fig. XI.7. It is not hard to see that

$$\phi = A(x^2 - y^2)$$

is a solution of $\nabla^2\phi = 0$ that has the same velocity components as

$$\psi = 2Axy$$

and so hyperbolic streamlines as shown. We met this flow earlier as a

Fig. XI.7. Flow against a plane.

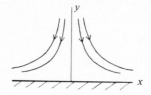

Fig. XI.8. Flow against a plane with a semicircular bump.

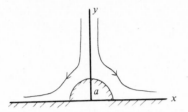

description of a stagnation point, and now we see that an irrotational flow against a wall must give exactly this. Notice that in reality at high Reynolds numbers there must be boundary layers along the wall; these have no reason to separate, and so the description by a potential should be quite successful.

Now add a semicircle to the wall, as shown in fig. XI.8. The flow will be disturbed near the origin, but at large distances we still expect

$$\phi \sim A(x^2 - y^2) \text{ as } r \to \infty.$$

(This is read as ϕ is asymptotically equal to $A(x^2 - y^2)$ as $r \to \infty$, and means that ϕ and $A(x^2 - y^2)$ differ by terms of smaller size for all large values of r.) We shall solve the problem

$$\begin{cases} \nabla^2 \phi = 0 \text{ for } r > a, \ 0 < \theta < \pi, \\ \partial\phi/\partial n = 0 \text{ on the wall and semicircle,} \\ \phi \sim Ar^2 \cos 2\theta \text{ as } r \to \infty. \end{cases}$$

Since our previous theory does not prove the uniqueness of the solution we derive, we really should consider proving another uniqueness theorem, but we shall omit this.

(b) Separation of variables

The method of solution is the important method of 'separation of variables'. Assume a solution in the form of a product

$$\phi = R(r) \, H \, (\theta).$$

Now $\nabla^2 \phi$ in plane polar coordinates has the form

$$\frac{\partial^2 \phi}{\partial r^2} + \frac{1}{r}\frac{\partial \phi}{\partial r} + \frac{1}{r^2}\frac{\partial^2 \phi}{\partial \theta^2},$$

and so Laplace's equation becomes

$$R''(r)H(\theta) + r^{-1}R'(r)H(\theta) + r^{-2}R(r)H''(\theta) = 0,$$

where dashes denote derivatives. This may be rewritten as

$$\{r^2 R''(r) + rR'(r)\}/R(r) = -H''(\theta)/H(\theta).$$

In this equation the left side is independent of θ, and the right side is independent of r. Hence each side must be independent of both, and so each is a constant. There are three essentially different choices for this constant:

(i) positive;

(ii) zero;

(iii) negative;

and each should be investigated in turn. In fact only the first one will give a solution in the present case, so the investigation of the other two will be left as an exercise.

We take the constant to be k^2, then, and hence we have

$$\begin{cases} H''(\theta) = -k^2 H(\theta), \\ r^2 R''(r) + r R'(r) - k^2 R(r) = 0. \end{cases}$$

The variables have been separated, and we are left with two ordinary differential equations. The boundary conditions are easily put in terms of R and H:

$$\begin{cases} \text{on } r = a \text{ for } 0 < \theta < \pi, \ R'(a)H(\theta) = 0; \\ \text{on } \theta = 0 \text{ for } r > a, \ R(r)H'(0) = 0; \\ \text{on } \theta = \pi \text{ for } r > a, \ R(r)H'(\pi) = 0; \\ \text{as } r \to \infty, \ R(r)H(\theta) \sim Ar^2 \cos 2\theta. \end{cases}$$

It is evident that the solution for H is in terms of $\cos k\theta$ and $\sin k\theta$, and we need

$$H'(0) = H'(\pi) = 0,$$

since it is unreasonable to take $R(r) = 0$. The condition $H'(0) = 0$ requires us to take

$$H(\theta) = B \cos k\theta,$$

and then $H'(\pi) = 0$ means that we must have

$$Bk \sin k\pi = 0.$$

The solutions of this other than $B = 0$ and $k = 0$ are

$$k = \text{non-zero integer, say } n.$$

These are the eigenvalues of the equation and boundary conditions for $H(\theta)$.

Having now determined that the solutions for H are

$$\text{constant} \times \cos n\theta, \text{ integer } n$$

we can return to the problem for $R(r)$. It is

$$\begin{cases} r^2 R''(r) + rR'(r) - n^2 R(r) = 0, \\ R'(a) = 0. \end{cases}$$

The equation is of standard (Euler, homogeneous) type and has solutions like r^α, where

$$r^2\alpha(\alpha - 1)r^{\alpha-2} + r\alpha r^{\alpha-1} - n^2 r^\alpha = 0.$$

This requires $\alpha = \pm n$, and so

$$R(r) = Cr^n + Dr^{-n};$$

to fit the boundary condition we must take the combination

$$\text{constant} \times \{(r/a)^n + (a/r)^n\}.$$

(c) The solution for the example
We have now reached the stage that

$$\text{constant} \times \{(r/a)^n + (a/r)^n\} \cos n\theta$$

satisfies the equation and the boundary conditions on semicircle and wall for any constant and any integer n. It is clear that we do not need to look beyond

$$\begin{cases} \text{constant} = Aa^2 \\ n = 2 \end{cases}$$

to fit the condition at $r \to \infty$. Hence the solution of the problem is

$$\phi = A\{r^2 + a^4/r^2\} \cos 2\theta.$$

However the problem as posed is not in fact uniquely determined, because

$$\phi = A\{r^2 + a^4/r^2\} \cos 2\theta + E\{r/a + a/r\} \cos \theta$$

is also a solution for any value of E. The last term corresponds to a stream parallel to the x-axis being disturbed by the semicircle of radius a. The problem could have been properly set up by asking for ϕ to be a function symmetric about the y-axis, or by demanding that

$$\phi - Ar^2 \cos 2\theta = O(1) \text{ at most}$$

as $r \to \infty$.

The solution above satisfies $\nabla^2\phi = 0$, but this is only the relevant equation for the given region on the assumption that boundary layers are thin and do not separate. It is in fact rather likely that the boundary layers will separate. At the points $r = a$, $\theta = 0$, $\frac{1}{2}\pi$, π (see fig. XI.9), there must be

Fig. XI.9. Relative pressures on the semicircle.

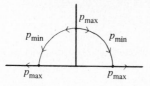

stagnation points, and hence pressure maxima, by Bernoulli's formula. But at $r = a$, $\theta = \frac{1}{4}\pi$, $\frac{3}{4}\pi$ the velocity will not be zero, and so p will be smaller. Hence there will be a rising pressure towards the plane, and so a tendency for reverse flow and separation.

(d) *The method in general*

The method of separation of variables is very powerful when the boundary conditions are on reasonably simple geometrical shapes. The most important cases are as follows.

(i) Flat boundaries, cartesian coordinates. In this case the functions are sines, cosines and exponentials or hyperbolic functions. We shall find these solutions in the chapters on sound and water waves.

(ii) Circular boundaries in two dimensions. As we have seen we can get sines and cosines for the angular variable, but we may also find a term in $k\theta$ (leading to a circulation) if the region is not simply connected. The terms in r are positive and negative powers, with a term in $K \ln(r/a)$ if there is an outflow from (the region near) the origin.

(iii) Cylindrical polar coordinates. The angular dependence is as in (ii), and the axial dependence as in (i), but the radial dependence now brings in Bessel functions. These solutions are also appropriate to the chapters on waves, and we leave them till then.

(iv) Spherical polar coordinates. The radial dependence is again in powers of r, and the angular dependences are usually in terms of polynomials in powers of $\cos \theta$ and sines or cosines of multiples of λ. We go on to an example of this type in the next section.

The method of separation of variables is not always as simple as it was in the example above. Often the method will produce an infinite set of suitable functions, and the correct solution requires the determination of the coefficients in an infinite series. For example, in the above problem we could have written

$$\phi = \sum_{n=1}^{\infty} c_n \{(r/a)^n + (a/r)^n\} \cos n\theta$$

as the appropriate solution of $\nabla^2 \phi = 0$ and the conditions on the wall. In that case the asymptotic condition enabled us to reject almost all the c_n as zero. In other cases we may have to keep them all.

5. Separation of variables for an axisymmetric flow: Legendre polynomials

(a) A model flow

When a puff of hot gas rises through still air, it takes up a roughly spherical shape, and air is sucked into the gas near the rear of the sphere. Exactly the same sort of behaviour is seen when a blob of coloured liquid falls through water, and there are excellent photographs of this in Batchelor's text, plate 20. The flow is sketched in fig. XI.10. This is really a very complicated flow, with interaction between the inner ring vortex and the outer flow; we shall model it in the simplest possible way to discuss the flow round the outside of the puff or blob, which should certainly be irrotational because the outer air or water is initially at rest.

We must build into our model:

(i) that there is an inflow at the back;

(ii) that the puff expands because of this;

(iii) that there is a flow past the puff.

What we shall do is regard the puff as a sphere with a fixed centre which at time t has radius a. At this radius a we impose a boundary condition to

Fig. XI.10. The relative motion for a rising puff.

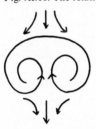

Fig. XI.11. The model flow for the rising puff.

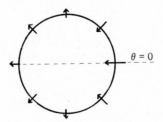

$\theta = 0$

represent (i) and (ii) above:

$$(\partial \phi / \partial r)_{r=a} = f(\theta),$$

where $f(\theta)$ is chosen so that there is an inflow at the back, but no total inflow through the whole sphere, as in fig. XI.11. In addition we need

$$\phi \sim Ur \cos \theta,$$

which is the potential for a uniform stream in spherical polar coordinates.

(b) *The mathematical model*

The mathematical model we solve will be as follows.
The equation for $r > a$ is

$$\nabla^2 \phi = 0.$$

The condition at large values of r is

$$\phi \sim Ur \cos \theta.$$

The condition on $r = a$ we take to be

$$(\partial \phi / \partial r)_{r=a} = \tfrac{1}{5} u(1 - 3 \cos \theta - 3 \cos^2 \theta)$$

because this is a reasonably simple function satisfying conditions (i) and (ii) above: it gives an inflow u at $\theta = 0$, and no *total* inflow into the sphere because

$$\int_0^\pi (1 - 3 \cos \theta - 3 \cos^2 \theta) \sin \theta \, d\theta = 0.$$

Notice that we are not trying to model the flow for $r < a$, which will certainly not be irrotational. And also that we have produced a steady model if we take

$$a = \text{constant}.$$

The original flow is unsteady, partly because the size of the puff increases with time. We could, but we shall not, model this by choosing something like

$$a(t) = \tfrac{1}{5} ut,$$

which matches the radial velocity at $\theta = \pi$ and $\theta = \pm \tfrac{1}{2}\pi$.

(c) *Separation of variables*

The method we shall use is, again, that of separation of variables; naturally we operate this time in spherical polar coordinates r, θ, λ. Now the problem we have chosen is one with an axis of symmetry, and no swirl velocity round this axis, so we expect to get

$$\phi(r, \theta)$$

as a solution. This makes the mathematics somewhat easier – the full solution by separation of $\nabla^2 \phi = 0$ in spherical polar coordinates is available in larger texts. So we shall assume a solution of the form

$$\phi(r, \theta) = f(r)g(\theta)$$

for the equation

$$\frac{1}{r^2}\frac{\partial}{\partial r}\left(r^2\frac{\partial\phi}{\partial r}\right) + \frac{1}{r^2 \sin\theta}\frac{\partial}{\partial\theta}\left(\sin\theta\frac{\partial\phi}{\partial\theta}\right) = 0,$$

which is the appropriate form of $\nabla^2 \phi = 0$.

On substitution, we may rearrange the equation as

$$\frac{1}{f(r)}\frac{d}{dr}\left(r^2\frac{df}{dr}\right) = -\frac{1}{g(\theta)\sin\theta}\frac{d}{d\theta}\left(\sin\theta\frac{dg}{d\theta}\right).$$

The right side is independent of r, the left side of θ, so each side can only be a constant. As before, this constant can be positive, zero, or negative; and only when the constant is of suitable sign can a physically appropriate solution be found. Now the solution of

$$\frac{d}{d\theta}\left(\sin\theta\frac{dg}{d\theta}\right) = -\text{constant} \times g(\theta)\sin\theta$$

is quite an involved task, which we cannot undertake here – the details are readily available in texts on ordinary and partial differential equations. Here we will be content with quoting the form of the solution: the equation for θ has a solution continuous for all θ, $0 \leqslant \theta \leqslant \pi$, if the constant has the value

$$n(n + 1)$$

where n is a non-negative integer. The solution is then a polynomial of degree n in powers of $\cos\theta$, usually written as

$$P_n(\cos\theta),$$

and known as a Legendre polynomial. The first few of them are

$$P_0(\cos\theta) = 1,$$
$$P_1(\cos\theta) = \cos\theta,$$
$$P_2(\cos\theta) = \tfrac{1}{2}(3\cos^2\theta - 1),$$
$$P_3(\cos\theta) = \tfrac{1}{2}(5\cos^3\theta - 3\cos\theta).$$

Their most important properties from our present point of view are:

(i) when $\theta = 0$, $P_n(\cos\theta) = 1$;

(ii) $\int_0^\pi P_m(\cos\theta)\,P_n(\cos\theta)\sin\theta\,d\theta = 0$ if $m \neq n$;

(iii) $\int_0^\pi \{P_n(\cos\theta)\}^2 \sin\theta\,d\theta = 2(2n + 1)^{-1}$;

(iv) any reasonable function of θ may be put as

$$F(\theta) = \sum_{n=0}^{\infty} c_n P_n (\cos \theta).$$

These properties are rather similar to those of $\cos n\theta$, which came into the solution of $\nabla^2 \phi = 0$ in §4 above. In particular, the coefficients c_n in (iv) may be found by the same sort of method as in Fourier series: you multiply both sides of the expression

$$F(\theta) = \sum_{n=0}^{\infty} c_n P_n (\cos \theta)$$

by $P_m (\cos \theta) \sin \theta$ and integrate from 0 to π, using (ii) and (iii) to evaluate the integrals. The result is

$$c_m = \tfrac{1}{2}(2m + 1) \int_0^{\pi} F(\theta) \, P_m(\cos \theta) \sin \theta \, d\theta,$$

for $m \geqslant 0$. Hence if $F(\theta)$ is known, the coefficients can all be calculated.

The Legendre polynomials have many more properties than the ones quoted here. Larger texts give full details. Larger texts also explain why it is that solutions of physically relevant equations like $\nabla^2 \phi = 0$ always turn out to have properties rather like (ii), (iii) and (iv).

When the separation constant has the value $n(n + 1)$, the equation for $f(r)$ becomes

$$r^2 f''(r) + 2rf'(r) - n(n + 1) \, f(r) = 0$$

and this has solutions

$$Ar^n + B/r^{n+1}$$

for any constants A and B.

We have therefore derived solutions of $\nabla^2 \phi = 0$ in spherical polar coordinates which have the form

$$(Ar^n + Br^{-n-1})P_n(\cos \theta)$$

for any integer $n \geqslant 0$. Since the equation for ϕ is linear, we may add solutions of this type and still have a solution. Hence the general separation solution of $\nabla^2 \phi = 0$ for this case (spherical polars, axisymmetric, $0 \leqslant \theta \leqslant \pi$) is

$$\phi(r, \theta) = \sum_{n=0}^{\infty} (A_n r^n + B_n r^{-n-1})P_n(\cos \theta),$$

where the A_n and B_n are constants which must be chosen so that the boundary conditions are satisfied.

(d) Application to the model

It seems most sensible to put the boundary conditions in terms of the Legendre polynomials $P_n(\cos \theta)$. In the model that we set off with, this is quite easily achieved. The conditions are

$$\phi \sim Ur P_1(\cos \theta)$$

because

$$P_1(\cos \theta) = \cos \theta,$$

and

$$(\partial \phi / \partial r)_{r=a} = -(2u/5)P_2(\cos \theta) - (3u/5)P_1(\cos \theta)$$

because

$$P_2(\cos \theta) = \tfrac{1}{2}(3 \cos^2 \theta - 1).$$

Our problem is now to choose constants A_n and B_n in the general solution

$$\phi = \sum_{n=0}^{\infty} (A_n r^n + B_n r^{-n-1})P_n(\cos \theta),$$

so that

$$\phi \sim Ur P_1(\cos \theta) \text{ as } r \to \infty$$

and

$$(\partial \phi / \partial r)_{r=a} = -(2u/5)P_2(\cos \theta) - (3u/5)P_1(\cos \theta).$$

It is clear that we shall not need any terms with $n > 2$, because no term in the boundary conditions has $n > 2$: this may be proved formally, but is sufficiently obvious. Moreover

$$A_2 = 0, \ A_1 = U$$

are needed to match the uniform stream at large values of r. The remaining condition is then

$$\left\{ \frac{\partial}{\partial r}(UrP_1 + A_0 + B_0 r^{-1}P_0 + B_1 r^{-2}P_1 + B_2 r^{-3}P_2) \right\}_{r=a}$$
$$= -(3u/5)P_1 - (2u/5)P_2.$$

This is

$$-(B_0/a^2)P_0(\cos \theta) + (U - 2B_1/a^3 + 3u/5)P_1(\cos \theta)$$
$$+ (-3B_2/a^4 + 2u/5)P_2(\cos \theta) = 0.$$

It is evident that one way of satisfying this equation is to equate the coefficients of each of P_0, P_1 and P_2 to zero. That this is the *only* way of satisfying this equation for all θ in $[0, \pi]$ may be proved by the device of multiplying the equation by

$$P_m(\cos \theta)\sin \theta$$

where m takes values 0, 1, 2 in turn, and integrating from $\theta = 0$ to π. The formula (c) (ii) above then proves the result.

The inner boundary condition has thus given us the three equations

$$\begin{cases} B_0 = 0, \\ U - 2B_1/a^3 + 3u/5 = 0, \\ -3B_2/a^4 + 2u/5 = 0. \end{cases}$$

The solution to our mathematical model for the puff is therefore

$$\phi(r, \theta) = Ur\cos \theta + \tfrac{1}{2}(a^3/r^2)(U + 3u/5)P_1(\cos \theta) \\ + (2ua^4/15r^3)P_2(\cos \theta),$$

where we have taken $A_0 = 0$ without any real loss of generality.

Notice that if we put $u = 0$ we get the potential

$$\phi = Ur\cos \theta + \tfrac{1}{2}Ua^3 \cos \theta/r^2,$$

which is appropriate for the flow of a stream U past a rigid sphere of radius a.

The term in $P_2(\cos \theta)$ is often called a 'quadrupole' term: it corresponds to an arrangement of two dipoles which almost cancel each other out, or to two sources and two sinks, arranged as shown in fig. XI.12. In axisymmetric flow these dipoles can only lie along the line $\theta = 0$, but in more general flows you could have other arrangements of the sources and sinks that form them, for example those in fig. XI.13.

Fig. XI.12. A quadrupole of sources and sinks along a line, equivalent to opposite dipoles.

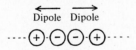

Fig. XI.13. Other quadrupole configurations.

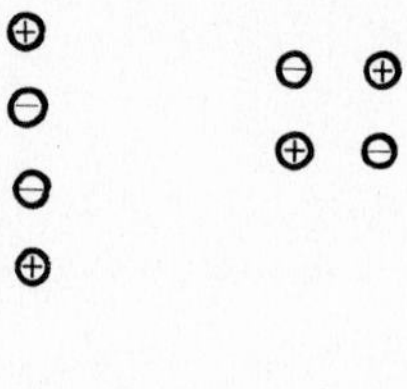

(e) *The force on the sphere in steady flow*

The solution found in (d) for the case $a =$ constant enables us to calculate the force on the puff of gas for this model. We use Bernoulli's equation to derive the pressure:

$$p_0 + \tfrac{1}{2}\rho U^2 = p + \tfrac{1}{2}\rho v^2$$

relates quantities at large distances to those on the sphere $r = a$, where

$$\mathbf{v} = \nabla\phi.$$

On $r = a$ we find that

$$\nabla\phi = \{\tfrac{1}{5}u(1 - 3\cos\theta - 3\cos^2\theta), -\tfrac{3}{2}U\sin\theta$$
$$- \tfrac{3}{10}u\sin\theta - \tfrac{2}{5}u\cos\theta\sin\theta, 0\}$$

and so

$$v^2 = \tfrac{1}{25}u^2(1 - 3\cos\theta - 3\cos^2\theta)^2 + (\tfrac{3}{2}U\sin\theta + \tfrac{3}{10}u\sin\theta$$
$$+ \tfrac{2}{5}u\cos\theta\sin\theta)^2.$$

The force on the sphere can only be in the direction $\theta = 0$, because the flow is axisymmetric. Hence the force in this direction is

$$-\int p\cos\theta\, dS = -\int_0^\pi p\cos\theta \times 2\pi a^2 \sin\theta\, d\theta,$$

taking the area dS to be a ring on the sphere $r = a$ at angle θ, as in fig. XI.14. This integral is easily calculated, because

$$\begin{cases} \displaystyle\int_0^\pi \cos^{2n+1}\theta \sin\theta\, d\theta = 0, \\[2ex] \displaystyle\int_0^\pi \cos^{2n}\theta \sin\theta\, d\theta = 2/(2n+1). \end{cases}$$

so most of the terms in p give a zero integral and the force is calculated to be

$$\frac{24\pi\rho a^2 u}{25}(\tfrac{1}{3}U + \tfrac{1}{5}u).$$

Fig. XI.14. The surface element for the force calculation.

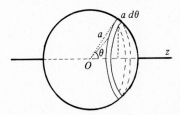

This is the force for the *steady* solution with $a = $ constant. We shall see later that an extra term comes in when $\partial\phi/\partial t$ is non-zero, and the calculation of the force for this model when $a(t) = \frac{1}{5}ut$ is left to the exercises. In the real rise of a puff of hot gas it is also unlikely that U is a constant, as you would expect the puff to slow down as it gets bigger and relatively cooler.

If you take $u = 0$ in this result, you get zero force, the correct result for steady potential flow in this case. But if you take $U = 0$, you still get a force on the sphere, because the velocities on one side are higher than those on the other.

We should note, before we abandon this problem, the smallness of the wake that is shown by the photographs. This flow has little tendency to separate for two reasons: firstly, the boundary layer is very weak because there is no solid surface at which a no-slip condition has to be imposed; and secondly, much of the boundary layer is sucked into the interior by the inflow at the rear. Such a 'boundary layer suction' has been used as a practical method of preventing the growth and separation of boundary layers on aircraft wings.

6. Two unsteady flows

The potential flows we have solved so far have been steady; let us now consider two unsteady ones.

(a) Radial oscillations of a bubble

Bubbles arise in a liquid for three main reasons. Firstly, some unrelated gas may be present in the liquid. Secondly, the local pressure in a rapid fluid motion may be seriously less than the vapour pressure of the liquid, and a cavity filled with vapour from the liquid may form at that point – this is the important phenomenon known as cavitation. Thirdly, local heating may cause the vapour pressure to rise above the local liquid pressure and a cavity of vapour to form by boiling. We shall discuss the

Fig. XI.15. Radial motion outside a bubble of gas in a liquid.

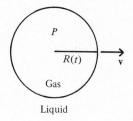

liquid motion outside a spherical bubble of radius $R(t)$. The inner motion in the bubble will not be discussed: we just assume a uniform pressure P inside it. The situation is sketched in fig. XI.15.

Outside the bubble, the motion is taken to be radial:

$$\mathbf{v} = v(r, t)\hat{\mathbf{r}}.$$

Now such a motion is inevitably irrotational, because

$$\nabla \times \{v(r, t)\hat{\mathbf{r}}\} = 0.$$

So there must be a potential $\phi(r, t)$, with

$$\nabla^2 \phi = 0$$

in the liquid. The only spherically symmetric solution of this equation (i.e. independent of θ and λ) is

$$\phi = A(t)/r.$$

It is, in fact, a special case of the axisymmetric solution in §5 above. We may find $A(t)$ in terms of $R(t)$, because the liquid's velocity at the interface $R(t)$ must be just dR/dt. Hence

$$v(R, t) = (\partial\phi/\partial r)_{r=R} = \dot{R}(t).$$

That is $-A(t)/R^2 = \dot{R}$ (where the dot denotes a time derivative) and so

$$A(t) = -\{R(t)\}^2 \dot{R}(t),$$

which gives us

$$\phi(r, t) = -\{R(t)\}^2 \dot{R}(t)/r.$$

We may find an equation for $R(t)$ from the radial equation of motion of the liquid, which is

$$\partial v/\partial t + v \,\partial v/\partial r = -\rho^{-1} \partial p/\partial r,$$

from Euler's equation. Use $v = \partial\phi/\partial r$ to convert this equation to

$$\frac{\partial}{\partial t}\left(\frac{R^2\dot{R}}{r^2}\right) + v\frac{\partial}{\partial r}\left(\frac{R^2\dot{R}}{r^2}\right) = -\frac{1}{\rho}\frac{\partial p}{\partial r}$$

or

$$\ddot{R}R^2/r^2 + 2R\dot{R}^2/r^2 - 2R^4\dot{R}^2/r^5 = -\rho^{-1}\partial p/\partial r.$$

This rather unpromising equation may be integrated from $r = R(t)$ to $r = \infty$, by making some assumptions about the pressure at these distances – we assume

$$\begin{cases} p \to p_0 \text{ as } r \to \infty, \\ p = \text{constant} \times (4\pi R^3/3)^{-\gamma} = KR^{-3\gamma} \text{ at } r = a. \end{cases}$$

This represents the free oscillations of a bubble of gas which behaves isentropically (note that surface tension has been omitted). The integration gives

$$[-\ddot{R}R^2/r - 2R\dot{R}^2/r + R^4\dot{R}^2/2r^4]_R^\infty$$
$$= -p_0/\rho + KR^{-3\gamma}/\rho,$$

or

$$R\ddot{R} + 3\dot{R}^2/2 = AR^{-3\gamma} - B,$$

where A and B are constants.

We continue the integration by putting

$$\frac{d}{dt}\left(\frac{dR}{dt}\right) = \frac{dR}{dt}\frac{d}{dR}\left(\frac{dR}{dt}\right),$$

so that

$$R\dot{R}\,d\dot{R}/dR + 3\dot{R}^2/2 = AR^{-3\gamma} - B.$$

Now put $Z = \frac{1}{2}\dot{R}^2$ to derive

$$R\,dZ/dR + 3Z = AR^{-3\gamma} - B.$$

This too can be integrated, as it can be put as

$$(d/dR)(R^3 Z) = AR^{2-3\gamma} - BR^2.$$

Thus $R^3 Z = \frac{1}{3}A(1-\gamma)^{-1}R^{3(1-\gamma)} - \frac{1}{3}BR^3 + C = f(R)$ say, and so

$$dR/dt = \{2f(R)/R^3\}^{1/2}.$$

This can be finally expressed as an integral,

$$\int \frac{dR}{F(R)} = \int dt,$$

but it cannot be evaluated in elementary terms.

Oscillations of bubbles like this occur frequently in small streams which develop air bubbles as they flow over stones, and the oscillations contribute to the pleasant sound of such streams. Similar oscillations occur in bubbles formed when a drop falls into water, though in both the cases the motion will not often be quite as simple as that described above. To get some idea of how the frequency of the oscillation depends on the size of the bubble, assume a small oscillation:

$$R(t) = a + x(t),$$

where a is constant and $|x/a| \ll 1$. Substitute into the equation for $\ddot{R}$

and neglect terms of size x^2 and smaller:

$$a\ddot{x} + 0 \doteq A(a + x)^{-3\gamma} - B$$
$$\doteq Aa^{-3\gamma}(1 - 3\gamma x/a) - B.$$

Now suppose that $R = a$ is the equilibrium radius of the bubble, so that

$$0 = Aa^{-3\gamma} - B.$$

The equation for x reduces to

$$a\ddot{x} + 3\gamma Aa^{-3\gamma-1}x = 0,$$

which is a linear oscillation equation with (radian) frequency

$$(3\gamma Aa^{-3\gamma-2})^{1/2}.$$

But the constant A is just $p_0 a^{3\gamma}/\rho$, so the radian frequency is $(3\gamma p_0/\rho a^2)^{1/2}$, and this can be evaluated for various sizes of bubbles, taking p_0 to be atmospheric pressure and ρ to be the density of water. The result is that the frequency in cycles per second is approximately 3.26/(radius in metres); thus a bubble of radius 10^{-2} m oscillates at about 326 Hz.

(b) *Flow down a pipe*

We considered the flow out of a reservoir and through a pipe in Chapter X, where we assumed that the flow was almost steady at any time because the exit pipe was so narrow compared with the reservoir radius. Now consider the opposite case, where the exit radius is not much smaller than the radius of the main pipe in fig. XI.16. Assume, however, that the main pipe length L is much greater than the contracting length l, and that the contraction is slow enough and smooth enough that we may consider the flow to be almost parallel to the axis, which is vertical.

Fig. XI.16. Flow out of a narrowing tube.

Take coordinate y up the axis from the exit, and let the cross-sectional area at height y be $A(y)$. Then if the flow rate out of the exit is $Q(t)$, the downward velocity at height y must be

$$Q(t)/A(y),$$

where we assume a uniform velocity at all points of the cross-section, as boundary layers will be thin and there can be no vorticity elsewhere if the flow starts from rest.

Because the flow started from rest and the boundary layers will not have very long to diffuse vorticity outwards, a potential will exist: it must be

$$\phi(y, t) = -\int_0^y \{Q/(t)/A(\eta)\}\, d\eta$$

so that the (upward) velocity $\partial\phi/\partial y$ has the correct value.

Evidently to solve this problem we must find an equation for $Q(t)$ from the vertical equation of motion. It is in fact easiest to do some general theory before we carry this out.

7. Bernoulli's equation for unsteady irrotational flow

(*a*) *Derivation of the equation*

In the last chapter we derived an integral of the equation of motion for steady motion of an incompressible fluid:

$$p + \tfrac{1}{2}\rho v^2 + \rho\Phi$$

was constant along a streamline that avoided regions where viscosity was important. We now construct a similar integral for unsteady irrotational flow, firstly for an incompressible fluid and then for one in which $\rho = f(p)$.

The equation of motion is taken to be

$$\partial \mathbf{v}/\partial t + \mathbf{v}\cdot\nabla\mathbf{v} = -\rho^{-1}\nabla p - \nabla\Phi,$$

where Φ is the potential for the body forces, and where we are now assuming that

$$\mathbf{v} = \nabla\phi.$$

As before, we use the identity

$$\mathbf{v}\cdot\nabla\mathbf{v} = \nabla(\tfrac{1}{2}v^2) - \mathbf{v} \times (\nabla \times \mathbf{v}),$$

but because we now have irrotational flow, the last term is zero. Hence we may rewrite the equation of motion as

$$\partial/\partial t(\nabla\phi) + \nabla(\tfrac{1}{2}v^2) = -\rho^{-1}\nabla p - \nabla\Phi.$$

Finally we use the constancy of ρ and the fact that $\partial/\partial t(\nabla\phi) = \nabla(\partial\phi/\partial t)$

to obtain

$$\nabla\{p/\rho + \tfrac{1}{2}\mathbf{v}^2 + \Phi + \partial\phi/\partial t\} = 0.$$

This may be integrated to give

$$p/\rho + \tfrac{1}{2}\mathbf{v}^2 + \Phi + \partial\phi/\partial t = F(t)$$

for some function $F(t)$, which may usually be determined from the boundary conditions of the problem. Alternatively, $F(t)$ may be taken into the potential ϕ by defining

$$\phi_1 = \phi - \int_0^t F(s)ds;$$

then ϕ_1 and ϕ give the same velocities but

$$\partial\phi_1/\partial t = \partial\phi/\partial t - F(t),$$

so that

$$p/\rho + \tfrac{1}{2}\mathbf{v}^2 + \Phi + \partial\phi_1/\partial t = 0.$$

This is the Bernoulli integral for unsteady irrotational motion of fluid of constant density, and it applies to all parts of the fluid which are in irrotational motion, i.e. away from boundary layers, wakes, eddies and so on.

In cases when the density is not constant, but we may still put

$$\rho = f(p),$$

then the term in p/ρ is replaced (as we have seen before) by

$$\int dp/\rho.$$

In this case the Bernoulli integral is

$$\int dp/f(p) + \tfrac{1}{2}\mathbf{v}^2 + \Phi + \partial\phi_1/\partial t = 0$$

throughout the irrotational flow and in regions where $\rho = f(p)$ is a good approximation.

(b) Example: flow down a pipe

We may now complete the calculation of flow down the pipe of changing cross-section. We suppose the area contracts from a value A_1 in most of the pipe to an exit value A_0, and that the water height starts at $l + L$, and at time t is $H(t)$, as indicated in fig. XI.17. Then the water speed in the upper part of the pipe is

$$dH/dt = -Q(t)/A_1;$$

Fig. XI.17. Definition sketch for flow out of a narrowing tube.

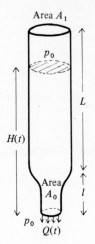

and at the exit the speed is

$$Q(t)/A_0 = -(A_1/A_0)dH/dt.$$

Bernoulli's integral can be applied to this unsteady potential flow (outside the boundary layers):

$$p/\rho + \tfrac{1}{2}v^2 + \Phi + \partial\phi/\partial t = F(t).$$

Now we know that $p = p_0$ at the upper water surface and at the exit (near enough), and for this flow

$$\begin{cases} \Phi = gy, \\ \phi = -\displaystyle\int_0^y \{Q(t)/A(\eta)\}\, d\eta. \end{cases}$$

So at the upper surface we get

$$F(t) = p_0/\rho + \tfrac{1}{2}(dH/dt)^2 + gH - (dQ/dt)\int_0^H \{A(\eta)\}^{-1} d\eta;$$

and at the exit we obtain similarly

$$F(t) = p_0/\rho + \tfrac{1}{2}(A_1/A_0)^2 (dH/dt)^2.$$

(Notice that $\partial\phi/\partial t = -(dQ/dt)\int_0^y \{A(\eta)\}^{-1}\, d\eta$, y is held constant during this differentiation.)

It is easy now to eliminate $F(t)$ and dQ/dt to get an equation for $H(t)$:

$$\tfrac{1}{2}(1 - A_1^2/A_0^2)(dH/dt)^2 + gH + A_1(d^2H/dt^2)\int_0^H \{A(\eta)\}^{-1}\, d\eta = 0.$$

There is still one awkward term in this equation,

$$\int_0^H \{A(\eta)\}^{-1} d\eta.$$

We can approximate this rather than calculate it exactly from the shape of the contraction. The easiest approximation would be to take $A(\eta) = A_1$ for the whole distance, which is reasonable if $L \gg l$. A better approximation might be to take

$$A(\eta) = A_0 + (\eta/l)(A_1 - A_0)$$

for $0 \leqslant \eta \leqslant l$, and $A(\eta) = A_1$ thereafter. But we will use the easiest one, to give

$$\int_0^H \{A(\eta)\}^{-1} d\eta \doteqdot H/A_1.$$

The equation for $H(t)$ now reduces to

$$H\, d^2H/dt^2 + \tfrac{1}{2}(1 - A_1^2/A_0^2)(dH/dt)^2 + gH(t) = 0.$$

This may be solved without too much difficulty by writing

$$d^2H/dt^2 = \dot{H}\, d\dot{H}/dH$$

(as was done for the bubble problem above), and then by using $Z = \tfrac{1}{2}\dot{H}^2$. The equation becomes

$$dZ/dH - cZ/H = -g$$

where $c = (A_1^2/A_0^2 - 1)$. This linear equation may be solved for Z in terms of H, and finally the variables separate to give an integral involving H equal to t. The details are left as an exercise, as it is all so similar to the bubble problem of §6(a); the solution is, anyway, not particularly simple in structure, and does not give any new insights.

(c) *Example: bubble oscillation*

It is worth returning to the bubble problem to see that Bernoulli's integral would have made it slightly easier. We had

$$\begin{cases} \phi(r, t) = -\{R(t)\}^2 \dot{R}(t)/r \\ v(r, t) = \{R(t)\}^2 \dot{R}(t)/r^2 \end{cases}$$

with $p \to p_0$ as $r \to \infty$ and $p = P$ at $r = R(t)$, where P was given in terms of R. The Bernoulli integral (with body force neglected) is then

$$p/\rho + \tfrac{1}{2}v^2 + \partial\phi/\partial t = F(t),$$

and using the two ends $r \to \infty$ and $r = R$ we get

$$p_0/\rho = P/\rho + \tfrac{1}{2}\dot{R}^2 - R\ddot{R} - 2\dot{R}^2$$

directly, where we have used

$$\partial\phi/\partial t = - R^2\ddot{R}/r - 2R\dot{R}^2/r.$$

Naturally, this is the same answer as before.

8. The force on an accelerating cylinder

(a) *The potential for the motion*

We used Bernoulli's integral before to calculate the force on a body, by integrating the pressure round the surface. We shall now do this for an unsteady problem for which a potential flow is a reasonably good model. The flow is that round a circular cylinder which accelerates from rest, sketched in figs. XI.18 and 19. This is adequately represented by a potential flow until the secondary flows develop and separation starts. If the acceleration is part of a small amplitude oscillation, then potential theory is a good approximation for all times, as no wake will develop.

It is convenient to choose axes which are at rest relative to the fluid at large distances, so that $\mathbf{v} \to 0$ as $r \to \infty$, and we take $\theta = 0$ along the direction of motion. However, such axes do get left behind by the cylinder as it accelerates away. And it is inconvenient to apply boundary conditions on the circle

$$\{x - (x_0 + Ut)\}^2 + \{y - y_0\}^2 = a^2.$$

So we choose to use *fixed* axes which are instantaneously at the centre of

Fig. XI.18. Definition sketch for potential flow past an accelerating cylinder.

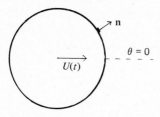

Fig. XI.19. Polar coordinates measured from a fixed origin coinciding with the centre.

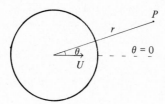

the circle $r = a$ – this is perfectly fair since there is always such a set of axes and the potential flow is determined at each instant by the conditions then and not by its past.

We need to solve $\nabla^2\phi = 0$ for $r > a$, then, and the boundary conditions are that

$$\nabla\phi \to 0 \text{ as } r \to \infty$$

and

$$\partial\phi/\partial r = U\cos\theta \text{ on } r = a,$$

because this is the outward velocity at a point on the cylinder. We have already seen the general method of solution of $\nabla^2\phi = 0$ by separation of variables in terms of r, θ. The full solution is

$$\phi = A\theta + B\ln(r/a) + \sum_{n=1}^{\infty}(C_n r^n + D_n r^{-n})\cos(n\theta + \alpha_n)$$

for integers n and constants A, B, C_n, D_n, α_n. When we fit the boundary values, it is clear that the only term remaining is

$$\phi = -Ua^2\cos\theta/r,$$

assuming that there is no circulation ($A = 0$) and no source term ($B = 0$). This solution is anyway to be expected, from the one we had previously for a cylinder in a stream: we have 'subtracted the stream'.

(b) Dependence on t

In this solution it is obvious that U depends on t. It is less obvious that r and θ also depend on t when a fixed point P is considered – as the cylinder moves along and we use new axes at the centre, r decreases and θ increases in fig. XI.19. Thus this solution ϕ is unsteady *even* when U is constant, because of the variation of r and θ with t. As we need $\partial\phi/\partial t$ in Bernoulli's integral, we must calculate $\partial r/\partial t$ and $\partial\theta/\partial t$ for this system of axes. We may do this most easily by fixing the origin O and letting P move

Fig. XI.20. The relative motion of field point and origin.

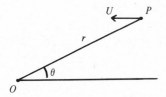

in the direction shown in fig. XI.20 at speed U. Then clearly

$$\begin{cases} \dot{r} = - U \cos \theta, \\ r\dot{\theta} = U \sin \theta, \end{cases}$$

are the velocity components of P in polar coordinates. So we calculate $\partial\phi/\partial t$ as

$$- \dot{U}(t)a^2 \cos \theta/r + Ua^2\dot{\theta} \sin \theta/r + \dot{r}Ua^2 \cos \theta/r^2$$
$$= - \dot{U}a^2 \cos \theta/r - U^2a^2 \cos 2\theta/r^2.$$

Notice that what we have calculated can be expressed as

$$\partial\phi/\partial t + \dot{r}\,\partial\phi/\partial r + \dot{\theta}\,\partial\phi/\partial\theta$$

(where now $\partial\phi/\partial t$ means, keep r and θ constant); and on using the values of $\dot{r}$ and $\dot{\theta}$ this is

$$\partial\phi/\partial t - \mathbf{U}\cdot\nabla\phi$$

where it is now a negative sign because P is moving at $- \mathbf{U}$ with respect to the axes.

If you prefer you may write x and y for P explicitly in terms of t as

$$x = x_0 - Ut, \quad y = y_0;$$

then, for example,

$$r = \{(x_0 - Ut)^2 + y_0^2\}^{1/2}$$

so that, as before,

$$\partial r/\partial t = - U \cos \theta.$$

(c) *The force in terms of added mass*

The pressure on the cylinder is calculated from Bernoulli's integral

$$p/\rho + \tfrac{1}{2}\mathbf{v}^2 + \partial\phi/\partial t = F(t), \text{ neglecting gravity.}$$

On the surface we have

$$\mathbf{v} = \nabla\phi$$
$$= (Ua^2 \cos \theta/r^2, \, Ua^2 \sin \theta/r^2);$$

and at infinity we have $p = p_0$ and $\phi \to 0$. Thus on the surface

$$p = p_0 - \tfrac{1}{2}\rho U^2 + \rho\dot{U}a \cos \theta + \rho U^2 \cos 2\theta.$$

The force opposing the motion of the cylinder is clearly along the line $\theta = 0$, by symmetry, and it has the value (per length of the cylinder)

$$\int_{-\pi}^{\pi} p \cos \theta \, a \, d\theta.$$

Fig. XI.21. The force element on the cylinder.

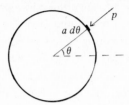

(See fig. XI.21). This integral evaluates to

$$\pi \rho a^2 \dot{U}$$

per length. This is dimensionally a mass-per-length times an acceleration. It is physically due to the acceleration of fluid near the cylinder by the motion of the cylinder. This term is often called an added mass or a virtual mass term.

The calculation above can be done much more rapidly via the kinetic energy T of the motion. This is

$$T = \tfrac{1}{2}\rho \int_V (\nabla\phi)^2 \, dV,$$

where V is outside the cylinder. We may convert this by means of the divergence theorem and $\nabla^2\phi = 0$, as in §2 above, to

$$T = -\tfrac{1}{2}\rho \int_S \phi \, \partial\phi/\partial n \, dS,$$

where $\partial/\partial n$ is *into* V. For the given ϕ this has the value (per length)

$$\tfrac{1}{2}\rho \int_{-\pi}^{\pi} Ua\cos\theta \times U\cos\theta \times a \, d\theta = \tfrac{1}{2}\pi\rho a^2 U^2.$$

This is the kinetic energy for a motion along the x-axis, and the associated force is (from Newton's law, or Lagrange's version of it)

$$\frac{d}{dt}\left(\frac{\partial T}{\partial \dot{x}}\right) = \frac{d}{dt}\left(\frac{\partial T}{\partial U}\right)$$
$$= \pi\rho a^2 \dot{U}$$

as before.

This kind of kinetic energy argument is not always available; for example if there were a circulation round the cylinder, the kinetic energy would be infinite. We may deduce from this that a truly two-dimensional vortex motion

$$\mathbf{v} = A\hat{\boldsymbol{\theta}}/r$$

outside a cylinder cannot be set up, as it would take infinite time to assemble enough energy. There must always be an outer limit to any vortex motion of this type, and this outer boundary condition may be important.

The calculation of added masses for accelerating cylinders has recently become important, with the interest in the extraction of energy from surface waves by submerged oscillating devices. However the calculations are rendered somewhat more difficult when there is a free surface to the liquid in which the cylinder moves.

9. D'Alembert's paradox

(a) The potential for a moving body

In many of the cases for which we have calculated the force on a body, it has turned out to be zero. It is worth proving a general result on forces in potential flow, firstly because it saves a lot of detailed calculation and secondly because the generality of the method is interesting. The result, often known as d'Alembert's paradox, is that there is *no* force on a steadily moving finite rigid body in three dimensions. It is, of course, no longer a paradox, as we have seen that drag forces on a body arise quite easily through the separation of the flow from the surface of the body. There are related results in two dimensions and when the body accelerates in a simple fashion or expands as it moves.

The general method is to use Bernoulli's integral to calculate the pressure on the body; and to convert many of the surface integrals over the body by means of the divergence theorem to integrals 'at infinity' (i.e. over a sphere of large radius R, where $R \to \infty$ eventually) which can be shown to be vanishingly small.

Outside the body, as shown in fig. XI.22, the potential satisfies

$$\nabla^2 \phi = 0,$$

and the boundary condition on S is

$$\partial \phi / \partial n = \mathbf{U} \cdot \mathbf{n}.$$

Moreover, we require $\nabla \phi \to 0$ as $r \to \infty$ and also that there shall be no net outflow from near the body, i.e. no source term in the solution for ϕ. But

Fig. XI.22. Definition sketch for §9(a).

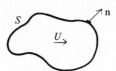

we may solve $\nabla^2 \phi = 0$ by separation of variables, and if we do this and impose the given conditions at infinity, we are left with terms such that

$$\begin{cases} \phi = O(r^{-2}) \text{ as } r \to \infty, \\ \nabla\phi = O(r^{-3}) \text{ as } r \to \infty. \end{cases}$$

The detailed calculation required to prove this is not done here, because we did not do a full separation of Laplace's equation in spherical polars; but the axisymmetric solution quoted in §5(c) is enough to give a good idea of what is needed.

The motion of the body is steady, so $\dot{U} = 0$; but this does not mean that ϕ does not change with time. As explained in §8, the coordinates of a fixed point change, because the axes at the centre of the moving body are continually having to be changed. Since $\dot{U} = 0$, in this case we get

$$\partial\phi/\partial t = -\mathbf{U}\cdot\nabla\phi$$

from §8.

(b) The force in a simple form

The force on the body is $\mathbf{F}$, where

$$F_i = \int_S p\, dS_i,$$

and where dS is *into* the body, as in fig. XI.23. Using Bernoulli's integral and neglecting the body force and assuming that $p \to p_0$ at infinity, this converts to

$$F_i = -\rho \int_S \dot\phi\, dS_i - \tfrac{1}{2}\rho \int_S \mathbf{v}^2\, dS_i + \int_S p_0\, dS_i.$$

Now $\int_S p_0\, dS_i = 0$, from the divergence theorem, and we substitute for $\partial\phi/\partial t$ from above:

$$F_i = \rho U_j \int_S (\partial\phi/\partial x_j)\, dS_i - \tfrac{1}{2}\rho \int_S \mathbf{v}^2\, dS_i,$$

$$= \rho U_j \int_S v_j\, dS_i - \tfrac{1}{2}\rho \int_S (\nabla\phi)^2\, dS_i.$$

Fig. XI.23. Definition sketch for §9(b).

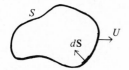

The term in $(\nabla\phi)^2$ can be converted to a volume integral by using a version of the divergence theorem:

$$\int_S (\nabla\phi)^2 \, dS_i = \int_V \partial/\partial x_i (\nabla\phi)^2 dV - \int_\infty (\nabla\phi)^2 dS_i$$

where the normals to the surfaces are taken to be out of the fluid, i.e. out of V. The integral at infinity is $O(r^{-4})$ because $\nabla\phi$ is $O(r^{-3})$ and dS is $O(r^2)$, and hence it vanishes as $r \to \infty$. The integral over the fluid can be rearranged by using the identity

$$\nabla(v^2) = 2\mathbf{v}\cdot\nabla\mathbf{v} + 2\mathbf{v} \times \boldsymbol{\omega}$$
$$= 2\mathbf{v}\cdot\nabla\mathbf{v}$$

because the flow is irrotational; and further

$$(\mathbf{v}\cdot\nabla\mathbf{v})_i = v_j \partial v_i/\partial x_j$$
$$= \partial/\partial x_j (v_i v_j)$$

because $\nabla\cdot\mathbf{v} = \partial v_j/\partial x_j = 0$ for an incompressible fluid. So we can replace $\partial/\partial x_i (\nabla\phi)^2$ by

$$2\partial/\partial x_j (v_i v_j).$$

We may then reuse the divergence theorem on

$$\rho \int_V \partial/\partial x_j (v_i v_j) dV$$

to get

$$\rho \int_S v_i v_j \, dS_j,$$

together with a term at infinity which is easily shown to be vanishingly small. So we have shown that

$$\tfrac{1}{2}\rho \int_S (\nabla\phi)^2 \, dS_i = \rho \int_S v_i v_j \, dS_j$$
$$= \rho U_j \int_S v_i \, dS_j$$

because of the boundary condition on S.

If we use this result in the formula for $\mathbf{F}$, we get

$$F_i = -\rho U_j \int_S (v_i \, dS_j - v_j \, dS_i).$$

Fig. XI.24. A flow with zero force in potential theory.

(c) *General results on the force*

It follows immediately that $\mathbf{F} \cdot \mathbf{U} = 0$, because this is

$$F_i U_i = -\rho \int_S (v_i U_i U_j dS_j - U_j v_j U_i dS_i)$$

and the integrand is clearly zero. So we have shown that there is no force component parallel to $\mathbf{U}$.

We now reprocess $\int_S (v_i dS_j - v_j dS_i)$ in a very similar fashion to prove that it is identically zero provided ϕ is small enough at large distances. From a version of the divergence theorem

$$\int_S (v_i dS_j - v_j dS_i) = \int_V (\partial v_i / \partial x_j - \partial v_j / \partial x_i) \, dV - \int_\infty (v_i dS_j - v_j dS_i).$$

This volume integral has a component of $\nabla \times \mathbf{v}$ as its integrand, which is zero. And the integral at infinity is zero because $\mathbf{v} = O(r^{-3})$ and $dS = O(r^2)$.

So we have finally proved, with rather more fiddling with integrals than is pleasant, that $\mathbf{F} = 0$, there is no force on the body. The relevance of this result is to flows that do not separate and are truly irrotational: there are few of these, but enough to make the result useful. For example, the symmetrical streamlined strut sketched in fig. XI.24 has a narrow wake and a flow round it which is almost potential flow. The drag force on it is indeed small. Similarly a smallish spherical bubble rising steadily through a liquid has almost no pressure force on it, as there is no boundary layer separation and flow round it which is quite close to potential flow.

The corresponding result in two dimensions is very similar, except that a circulation may exist round a body, leading to $\nabla \phi = O(r^{-1})$ at infinity. This, with $dS = O(r)$ as $r \to \infty$, leads to a non-zero force, the Magnus force perpendicular to the direction of motion. The details are left as an exercise. The two-dimensional case is anyway most easily done by the methods of Chapter XVI.

Exercises

1. Concentric cylinders of radii a and $2a$ contain a liquid at rest and a radial barrier as shown in fig. XI.25. The barrier accelerates from rest. Explain

Fig. XI.25. Sketch for Q1 and Q2.

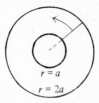

$r = a$

$r = 2a$

why the motion must be initially a potential flow, and hence why it cannot be rigid body rotation $\mathbf{v} = \Omega r \hat{\theta}$.

2. What boundary conditions should be imposed on the potential ϕ for Q1 above? Prove that there is a unique solution for ϕ in this problem. [Don't try to solve for ϕ: it is hard.]

3. Show from $\nabla^2\phi = 0$ that $\int_S \partial\phi/\partial n \, dS = 0$ for any closed surface S in the fluid, and hence that ϕ cannot have an extreme value at an interior point of the fluid. Show that $\partial\phi/\partial x$ satisfies Laplace's equation and so cannot have an extreme value at an interior point of the fluid. Deduce that the speed cannot have a maximum value at an interior point.

4. Show that $\phi = A(x^2 - y^2)$ satisfies $\nabla^2\phi = 0$ and that $\psi = 2Axy$ gives the same velocity field. Find a polynomial of degree three in x and y which satisfies $\nabla^2\phi = 0$, and determine the corresponding stream function ψ. Find typical polynomials of degree two or less in x, y, z that satisfy Laplace's equation, and express them in terms of spherical coordinates r, θ, λ.

5. The equation $\nabla^2\phi = 0$ is to be solved with the conditions

$$\begin{cases} \partial\phi/\partial x = 0 & \text{on} \quad x = \pm a, \\ \partial\phi/\partial y = 0 & \text{on} \quad y = 0, \end{cases}$$

with the solution being required in

$$\begin{cases} -a < x < a, \\ \ \ 0 < y < a. \end{cases}$$

The situation is sketched in fig. XI.26. Use separation of variables to derive the solution

$$\phi(x, y) = \sum_{n=0}^{\infty} c_n \cosh(n\pi y/a) \cos(n\pi x/a)$$

$$+ \sum_{n=0}^{\infty} d_n \cosh(2n + 1)\pi y/2a \sin(2n + 1)\pi x/2a.$$

Calculate the coefficients d_n when also

$$\partial\phi/\partial y = Ax \quad \text{on} \quad y = a.$$

Interpret your solution in terms of motion of an ideal fluid. Show that ϕ is quite well approximated by the term with $n = 0$.

Fig. XI.26. Sketch for Q5.

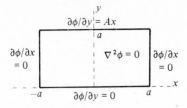

Fig. XI.27. Sketch for Q7.

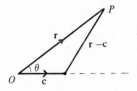

6. Derive the general form of separation solution

$$\phi(r, \theta) = A\theta + B \ln(r/a) + \sum_{n=0}^{\infty} (C_n r^n + D_n r^{-n}) \cos(n\theta + \alpha_n)$$

for the equation $\nabla^2 \phi = 0$ and the region

$$\begin{cases} 0 < r < \infty, \\ -\pi < \theta \leqslant \pi, \end{cases}$$

with continuity of ϕ at the cut $\theta = -\pi$ except for the term representing a circulation.

7. A point source at a point $\mathbf{c}$ on the axis $\theta = 0$ of spherical polar coordinates has potential

$$\phi = A/|\mathbf{r} - \mathbf{c}|$$

(see fig. XI.27). But this is an axisymmetric situation, and so has a separation solution

$$\phi = \sum_{n=0}^{\infty} (a_n r^n + b_n r^{-n-1}) P_n(\cos \theta).$$

Expand $|\mathbf{r} - \mathbf{c}|^{-1}$ for $r > c$ to determine the coefficients b_n for $n = 0, 1, 2, 3$.

8. Water is at rest in a conical funnel (see fig. XI.28) of angle $\frac{1}{2}\pi$ (so that the walls are $\theta = \frac{1}{4}\pi$ when the axis is $\theta = 0$), when the lower end is suddenly opened. Show that a potential flow is to be expected at later times, and set up the mathematical problem to be solved in the upper conical part in terms of a sink at the vertex. Outline the steps in the solution, but do not carry it out.

9. A sphere accelerates from rest along the axis $\theta = 0$, with speed $U(t)$ at

Fig. XI.28. Sketch for Q8.

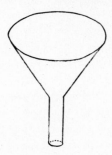

time t. The pressure at large distances is p_0. Determine the greatest value of dU/dt such that the pressure is positive over the whole sphere when $U = 0$. [Neglect gravity.]

10. A cylinder of radius a is moved so that its axis describes a small circle of radius $b(<a)$ at speed Ωb. Assuming that potential flow is a good model, determine the force on the cylinder from the fluid.

11. Calculate the force needed to accelerate a sphere of radius a and mass M from rest through a fluid. Hence calculate the initial upward acceleration of a spherical bubble through a liquid, neglecting viscous forces.

12. In the model of §5, take the radius of the sphere to be

$a(t) = ut/5$

where u is a constant velocity. Calculate the extra force on the sphere because now

$\partial\phi/\partial t \neq 0$.

13. Solve the equation for $H(t)$ as far as is possible for the flow of liquid down a pipe with a contraction near the exit. Verify that the neglect of unsteadiness in the case $A_1 \gg A_0$ is a reasonable approximation.

14. A line source lies at distance $2a$ from the axis of a cylinder of radius a, and parallel to the axis, as in fig. XI.29. Give an expression for ϕ in terms of images, and express ϕ at large distances in the form

$\phi = A \ln(r/a) + B + Cr^{-1} \cos\theta + O(r^{-2})$.

Calculate the force on the cylinder, by integrating the pressure round the surface of the cylinder, and explain why this does not contradict d'Alembert's paradox.

Fig. XI.29. Definition sketch for Q14.

$r = a$

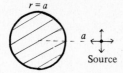

Source

Fig. XI.30. Definition sketch for Q15.

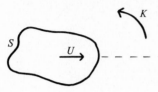

15. The solution ϕ in two dimensions outside the surface S which moves steadily along the line $\theta = 0$ has the form

$$\phi = K\theta/2\pi + O(r^{-1})$$

for large r (see fig. XI.30). Use the methods of §9 to find the force on the surface S.

References

(a) Texts mainly on potential flow have been written for a century; many treat it as the major aspect of fluid dynamics. The best of the older texts is *Hydrodynamics*, H. Lamb, C.U.P.; it was written in 1879 but was so good as to be brought out in several editions up to 1945.

(b) *An Introduction to Fluid Dynamics*, G. K. Batchelor, C.U.P. 1967, gives a more thorough modern approach to potential flow than is possible here.

(c) Other texts written from a mathematical point of view tend to follow Lamb in material. Of these it may be worth looking at
  (i) *Theoretical Hydrodynamics* (2nd edn), L. M. Milne-Thomson, Macmillan 1949
  (ii) *Textbook of Fluid Dynamics*, F. Chorlton, Van Nostrand 1967.
  These texts are more approachable than Lamb, and give much more detail than is possible here.

(d) There are two film loops that give the Hele–Shaw visualisation of two-dimensional potential flow. These have been mentioned earlier in connection with images and Rankine bodies. They are not a major help for this chapter. They are 39.5117 *Hele–Shaw Analogue to Potential Flow Part I* (FM-80), 39.5118 *Hele–Shaw Analogue to Potential Flow Part II* (FM-81).

XII

Sound waves in fluids

1. Background

The basic facts of the propagation and generation of sound are well known. Sound is a pressure disturbance in air (or water) which travels at a rather high speed. Sound is generated in regions of fluid motion; sometimes a motion of solid boundaries is involved, as in a drum; sometimes not, as in a flute. Sound is received in our ears (or by a microphone) where the pressure oscillations are converted into electrical signals.

The speed of sound is typically around $330\text{--}40$ m s^{-1} in still air near ground level, and around $1400\text{--}50$ m s^{-1} in water. Sound also travels through solid materials in the form of elastic waves; we cannot do the theory for such waves here, though it is of the same general type as what we shall do. Often the speed of sound in solids is much higher than the speed of sound in air.

The pressure fluctuations in sound in air are typically between 10^{-4} and 1 N m^{-2}, so that the ratio of pressure fluctuations to ambient pressure in the atmosphere lies between

$$10^{-9} \quad \text{and} \quad 10^{-5}.$$

The smallness of this quantity is one of the basic facts in the simple theory of sound. It is quite possible to get much higher pressure fluctuations, in explosions or in jet engines, so that the ratio of pressures is not small: the simple theory of sound cannot apply to such cases.

The frequencies of audible sound waves lie between about 20 Hz and 2×10^4 Hz. Below about 20 Hz, oscillations are felt by the body rather than heard by the ear, and so may not be classed as sound waves, though they are still described by the same mathematics – sometimes the word infrasound is used. Similarly, oscillations above 2×10^4 Hz cannot be heard by everyone; again, the mathematics is mainly the same, and the word ultrasound is used.

The corresponding wavelengths are, in air,

$$\begin{cases} 16.5 \text{ m for } 20 \text{ Hz}, \\ 1.65 \text{ cm for } 2 \times 10^4 \text{ Hz}. \end{cases}$$

These are relatively large. The musical note E above middle C is within the range of most human singing voices (at the top of the bass range and the bottom of the soprano); it has a frequency of 330 Hz, and so a wavelength of about 1 m; this is considerably larger than the human head from which it is produced. We shall see that it is quite typical for the wavelengths produced to exceed the size of the production region by quite a wide margin.

In this text we shall stick to the easiest aspects of the theory of sound. Thus we shall not be able to cover the (very interesting) topics of:

   (i) refraction by variations in density or velocity in the air in which sound propagates;
  (ii) diffraction of sound waves by boundaries and the formation of sound shadows;
 (iii) generation of sound by general fluid motions.

We shall concentrate on the basic model for sound waves, which is a small disturbance to a region of air which is uniform and at rest before the small disturbance reaches it.

2. The linear equations for sound in air

(a) Pressure and density in a sound wave

Sound waves typically travel through still air away from boundaries. The circulation round any circuit in the still air is zero; consequently there is no vorticity in sound waves provided it is fair to assume $\rho = f(p)$. Now the length scale of sound waves is rather large, and the time scale is rather short, and so there is no time for heat conduction to be noticeable over so large a scale; hence we may assume that there is effectively no heat flow. The flow is not much affected by viscosity, and is effectively reversible, and so we have isentropic conditions: that is, $\rho = f(p)$ is fair since $S = $ constant; and in particular for air

$$p = k\rho^{\gamma}$$

is the appropriate relation. And as a consequence $\mathbf{v} = \nabla\phi$. For sound in water you reach the same conclusion, but the relation between p and ρ will be different – see Chapter VII §1 for a model. Naturally, neither of these equations is exactly true; viscosity and so vorticity must enter near boundaries, brought in by the no-slip condition; and heat conduction, though small, will eventually be important. But the basic theory, derived from these equations, will be enough for the present.

We shall write

$$\begin{cases} p = p_0 + p', \\ \rho = \rho_0 + \rho', \end{cases}$$

where p_0, ρ_0 are the values in the still air. We are assuming that

$$p'/p_0, \rho'/\rho_0$$

are very small, so that we will frequently neglect second and higher powers of these small numbers. For example, the adiabatic relation

$$p = k\rho^\gamma$$

is

$$p_0 + p' = (p_0/\rho_0^\gamma)(\rho_0 + \rho')^\gamma,$$

on deriving the value of k from the state of rest, and this can be approximated as

$$p_0 + p' = \rho_0^\gamma(1 + \gamma\rho'/\rho_0)(p_0/\rho_0^\gamma)$$

or

$$p' = (\gamma p_0/\rho_0)\rho'.$$

An alternative view of this formula is that it is a Taylor approximation

$$\rho = \rho_0 + (p - p_0)f'(p_0) + O(p - p_0)^2,$$

i.e.

$$\rho' \doteqdot p'f'(p_0),$$

which shows that ρ' is in general proportional to p' for any fluid with $\rho = f(p)$.

The velocity $\mathbf{v}$ is also a small quantity, because it is driven by the changes in pressure. Exactly in what sense it is small will appear later in the theory. Similarly ϕ will be small.

(b) Pressure and velocity potential in a sound wave

Bernoulli's integral applies, in the form appropriate for unsteady irrotational compressible flow:

$$\int \rho^{-1}dp + \tfrac{1}{2}\mathbf{v}^2 + \partial\phi/\partial t = F(t).$$

Notice that we assume that gravity is of no real importance in the basic theory of sound. Now in the undisturbed air, everything is steady, and so $F(t)$ can only be a constant: and we evaluate the integral by using the approximate relation

$$\rho = \rho_0(1 + \rho'/\rho_0)$$
$$\doteqdot \rho_0(1 + p'/\gamma p_0);$$

moreover, we neglect $\frac{1}{2}v^2$ as a second power of a small quantity. So using

$$\rho^{-1} = \rho_0^{-1}(1 - p'/\gamma p_0),$$

and

$$dp = dp'$$

we get

$$\rho_0^{-1}p' + \partial\phi/\partial t = 0,$$

where the constant is zero as p' and ϕ vanish in still air, So an approximate relation between pressure and velocity is

$$p' = -\rho_0\partial\phi/\partial t.$$

This, with $\mathbf{v} = \nabla\phi$, gives the relations between the observables p', $\mathbf{v}$ and the mathematical variable ϕ in the simple theory of sound.

(c) The equation for the potential

We have still to satisfy the mass-conservation equation

$$\partial\rho/\partial t + \nabla\cdot(\rho\mathbf{v}) = 0.$$

We approximate this by neglecting $\rho'\mathbf{v}$ as a second order quantity. This gives

$$\partial\rho'/\partial t + \rho_0\nabla\cdot\mathbf{v} = 0$$

or

$$\partial\rho'/\partial t + \rho_0\nabla^2\phi = 0.$$

Now since $\rho' \doteqdot (\rho_0/\gamma p_0)p'$, we may put this as

$$\partial p'/\partial t = -\gamma p_0\nabla^2\phi.$$

But we have already derived another relation between p' and ϕ, so we can eliminate p' to get

$$\partial^2\phi/\partial t^2 = (\gamma p_0/\rho_0)\nabla^2\phi.$$

This is the basic equation for the simple theory of sound; it has the same form as the equation that comes into many physical applications in which waves propagate. The constant

$$\gamma p_0/\rho_0 = c^2$$

has the dimensions of a speed squared: taking

$$\gamma = 1.4, \, p_0 = 1.01 \times 10^5 \, \text{N m}^{-2}, \, \rho_0 = 1.23 \, \text{kg m}^{-3},$$

which are figures for the standard atmosphere at sea level gives

$$c = 339 \, \text{m s}^{-1},$$

which is quite closely equal to the observed speed of sound in dry air at sea level at 288 K.

Since $p' = -\rho_0 \partial\phi/\partial t$, p' also satisfies a wave equation, with the same wave speed; and consequently $\rho' = (\rho_0/\gamma p_0)p'$ also satisfies this wave equation. It really does not matter which of ϕ, p', ρ' the theory is put in terms of; we choose to use ϕ mainly to emphasize that this is a part of potential theory, though the basic equation is no longer $\nabla^2\phi = 0$.

(d) Boundary and initial conditions

The wave equation has some rather different conditions imposed on it, because it is associated with finite speeds rather than the instantaneous propagation of $\nabla^2\phi = 0$.

(i) As before, at a solid boundary the normal velocity of boundary and fluid must be equal, and so

$$\partial\phi/\partial n = \mathbf{U}\cdot\mathbf{n}$$

there. As before, there may be a slip velocity tangential to the wall, which will be taken up by a boundary layer in reality.

(ii) Quite often the boundary condition at a narrow opening is well modelled by

$$p' = 0$$

at or near the opening. The sound in a room does not leak out through a slightly open window; draughts (governed more nearly by $\nabla^2\phi = 0$) rush happily through narrow gaps, but sound waves of scale larger than the slit tend to be reflected rather than pass through, and this is associated with a small (almost zero) value of p' at the opening.

(iii) Because the wave equation involves time derivatives of second order, there will need to be initial conditions on both ϕ and $\partial\phi/\partial t$; equivalently, initial conditions are needed on $\mathbf{v}$ and p, as we then know the motion and the initial acceleration (given by the initial value of ∇p).

(iv) We may need to have some sort of boundary condition at large distances. This is now of a rather different kind from the ones appropriate for $\nabla^2\phi = 0$. Since the wave equation is to do with propagation, we need to know whether energy is being sent in from infinity, or whether all energy propagation is outwards at infinity. This is called a radiation condition, and we shall see how to impose it on ϕ later.

(v) At non-rigid boundaries, we shall have to set up an equation of motion, in which one of the forces is derived from the pressure in the sound wave.

3. Plane sound waves

(a) The wave velocity

The simplest solutions of the wave equation involve only one space dimension, and should be familiar from elementary work on waves on flexible strings under tension. In any case it is easily verified that

$$\phi = A \cos k(x - ct)$$

is a solution of

$$c^2 \nabla^2 \phi = \partial^2 \phi / \partial t^2$$

where c is the constant wave speed, and k is a constant called the wave number, and A is a constant called the amplitude. A more formal derivation of this result may have been met in calculus courses where it is often proved that the general solution of

$$c^2 \partial^2 \phi / \partial x^2 = \partial^2 \phi / \partial t^2$$

has the form

$$\phi = F(x - ct) + G(x + ct).$$

where F and G are *any* twice differentiable functions. We have merely chosen to put $G = 0$ for the moment and choose a special form for F.

This solution is called a plane wave, because at any time t_0, ϕ has a constant value on each plane $x = $ constant. The value of $k(x - ct)$ is called the phase of the wave, and so we may say that the surfaces of constant phase at time t_0 are the planes $x = $ constant. Let the phase have value s at $x = x_0$ and $t = t_0$, so that

$$s = k(x_0 - ct_0).$$

Now consider a later time $t_0 + \tau$: the phase still has the value s at the position

$$x_0 + c\tau$$

because

$$s = k\{x_0 + c\tau - c(t_0 + \tau)\}$$
$$= k(x_0 - ct_0).$$

This, of course, is true for any value of τ. The surface of constant phase s moves to the new position $x_0 + c\tau$ in time τ, and so is travelling at speed c. This justifies the use of the term wave speed for the constant c in the equation, and it shows that the theory is matching well to the experiments in which the wave speed is measured and is found to be very close to c.

(b) *Relations for a sinusoidal plane wave*
We have taken

$$\cos k(x - ct)$$

as the basic plane wave solution rather than

$$f(x - ct),$$

which is also a plane wave, for various reasons.

   (i) The periodic nature of cos allows us to define frequency and wavelength.

  (ii) Musical notes are periodic, and can be closely sinusoidal.

 (iii) Other functions can be built up from cosines and sines by Fourier series.

 (iv) The functions sin and cos are amongst the easiest to work with, especially if we put

$$\cos s = \mathcal{R}e\, e^{is},$$

where $\mathcal{R}e$ stands for 'take the real part of' and $i = \sqrt{-1}$. This is a help, because differentiating exponentials is so easy; the $\mathcal{R}e$ operation is hardly ever a trouble, because the equation and boundary conditions are linear, and the $\mathcal{R}e$ operator can be freely interchanged with the other operators that come in. For example,

$$\mathcal{R}e\, \partial\phi/\partial x = \partial/\partial x(\mathcal{R}e\,\phi),$$
$$\mathcal{R}e(\alpha\phi_1 + \beta\phi_2) = \alpha\,\mathcal{R}e\,\phi_1 + \beta\mathcal{R}e\,\phi_2.$$

Thus all the calculations may be done on

$$\phi = Ae^{ik(x-ct)} = A\exp\{ik(x - ct)\},$$

with the convention that it is the real part that is meant throughout. The amplitude A may now be a complex number; for example, to represent $\sin k(x - ct)$ we need

$$\mathcal{R}e\{-ie^{ik(x-ct)}\},$$

that is, we have chosen

$$A = -i.$$

Fig. XII.1. Wavelength and frequency in sinusoidal waves.

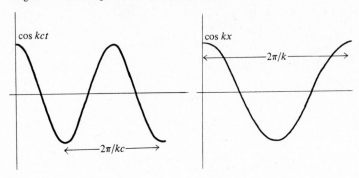

We choose therefore to discuss the plane wave

$$\phi = A \cos k(x - ct)$$

where A is real in the present context. Now take two points a distance $2\pi/k$ apart, and at the same time. Then ϕ has the same value because $\phi(s + 2\pi) = \phi(s)$. This is the justification for calling $2\pi/k$ the wavelength: a shift of $2\pi/k$ in the x-direction is the smallest one that returns you to an equivalent place.

In a similar fashion, a shift of $2\pi/kc$ in the time at a fixed point is the smallest one that gives an equivalent result, and so we call $2\pi/kc$ the period. Both these shifts are illustrated in fig. XII.1. The usual symbols for these quantities are

$$\begin{cases} \text{wavenumber } k, \text{ wavelength } 2\pi/k = \lambda, \\ \text{frequency } kc, \text{ period } 2\pi/kc = \tau \text{ or } T. \end{cases}$$

Notice that the frequency is measured in radians per second:

$$\begin{cases} \text{frequency (radians per second) } kc = \omega, \\ \text{frequency (cycles per second) } kc/2\pi = f. \end{cases}$$

Notice also that λ is not now a spherical polar angle. The relations

$$\lambda f = c, \, \omega/k = c, \, \lambda/\tau = c,$$

are often useful when you have a simple plane wave.

(c) *Plane waves in three dimensions*
A more general plane wave is

$$\phi = A \cos(k_1 x + k_2 y + k_3 z - kct)$$

where $k_1^2 + k_2^2 + k_3^2 = k^2$: check that this satisfies the wave equation. We may rewrite it as (the real part of)

$$\phi = A \exp\{i(\mathbf{k}\cdot\mathbf{x} - \omega t)\}$$

by introducing a wavenumber vector

$$\mathbf{k} = (k_1, k_2, k_3)$$

and defining

$$\omega = |\mathbf{k}|c = kc.$$

It is not hard to see that the planes of constant phase here are the planes

$$\mathbf{k}\cdot\mathbf{x} = \text{constant},$$

and that propagation is in the direction of $\mathbf{k}$ at speed c. Notice that $\mathbf{k}$ is not now the unit vector along the z-axis.

The velocity $\mathbf{v}$ of the air in this wave is

$$\mathbf{v} = \nabla\phi = (\partial\phi/\partial x, \partial\phi/\partial y, \partial\phi/\partial z)$$
$$= i\mathbf{k}A \exp\{i(\mathbf{k}\cdot\mathbf{x} - \omega t)\}$$

because $\mathbf{k}\cdot\mathbf{x} = k_1 x + k_2 y + k_3 z$. In other words, the air velocity is *along* the direction of propagation.

There is also a solution

$$\phi = A \exp\{i(-\mathbf{k}\cdot\mathbf{x} - \omega t)\}$$

to the wave equation. This merely corresponds to a change in the direction of travel of the wave. It may be rearranged as

$$\phi = A' \exp\{i(\mathbf{k}\cdot\mathbf{x} + \omega t)\}$$

if desired, by choosing a different amplitude, if only the real part is eventually needed:

$$A' = \bar{A},$$

the complex conjugate of A is the new amplitude.

The most general plane wave propagating in direction $\mathbf{k}$ is

$$\phi = f(\mathbf{k}\cdot\mathbf{x} - \omega t),$$

where $\omega = kc$ as before. So long as $\mathbf{k}\cdot\mathbf{x} - \omega t$ has the same value, you get the same value for

$$\phi = f(\mathbf{k}\cdot\mathbf{x} - \omega t).$$

Hence for a larger value of t, you get the same values of ϕ at larger values of $\mathbf{k}\cdot\mathbf{x}$. So the wave propagates in the direction of $\mathbf{k}$ without change of shape: it is merely moved along a distance c times the time interval, as you see in fig. XII.2. This means that (on this theory) you hear just the same at one place as at another, merely at a different time. In reality there are slight

Fig. XII.2. Preservation of the wave form in a plane wave.

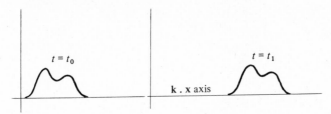

distortions due to non-linearity (which has been omitted) and due to diffusive effects (heat conduction and viscosity have both been neglected).

(d) *The size of the velocity and displacement*
Let us take the plane wave

$$\phi = A \exp\{ik(x - ct)\}.$$

Then from the equation in §2(b), we have

$$p = -\rho_0 \dot{\phi} = iA\rho_0 kc \exp\{ik(x - ct)\},$$

where we shall now usually drop the dash from p and ρ for convenience; so p is now the excess pressure. Also the air velocity is

$$v = \phi_x = iAk \exp\{ik(x - ct)\},$$

so that

$$v = p/\rho_0 c.$$

The quantity $\rho_0 c$ is often called the (specific) acoustic impedance, as it is the constant coming in the relation between the response (the velocity) and the forcing (the pressure). The term impedance really derives from electric circuit theory, where

'voltage = impedance × current'

i.e.

'forcing = impedance × response'.

This terminology is also used in mechanical oscillation theory, and because the theories are in parts so similar, experiments can sometimes be done on electrical networks instead of with sound waves.

Now we know that p has typical values $10^{-4} - 1\,\mathrm{N\,m^{-2}}$ in a sound wave. Hence the velocity of the air has typical values

$$3 \times 10^{-7} - 3 \times 10^{-3}\,\mathrm{m\,s^{-1}}.$$

In dimensionless form, all of p/p_0, ρ/ρ_0 and v/c have values in the range

$$10^{-9} - 10^{-5}.$$

We may now return to Bernoulli's equation and compare the sizes of (say) $\frac{1}{2}\mathbf{v}^2$ and $\partial\phi/\partial t$.

We have

$$\tfrac{1}{2}\mathbf{v}^2 = \tfrac{1}{2}iAk \exp\{ik(x-ct)\} \times c \times v/c$$

in this plane wave, and

$$\partial\phi/\partial t = -iAkc \exp\{ik(x-ct)\}.$$

Hence the ratio of sizes is $\frac{1}{2}v/c$, which is indeed tiny. This justifies that particular linearisation of the equations, at least over reasonable time scales (over very long times these very small terms may give a noticeable effect).

Finally, let us calculate the displacement ξ of the air in this sound wave. Since $\partial\xi/\partial t = v$ (this is a linearisation also), we have

$$\xi = -c^{-1}A \exp\{ik(x-ct)\} = iv/\omega.$$

Using the range of values given above for v and for ω, we find that the displacement of the air in a sound wave is typically

$$10^{-11}\,\text{m}-10^{-4}\,\text{m}.$$

Even loud low frequency sounds have *very* small displacements, and typical sounds have displacements that are around the mean free path of air molecules (10^{-7} m). This does not mean that continuum theory is no longer applicable, because the motions are organised over much larger scales laterally. An area of even as little as 10^{-6} m^2 moving with amplitude 10^{-9} m is covering a volume of 10^{-15} m^3, in which there will still be a very large number of molecules, about 3×10^{10}; and this is quite enough to average over.

(e) *Reflection of a plane wave*

So far we have discussed a plane wave travelling freely in space. Let us now put in a rigid boundary at $y=0$, and discuss the reflection of a plane wave which strikes $y=0$ at an angle θ as shown in fig. XII.3. We may approach this problem in two ways. We may say that, physically, we expect a reflected wave, and hence we will look for some sort of image wave to satisfy the boundary value

$$\partial\phi/\partial y = 0 \quad \text{on} \quad y=0.$$

Or we may treat the problem mathematically without any appeal to what we expect, by separating variables and imposing the condition that there

Fig. XII.3. Reflection of a plane wave at a rigid wall. Planes of constant phase in the three waves are indicated.

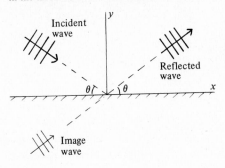

must be an incident wave from infinity, which we take to have the form

$$\phi = A \exp\{i(kx \cos\theta - ky \sin\theta - \omega t)\};$$

this is a plane wave travelling in the appropriate direction, downwards at θ to the x-axis.

Either method is acceptable for this problem; the separation of variables method is longer and needs more manipulation, but requires no physical insight. Since we shall have to do several separation solutions later, we choose to use an 'images' approach here. That is, we guess at a solution in terms of an image, and then verify that it satisfies all the required conditions. We then need a uniqueness theorem (which we shall not prove) to be convinced we have the correct solution.

The image we choose is a similar wave travelling upwards at θ to the x-axis. We choose it to have the same frequency, and the same wavelength $2\pi/k$. Hence we have

$$\phi_1 = B \exp\{i(kx \cos\theta + ky \sin\theta - \omega t)\}.$$

The changes from ϕ are that the amplitude B may not be equal to A, and the angle θ has been changed to $-\theta$ because of the different direction. We take $\phi + \phi_1$ to be our solution for $y > 0$. We verify that:

 (i) it satisfies $c^2 \nabla^2(\phi + \phi_1) = (\partial^2/\partial t^2)(\phi + \phi_1)$;

 (ii) there is appropriate inward radiation from infinity;

 (iii) $(\partial/\partial y)(\phi + \phi_1) = 0$ on $y = 0$, for all x and t.

Only the last condition needs any comment; it is

$$(-iAk \sin\theta + iBk \sin\theta) \exp\{i(kx \cos\theta - \omega t)\} = 0,$$

when we substitute in the forms of ϕ and ϕ_1. Clearly it is true for all x and t if $A = B$. Hence we have *the* solution to the problem.

Fig. XII.4. An echo will result from these almost plane waves.

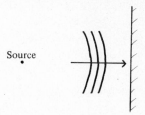

The simple reflection of a plane wave of sound from a plane rigid boundary gives an idea of how echoes may be formed: an almost plane sound wave is reflected from an almost plane wall of finite extent (as shown in fig. XII.4), so that a part of the emitted sound is received back at the source at a time $2 \times$ distance$/c$ later.

The slip velocity and pressure at the wall can be calculated in the model. The velocity along the wall is

$$\partial/\partial x(\phi + \phi_1) \quad \text{at} \quad y = 0$$

or

$$2iAk \cos \theta \exp\{i(kx \cos \theta - \omega t)\} ;$$

and similarly the pressure is

$$-\rho_0 \partial/\partial t(\phi + \phi_1) \quad \text{at} \quad y = 0$$

or

$$2iA\rho_0 \omega \exp\{i(kx \cos \theta - \omega t)\}.$$

This pressure is double the pressure in the incident wave and gives a periodic (both in x and t) force on the wall; physically, this is needed to reflect the compression wave. It should be noted that the wavelength of the pressure along the wall is not λ but $\lambda \sec \theta$, because the space variation is $k \cos \theta$ and not just k; similarly the apparent speed of this pressure disturbance along the wall is $\omega/(k \cos \theta) = c \sec \theta$. This exceeds c, and is not the real speed of the waves; it is just an effect of kinematics, of how the planes of constant phase intersect the wall.

The solution $\phi + \phi_1$ can be written as

$$2A \cos(ky \sin \theta) \exp\{i(kx \cos \theta - \omega t)\}.$$

This shows that the wave in the y-direction is a standing wave, i.e. not progressing forward or back. This is mathematically correct, but not physically helpful, as it does not distinguish between the incident wave which has to be supplied and the reflected wave created by the rigid wall.

Reflections of real sound waves in rooms and buildings do not always behave like this. In a cathedral, which has large flat hard expanses, sounds will echo around very much in accordance with this theory, being only gradually disorganised by irregular surfaces and lost through windows or degraded by diffusive processes. But in a concert hall the seats and audience are not hard, and absorb acoustic energy; the walls and ceiling may also be treated so as to remove energy (acoustic tiles, for example, or curtains), and sounds die away much more quickly. And in a small domestic room the losses are so great that sounds are gone almost at once, mainly because there are so many reflections per second that even a small loss at each reflection soon becomes effective.

4. Plane waves in musical instruments

(a) Modelling of the problem

Some major aspects of musical wind instruments can be fairly well described by rather simple mathematics. In particular the frequencies and the general quality of the emitted sounds can be found accurately enough, even though it is very hard to describe the processes which generate the sounds, and many of the practices of the skilled musician seem to be currently beyond mathematical description.

What we shall do is calculate plane wave solutions in tubes with parallel, straight sides – many instruments are very curvy and have changing cross-sections but this makes little difference to the results; we shall take

$$v = \partial\phi/\partial x = 0$$

at a closed end – it is not always obvious in a real instrument whether an end is open or closed; we shall take

$$p = -\rho_0\partial\phi/\partial t = 0$$

at an open end – even though doing this means that no work can be done there and so no radiation can result. Moreover we shall take solutions which represent standing waves, even though the evidence is that (at least in some instruments) there are considerable losses in the instrument to be made up by an input of energy from the player, and hence by a travelling wave down the tube. It may seem after all these comments that the theory is a poor one; actually it is surprisingly good, and not easy to improve so as to be much closer to reality.

(b) The flute and the clarinet

We shall carry through this theory for a flute; at least a modern flute has a straight cylindrical pipe, and the side holes are large enough that opening any one is nearly equivalent to terminating the pipe there. So we solve the system

$$\begin{cases} c^2 \partial^2 \phi / \partial x^2 = \partial^2 \phi / \partial t^2, \, 0 < x < l, \\ \partial \phi / \partial t = 0 \text{ at } x = 0 \text{ and } x = l, \end{cases}$$

because both ends of a flute are open. It is easily verified that

$$A_n \sin(n\pi x/l) \cos(n\pi ct/l + \alpha_n)$$

satisfies these requirements. Again, this solution may be derived mathematically by the method of separation of variables. Because of the linearity of the equation and boundary conditions, the general solution is

$$\phi = \sum_{n=1}^{\infty} A_n \sin(n\pi x/l) \cos(n\pi ct/l + \alpha_n),$$

where the constants A_n and α_n may be found so as to agree with any reasonable initial conditions.

The 'fundamental' or 'first partial' is the solution

$$A_1 \sin(\pi x/l) \cos(\pi ct/l + \alpha_1)$$

which has $n = 1$. The first overtone or second partial has $n = 2$ and so on. An experienced player can excite the mode $n = 1$ without generating an appreciably large amplitude of the other modes. In this mode, the pressure fluctuation is zero at each end, and maximal at the centre of the tube, while the velocity is zero at the centre and has maximum amplitude at the ends; fig. XII.5 illustrates this. The amplitude of the oscillation in p follows half a sine wave, as in fig. XII.6. The frequency of the oscillation is

$$\begin{cases} \pi c/l \text{ radians per second,} \\ c/2l \text{ Hz.} \end{cases}$$

Fig. XII.5. Velocities and excess pressures in the fundamental of the model flute.

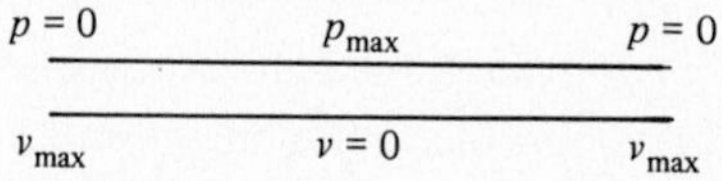

$p = 0$	p_{max}	$p = 0$
v_{max}	$v = 0$	v_{max}

Fig. XII.6. The pressure amplitude in the fundamental.

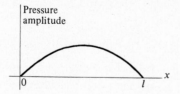

Take as example the note E just above middle C; this has frequency 330 Hz (based on A = 440 Hz), and with

$$c = 340 \text{ m s}^{-1}$$

we should have

$$l = c/660 \doteqdot 51 \text{ cm}.$$

Actual measurement gives about 46 cm: the errors come from 'end corrections' at each end to allow for p not being quite zero there, as it has to match into some flow outside the pipe.

The mode $n = 2$ can be excited quite purely by a slightly different blowing technique. The expected frequency is now just twice what it was before, which is what is observed. We now have $p = 0$ at the centre of the pipe, and amplitude

$$A_2 \sin 2\pi x/l$$

in the pipe (see fig. XII.7).

The mode $n = 4$ is excited by a further small change in blowing, with the opening of a hole near where $p = 0$ is required, as is shown in fig. XII.8. This gives a frequency four times the fundamental. A few higher modes than this can be used in the flute, and in brass instruments like the trumpet very high partials (high 'harmonics') are used.

The clarinet is another instrument with a mainly cylindrical bore, but it has an effectively closed end where the blowing is done. This can be shown to give solutions

$$\phi_n = B_n \cos\{\tfrac{1}{2}(2n+1)\pi x/l\} \cos\{\tfrac{1}{2}(2n+1)\pi ct/l + \beta_n\}.$$

Fig. XII.7. The mode $n = 2$ for the flute model.

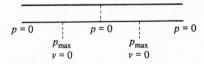

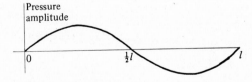

Fig. XII.8. An open hole assists the production of the mode $n = 4$.

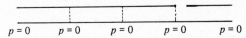

The fundamental is $n = 0$ with frequency

$c/4l$ Hz.

The same E now needs half the length compared with the flute, and the previous length of about 51 cm should give the next E below middle C. in fact the clarinet has rather small holes for opening, and this theory needs to be improved to get good agreement. However, the prediction is that the next harmonic has frequency $3c/4l$ Hz, followed by $5c/4l$ Hz; and this is indeed the ratio of frequencies that are observed in playing the clarinet. So the theory gives a moderate estimate of the actual frequencies, and gives the correct ratios between them. It does not tell you in what proportion the various frequencies occur when the instrument is blown, because it does not discuss the resonant interaction between lips, reed and air column (clearly rather hard to model successfully).

5. Plane waves interacting with boundaries

(*a*) *Transmission through a wall*

Sounds travel from one room to another in a building by a variety of mechanisms; one such is through the walls (think of some others). We shall give a very simple model for the transmission of a plane wave through a wall. We model the wall by a plane which has mass but no other properties. It moves in response to the pressure difference across it according to

$$\sigma\ddot{\eta} = p(y = 0 -) - p(y = 0 +)$$

where σ is the mass per area of the wall, $\eta(t)$ is its displacement, and $p(y = 0 -)$ means the pressure for y just less than 0.

For the sound we take, as sketched in fig. XII.9, an incident wave (in $y > 0$)

$$\phi_1 = A \exp\{i(-ky - \omega t)\},$$

Fig. XII.9. The plane waves of §5(*a*).

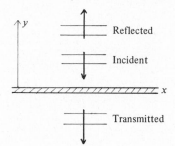

a reflected wave of the same frequency and hence the same wavelength (because the same wave speed)

$$\phi_2 = B \exp\{i(ky - \omega t)\},$$

and a transmitted wave (in $y < 0$)

$$\phi_3 = C \exp\{i(-ky - \omega t)\}.$$

We already have one condition at the wall, its equation of motion. We must also have that its velocity must equal the air velocity on each side, and hence

$$\dot{\eta} = \{\partial/\partial y(\phi_1 + \phi_2)\}_{y=0+} = (\partial\phi_3/\partial y)_{y=0-}.$$

That is,

$$\dot{\eta} = (-ikA + ikB)e^{-i\omega t} = -ikCe^{-i\omega t}.$$

Using $p = -\rho_0\dot{\phi}$ enables us to rewrite the equation of motion as

$$\sigma\ddot{\eta} = (i\omega\rho_0 C - i\omega\rho_0 A - i\omega\rho_0 B)e^{-i\omega t}.$$

It is not hard to eliminate η and solve for B and C in terms of A; the result is

$$\begin{cases} B = A(1 - 2i\rho_0/\sigma k)/(1 + 4\rho_0^2/k^2\sigma^2), \\ C = A - B, \\ \eta = (C/c)e^{-i\omega t}. \end{cases}$$

Take as a specific example

$$\begin{cases} A = 10^{-5}\ \text{m}^2\ \text{s}^{-1}, \\ \omega = 3.3 \times 10^3\ \text{s}^{-1}, \text{and so } k = 10\ \text{m}^{-1}, \\ \sigma = 250\ \text{kg m}^{-2}. \end{cases}$$

This is a moderately loud sound, and a moderately thick wall. Then

$$\begin{cases} 2\rho_0/\sigma k = 10^{-3}, \\ C \doteq 10^{-3}\, iA, \\ B \doteq A, \\ C/c \doteq 3 \times 10^{-6}\ \text{m}. \end{cases}$$

The transmitted sound has a considerably smaller amplitude as $|C/A| = 10^{-3}$, and there is a phase shift of $\frac{1}{2}\pi$ (as $\exp(\frac{1}{2}\pi i) = i$). The motion of the wall is virtually undetectable at about 3×10^{-6} m but for all that, it is the motion of the wall that has let the sound through. And the sound that has been transmitted is still easily heard, if the receiving room is itself quiet.

The ratio $2\rho_0/\sigma k$ which determines the amount of transmission may be written as

'mass of air in π^{-1} wavelengths/mass of wall'

(each per unit area). It may also be written as a ratio of impedances as

$2\rho_0 c/\sigma\omega$.

Thus long waves of low frequency get through the wall easily, and short waves cannot get through. Of course sounds from outside the building will come in at the thinnest point, such as a window (for which σ is quite low); a wide open window has $\sigma = 0$, and there is almost total transmission and no reflection.

A more realistic model of the wall would include some elasticity of the wall, and also some dissipation of the energy of the sound by the wall. Both of these are rather hard to treat thoroughly in a first course. However, since there are no energy losses in our model wall, there must be some result here on energy conservation; certainly the formula $A = B + C$ which we derived from the continuity of velocity at the wall *looks* like a conservation statement, but it cannot be about energy, which we would expect to be of second degree in velocities – we return to energy ideas later.

(b) Sound travelling parallel to an elastic wall

If we separate variables in

$$c^2\nabla^2\phi = \partial^2\phi/\partial t^2,$$

using cartesian coordinates x, y, so that

$$\phi = f(x)g(y)h(t),$$

we can derive quite easily

$$\begin{cases} f''(x) = \alpha f(x), \\ g''(y) = \beta g(y), \\ h''(t) = \gamma h(t), \end{cases}$$

where $c^2(\alpha + \beta) = \gamma$ and α, β, γ are constants. As above, these constants can be positive, negative, or zero; which value is appropriate depends on the assumed boundary values.

We shall discuss next a wave travelling along a model elastic boundary. That is, we are assuming that α and γ are negative constants, so as to get oscillations in x and t; and in particular we choose to have

$$f(x)h(t) = \exp\{i(kx - \omega t)\}.$$

This gives one choice of travelling wave, having frequency ω and wavenumber k along the x-direction. Now the y dependence is governed by β, and this must satisfy

$$c^2(-k^2 + \beta) = -\omega^2$$

because of our choice of x and t variation (and *not* $kc = \omega$).

If we choose β to be negative, then we get a solution that represents a wave travelling at an angle to the wall, as in §3(*e*) above. But we may sometimes find a solution with β positive; then $g(y)$ must be a *real* exponential, and we take

$$g(y) = Ae^{-ay}$$

so that ϕ is not infinite as $y \to \infty$. We go on to discuss such an example, where the sound wave only exists near the boundary, because the e^{-ay} factor causes it to vanish rapidly as y becomes large.

Consider, then, a sound wave propagating along a model elastic wall. Take the sound wave to be, for $y > 0$,

$$\phi = Ae^{-ay}e^{i(kx - \omega t)}$$

where

$$c^2(-k^2 + a^2) = -\omega^2;$$

and take the wall to have displacement $\eta(t)$ which satisfies an equation of motion

$$\sigma\ddot{\eta} + l^2\eta = -p(y = 0+).$$

This model of the wall now has a mass per area σ, but also a restoring stress $l^2\eta$ which is proportional to the displacement of the wall. As before, the velocity must be continuous at the wall, so that

$$\dot{\eta} = -aAe^{i(kx - \omega t)}.$$

And because the pressure in the sound wave is

$$p = -\rho_0\dot{\phi},$$

the equation of motion of the wall is

$$\sigma\ddot{\eta} + l^2\eta = -iA\rho_0\omega e^{i(kx - \omega t)}.$$

But

$$\eta = -i\omega^{-1}aAe^{i(kx - \omega t)},$$

and the equation of motion reduces to

$$ia\omega\sigma - ial^2/\omega = -i\rho_0\omega.$$

We are left with

$$a = \rho_0\omega^2/(l^2 - \sigma\omega^2)$$

and

$$c^2(-k^2 + a^2) = -\omega^2.$$

Suppose that the initial disturbance has a given frequency ω. Then if

$$l^2 > \sigma\omega^2$$

we find a positive value of a from the first of these equations, as is required to make ϕ decrease towards infinity; and we then find a positive value of k^2 from the second equation. Now $(l^2/\sigma)^{1/2}$ is the frequency of free oscillations of the wall, so the required condition is that the sound frequency should be less than the natural wall oscillation frequency. In this case a sound wave and an oscillation of the wall can run along together, at the same speed ω/k, with the sound not radiating away from the wall.

The modelling of the wall's motion here has been rather oversimple, and the effects are generally small for the interaction of a sound wave in air and an elastic wall. But the interaction between sound waves in water (which has a much higher density ρ_0), driven at some engine oscillation frequency, and the plates of a ship, could be more important.

Note that the speed of propagation ω/k is equal to $c(1 - a^2/k^2)^{1/2}$, which is less than the normal speed of sound. This is a value intermediate between that for the wall (which is zero in this model) and that for the air. Clearly it would be better to take a model for the wall in which waves could propagate at an appropriate speed before drawing general conclusions on wave speeds. But it does appear that the speed of disturbances in air need not always be c, depending on the boundary conditions.

(c) Sound propagating along a duct

Sound propagates rather well along large pipes and ducts, taking the noise from pumps and machinery to large distances; but speech is difficult down long corridors or tunnels. We may investigate these phenomena most easily by considering solutions

$$\phi = Y(y)Z(z)e^{i(kx - \omega t)}$$

which propagate along a square duct

$$\begin{cases} 0 \leqslant y \leqslant a, \\ 0 \leqslant z \leqslant a \end{cases}$$

with rigid walls, which is shown in fig. XII.10. Separation of variables leads to

$$\begin{cases} Y''(y) = \alpha Y(y), \\ Z''(z) = \beta Z(z), \end{cases}$$

Fig. XII.10. Definition sketch for §5(c).

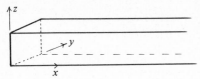

as before, with

$$c^2(\alpha + \beta - k^2) = -\omega^2$$

being required to satisfy the wave equation. The boundary conditions

$$\begin{cases} \partial\phi/\partial y = 0 \text{ on } y = 0 \text{ and } y = a, \\ \partial\phi/\partial z = 0 \text{ on } z = 0 \text{ and } z = a \end{cases}$$

for all x and t require us to impose

$$\begin{cases} Y'(0) = Y'(a) = 0, \\ Z'(0) = Z'(a) = 0. \end{cases}$$

These can only be satisfied by:

   (i) $Y(y)$ and $Z(z)$ both constant, when we get plane waves propagating down the duct;

  (ii) $\begin{cases} Y(y) = \cos(m\pi y/a), \\ Z(z) = \cos(n\pi z/a), \end{cases}$

      where m and n are non-zero integers;

  (iii) a combination of (i) and (ii).

The second case is the most interesting one, as it brings in new ideas: plane waves in ducts have in effect been dealt with in §4. We shall do the third case, as less demanding and still usefully new.

Take, as example, the solution

$$\phi = \cos(m\pi y/a)e^{i(kx - \omega t)}$$

which satisfies the wave equation provided

$$c^2(m^2\pi^2/a^2 + k^2) = \omega^2.$$

This is apparently a wave propagating down the duct at speed

$$\omega/k = c(1 + m^2\pi^2/a^2k^2)^{1/2}.$$

This is clearly greater than the speed of sound in air, which seems surprising as the walls are rigid. And indeed the speed can apparently be made indefinitely high by taking m very large. This result can be more readily understood if we remember that

$$\cos\theta = \tfrac{1}{2}(e^{i\theta} + e^{-i\theta}).$$

Fig. XII.11. Constituent travelling waves in the duct.

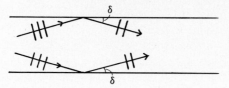

Thus the solution above can be written as

$$\phi = \cos(m\pi y/a)e^{i(kx - \omega t)}$$
$$= \tfrac{1}{2}\{\exp[i(m\pi y/a + kx - \omega t)] + \exp[i(-m\pi y/a + kx - \omega t)]\}.$$

That is, the kind of solution we have found may be regarded as the sum of two travelling waves, each with wave number

$$K = (m^2\pi^2/a^2 + k^2)^{1/2}$$

and travelling at angle

$$\delta = \pm\tan^{-1}(m\pi/ka)$$

with the direction of the duct (see fig. XII.11). This is because we may write ϕ in the form

$$\phi = \tfrac{1}{2}\exp\{i(Kx\cos\delta + Ky\sin\delta - \omega t)\}$$
$$+ \tfrac{1}{2}\exp\{i(Kx\cos\delta - Ky\sin\delta - \omega t)\}.$$

Now the speed of these waves is just

$$\omega/K = \omega k^{-1}(1 + m^2\pi^2/k^2a^2)^{-1} = c$$

from previous equations. So the correct wave speed is obtained for these constituent, slanting waves. The higher speed we seemed to get earlier was similar to that obtained in §3(e) above for the speed of surfaces of constant phase along a wall at an angle to **k**.

Let us examine one consequence of this theory. Consider a man standing in a long narrow corridor and trying to speak to another person at some distance down the corridor. The lowest frequencies he uses may be less than $\pi c/a$ radians per second or $\tfrac{1}{2}c/a$ Hz; so because we have shown that

$$\omega^2 = c^2(m^2\pi^2/a^2 + k^2)$$

for some integer m, we must have $m = 0$, and these waves travel down the corridor as plane waves. But some waves of a higher frequency will exist which will be of type $m = 1$ or $m = 2$. These will travel at speed c at an angle to the corridor, and so will have further to travel, and will take longer to reach the listener. So the low frequency part of the message travels at a higher speed than some of the high frequency part, and the message becomes incomprehensible. This variation of the speed with the wave frequency (or wavenumber) is called 'dispersion'. Plane sound waves are 'non-dispersive', but when they are confined and reflected in a 'wave guide' they become dispersive, and the original message becomes distorted.

Wave guides do not need to be physically visible as this one was. Temperature stratification in the sea (or air) can give the same reflective effect and confine waves to a horizontal section of the sea. A submarine can attempt to hide from listeners in such a wave guide: its sound signals (engine noise, flow noise) do not reach the sea surface.

(d) *Application to musical instruments*

We may now understand why musical wind instruments need to be narrow compared to their length. Take the flute model of a pipe open at both ends, but to make the mathematics easier consider a square section to the pipe with side a. The appropriate solutions of

$$c^2 \nabla^2 \phi = \partial^2 \phi / \partial t^2$$

are easily found to be

$$\phi = A \cos(m\pi y/a) \cos(n\pi z/a) \sin(s\pi x/l) e^{-i\omega t},$$

where m, n, s are integers and

$$c^2 \{ (m\pi/a)^2 + (n\pi/a)^2 + (s\pi/l)^2 \} = \omega^2.$$

The basic mode in such a flute is the plane wave, which has

$$\begin{cases} m = n = 0, s = 1, \\ \omega = \pi c/l. \end{cases}$$

So that the ratios of the frequencies are whole numbers or simple fractions, we ask for the next higher frequencies to be given by

$$m = n = 0, s = 2 \text{ and } 3 \text{ and } 4,$$

$$\omega = 2\pi c/l, 3\pi c/l, 4\pi c/l.$$

This needs

$$\pi/a > 4\pi/l$$

so that the transverse mode

$$\begin{cases} m = 1, n = s = 0, \\ \omega = \pi c/a \end{cases}$$

does not come into the set of frequencies that might be excited. That is, we need

$$a < \tfrac{1}{4}l.$$

In this model it would appear that we could keep a musical sequence of notes by making a a simple fraction of l. For a circular section this is not possible. And in practice $a \ll l$, so that no unmusical note can be excited.

6. Energy and energy flow in sound waves

(a) *Averages of products*

In the example of §5(c) we asked how fast the message is really propagated down the corridor. For a wave which is travelling at δ to the corridor, is the message speed $c \cos \delta$ or c or $c \sec \delta$? We need to ask how fast energy travels, and so in this section we discuss energy and its rate of flow.

Before we set off we need a minor theorem. Let

$$\begin{cases} x(t) = \mathscr{R}e\, Ce^{-i\omega t} \\ y(t) = \mathscr{R}e\, De^{-i\omega t}, \end{cases}$$

where C and D are complex constants. Then the average of $x(t)y(t)$ over one cycle has value

$$\tfrac{1}{4}(C\bar{D} + \bar{C}D)$$

where an overbar denotes a complex conjugate. This is easily shown: let $C = C_1 + iC_2$ and $D = D_1 + iD_2$, so that

$$\begin{cases} x(t) = C_1 \cos \omega t + C_2 \sin \omega t, \\ y(t) = D_1 \cos \omega t + D_2 \sin \omega t; \end{cases}$$

then the average of xy over a cycle is

$$\omega/2\pi \int_0^{2\pi/\omega} x(t)y(t)dt$$

which has value $\tfrac{1}{2}(C_1 D_1 + C_2 D_2)$, and this is just

$$\tfrac{1}{4}(C\bar{D} + \bar{C}D),$$

as required. We need this theorem because energies and rates of working are about products, and we are using real parts of complex exponentials to represent physical quantities.

(b) *Kinetic and potential energies*

The kinetic energy in a sound wave is clearly $\tfrac{1}{2}\rho \mathbf{v}^2$ per volume, and keeping only the leading term we get

$$\tfrac{1}{2}\rho_0 (\nabla \phi)^2 .$$

In a plane wave (choosing the axes suitably)

$$\phi = A \exp\{i(kx - \omega t)\},$$

so we have the velocity

$$\partial\phi/\partial x = ikA \exp\{i(kx - \omega t)\}.$$

Next average over a cycle, to get a representative energy density, and use the theorem above: the average kinetic energy density is

$$\tfrac{1}{4}\rho_0 k^2 |A|^2$$

per volume.

The 'potential' energy in a sound wave is due to compression, and so we must use the internal energy E of thermodynamics. Now sound waves have essentially no heat conduction in them and so

$$dE = \Delta W,$$

and as the work is derived from the local pressure and changes in volume we have

$$dE = \Delta W = -p\,dV, \text{ where } p \text{ is now total pressure,}$$
$$= -p\,d(\rho^{-1}), \text{ taking unit mass,}$$
$$= p\rho^{-2}\,d\rho.$$

This looks like a contradiction of statements in Chapter VII that E was independent of ρ. What we have here is $E(T)$, but in an isentropic change T is a function of ρ, so we may put E in terms of ρ. Now $E(\rho)$ is the energy per unit mass; the energy per unit volume is ρE. We shall expand E in a Taylor series about the equilibrium state ρ_0, using $\rho = \rho_0 + \rho'$; then we shall calculate ρE.

$$E(\rho) = E(\rho_0) + \rho'E'(\rho_0) + \tfrac{1}{2}\rho'^2 E''(\rho_0) + O(\rho'^3).$$

But we know $dE/d\rho = p\rho^{-2}$ and so

$$d^2E/d\rho^2 = \rho^{-2}dp/d\rho - 2p\rho^{-3}.$$

Evaluating these derivatives at $\rho = \rho_0$ and substituting gives us

$$E(\rho) = E(\rho_0) + \rho'p_0/\rho_0^2 + \tfrac{1}{2}\{\rho_0^{-2}(dp/d\rho)_0 - 2p_0\rho_0^3\}\rho'^2,$$

to second order in ρ'. Hence the energy per unit volume, which is $(\rho_0 + \rho')E(\rho)$, reduces to

$$\rho_0 E(\rho_0) + p_0\rho'/\rho_0 + \tfrac{1}{2}c^2\rho'^2/\rho_0,$$

because $(dp/d\rho)_0 = c^2$. Finally we choose to have $E(\rho_0) = 0$, as a reference level for the compressive energy. Thus the potential energy per volume is

$$p_0\rho'/\rho_0 + \tfrac{1}{2}c^2\rho'^2/\rho_0.$$

The first term in the potential energy averages to zero over a cycle, since ρ' is a periodic function in a plane wave. More generally, ρ' must average to zero over a large volume, as otherwise ρ_0 is not the appropriate undisturbed state. A fuller discussion of this linear term may be found in Lighthill's text *Waves in Fluids* (see the references). The second term may be averaged as before for a plane wave; we use

$$\begin{cases} \rho' = p'/c^2, \\ p' = -\rho_0 \dot{\phi} \end{cases}$$

(both of which are correct to first order) to get

$$\tfrac{1}{2} c^2 \rho'^2 / \rho_0 = \tfrac{1}{2} \rho_0 \dot{\phi}^2 / c^2,$$

with average value again

$$\tfrac{1}{4} \rho_0 k^2 |A|^2$$

because $\omega/c = k$ in this simple wave.

The total energy density in a plane wave of amplitude A is therefore

$$\tfrac{1}{2} \rho_0 k^2 |A|^2,$$

correct to second order in the amplitude. This is quadratic in the amplitude as one would expect. It also is quadratic in the wavenumber k (or frequency $\omega = kc$): high frequency waves have a larger energy for a given amplitude. The energy densities are not high in relative terms. Consider quite a loud sound with

$$\begin{cases} p' = 1 \text{ N m}^{-2}, \\ \omega = 10^3 \text{ radians/second.} \end{cases}$$

Then, because $p' = -\rho_0 \dot{\phi}$, we have

$$|A| = \rho_0^{-1} \omega^{-1}$$
$$= 0.8 \times 10^{-3} \text{ m}^2 \text{ s}^{-1}$$

and the energy density is about $4 \times 10^{-6} \text{ J m}^{-3}$. Compare this with air flowing at 1 m s^{-1}, which has a kinetic energy density of 0.6 J m^{-3}, vastly greater even for such a gentle breeze.

(c) *Example: the note from a wine bottle*

An example which can easily be done by energy methods is the natural frequency of a wine bottle. It is well known that a wine bottle (or other bottle) can be blown like a flute to give a pure note of wavelength long compared with the bottle. A typical French half bottle (illustrated in fig. XII.12) gave a note E below middle C, wavelength about 2 m, which is much more than the length of 23 cm.

Fig. XII.12. Definition sketch for §6(*c*). The values for *a*, *A*, *l*, *L* were 1.0, 2.7, 8, 23 cm, and *V* was 375 cm^3.

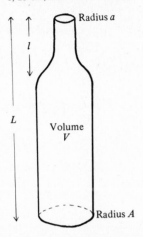

The sound is associated with air moving in and out of the neck of the bottle, and with the air in the bottle expanding and contracting. Take a coordinate x to describe the position at any time of a chunk of air in the neck: the rate of volume flow in and out is then

$$\pi a^2 \dot{x}$$

and the kinetic energy of the air in the neck is

$$\tfrac{1}{2}\pi \rho_0 a^2 l \dot{x}^2$$

approximately. The kinetic energy in the body of the bottle is much less, firstly because the velocity there must be reduced by a factor

$$a^2/A^2$$

so that mass flow is conserved where body and neck meet, and secondly velocity must reduce to zero at the base, bringing in a factor which we may estimate as

$$\text{(distance from base)}/(L - l).$$

On this basis the extra kinetic energy is about 1/12 of that for the neck. To allow for this and also for some 'added mass' of outer air moved by the air in the neck we will take

$$l' = 10 \text{ cm},$$

and use kinetic energy

$$\tfrac{1}{2}\pi \rho_0 a^2 l' \dot{x}^2.$$

Now when volume $\pi a^2 x$ leaves the neck, the density in the bottle reduces to

$$\rho_0(1 - \pi a^2 x/V),$$

and we take this as an effective uniform density in the bottle. This gives

$$\rho' = - \pi a^2 \rho_0 x/V$$

and potential energy per volume

$$\tfrac{1}{2}c^2(\pi a^2 \rho_0 x/V)^2/\rho_0.$$

So the total potential energy that we use is

$$\tfrac{1}{2}c^2 \pi^2 a^4 \rho_0 x^2/V.$$

For a simple oscillation with kinetic energy $\tfrac{1}{2}\alpha \dot{x}^2$ and potential energy $\tfrac{1}{2}\beta x^2$, it is in general true that its frequency is

$$(2\pi)^{-1}(\beta/\alpha)^{1/2} \text{ Hz.}$$

In the present case this gives a frequency of

$$157 \text{ Hz,}$$

which compares very well with the observed value of 165 Hz.

(d) Energy flow in a plane sound wave

Energy flow in a sound wave has two causes. Firstly, energy is carried along by the velocity **v** of the air in the wave: this effect must be of third order in the amplitude as energy is second order and air velocity is first order. And secondly, the air pressure p' does work at rate $p'\mathbf{v}$; this effect is of second order as each of p' and **v** is first order in the amplitude. Hence we only consider this second effect for small amplitude waves. So take a wave

$$\phi = A \exp\{i(\mathbf{k}\cdot\mathbf{x} - \omega t)\},$$

which has

$$\begin{cases} p' = i\rho_0 \omega A \exp\{i(\mathbf{k}\cdot\mathbf{x} - \omega t)\}, \\ \mathbf{v} = i\mathbf{k}A \exp\{i(\mathbf{k}\cdot\mathbf{x} - \omega t)\}. \end{cases}$$

The average rate of working is, by the theorem in (a) above,

$$\tfrac{1}{2}\rho_0 \omega \mathbf{k}|A|^2.$$

This is a rate of flow of energy which has direction **k** and is proportional to amplitude squared.

Now if we divide this rate of flow of energy by the previously derived energy density, we get a velocity, the velocity at which the energy is flowing.

Doing this, we get just

$$\omega \mathbf{k}/k^2,$$

which may be put as $\mathbf{c}$, where the direction is that of the wavenumber vector. Thus energy in a plane sound wave flows at speed c along the wavenumber vector.

Going back to the wave guide example in §5(c), we see that energy travels along the inclined plane waves at speed c, and so at speed $c \cos \delta$ down the corridor. This is the appropriate speed for that part of the message which is described by these waves.

The above calculations have been for plane waves of small amplitude in a uniform medium at rest. If these conditions are not met, energy flow may not be so simple; we shall discuss the easiest case of non-plane waves later.

(e) Loudness of a sound, decibels

A rate of flow of energy density should be described in terms of watts per square metre. However, another scale (and various subsidiary versions of it) are in common use. The reason for this is that physiological responses do not appear to follow a linear scale: ten times as many watts per square metre is not heard as ten times as loud a sound.

The current definition of this 'loudness' scale is that the number of decibels (db) in a noise is found from the formula

$$db = 120 + 10 \log_{10} (\text{W m}^{-2}).$$

So an energy density flow rate of 10^{-4} W m^{-2} is described as having a loudness of

$$(120 + 10 \log_{10} 10^{-4})db = 80 \text{ db}.$$

This is a loud noise. Ordinary speech at 10 m distance is about 10^{-8} W m^{-2} or 40 db. The quietest sound that can be heard is near 0 db (this used to be the basis of the decibel scale). At about 120 db or 1 W m^{-2}, the noise is so loud that pain starts to take over as the physiological response. And at about 200 db the pressure fluctuations are about equal to atmospheric pressure, well above the level at which structural damage can be caused.

The simple wave

$$\phi = A e^{i(kx - \omega t)}$$

corresponds to an energy flow rate

$$\tfrac{1}{2}\rho_0 \omega k |A|^2.$$

Using a frequency of 1000 Hz gives this the value $7 \times 10^4 |A|^2$, where A is measured in $m^2\ s^{-1}$. Thus the decibel value in terms of the $|A|$ value is, at *this* frequency,

$$db = 120 + 10 \log_{10} (7 \times 10^4 |A|^2).$$

For example, 40 db requires $|A| = 3.8 \times 10^{-5}\ m^2\ s^{-1}$.

There is a simple relation for *pressure* amplitudes, that is true for *all* frequencies: if

$$p' = B e^{i(kx - \omega t)}$$

then the energy flow rate is $\frac{1}{2}|B|^2/(\rho_0 c)$, so that in air

$$db = 120 + 10 \log_{10} (1.2 \times 10^{-3} |B|^2).$$

7. Sound waves in three dimensions

(a) *The spherically symmetric solution*

The study of plane sound waves is basic, and enough can be done to get a good idea of how sound behaves. But it is well worth looking at some of the easier three-dimensional solutions as well. The two-dimensional solutions are rather harder mathematically in some ways, and they will not be discussed here.

The basic three-dimensional solution of

$$c^2 \nabla^2 \phi = \partial^2 \phi / \partial t^2$$

is one in which ϕ depends only on r and t. By rearranging the wave equation you can show that such a spherically symmetric solution satisfies

$$c^2 \partial^2 / \partial r^2 \{r\phi(r, t)\} = \partial^2 / \partial t^2 \{r\phi(r, t)\}.$$

This is exactly like the one-dimensional equation, and so the general solution of it is

$$\phi(r, t) = r^{-1} \{F(r - ct) + G(r + ct)\}$$

for any twice differentiable functions F and G, just as in §3(a). These two terms correspond to an outward moving disturbance given by $r^{-1}F(r - ct)$ and an inward moving one given by $r^{-1}G(r + ct)$. Both these waves move without change of shape: when $t = 0$ you get shape $F(r)$, and when $t = \tau$ you get the same shape at the position $r = c\tau$. The r^{-1} factor merely reduces the amplitude of the sound as it spreads out further from its source. This preservation of shape enables speech to travel undistorted, as we saw for plane waves also.

Let us look at the outward radiation solution near $r = 0$, say at the sphere $r = \varepsilon$. The velocity is

$$\partial \phi / \partial r = -\varepsilon^{-2} F(\varepsilon - ct) + \varepsilon^{-1} F'(\varepsilon - ct),$$

and the volume flow through this small sphere is

$$4\pi\varepsilon^2(\partial\phi/\partial r)_{r=\varepsilon}.$$

This is approximately

$$-4\pi F(-ct)$$

on neglecting terms of size ε. So this outward wave requires a volume flow rate out from the origin of amount

$$-4\pi F(-ct),$$

i.e. there is a source of material at the origin.

This outward travelling solution corresponds to

$$p = c\rho_0 r^{-1} F'(r-ct)$$

and when r is large the outward velocity is

$$v_r = r^{-1}F'(r-ct) + O(r^{-2}).$$

From these we may calculate the rate at which energy flows out of a large sphere (of surface area $4\pi r^2$) surrounding this source flow: it is

$$4\pi r^2 c\rho_0 r^{-2}\{F'(r-ct)\}^2$$

on neglecting a term that is $O(r^{-1})$ as $r \to \infty$. So the energy outflow rate is

$$4\pi\rho_0 c\{F'(r-ct)\}^2$$

at large distances r. This is a radiation of energy provided that F is not constant, and hence this solution is the simple source solution of the wave equation in three dimensions.

The most obvious choice for F is a sinusoidal function. We take, as before, the real part of a complex exponential for ease in calculations:

$$\phi = Ar^{-1}e^{i(kr-\omega t)}.$$

This corresponds to a radiation of energy, on averaging over a cycle,

$$2\pi\rho_0\omega k|A|^2,$$

a formula a very like that derived for a plane wave.

It is not hard to calculate the averages of kinetic and potential (internal, compression) energy densities over a cycle; and provided we take r large we get the same results as before except for an extra r^{-2} factor. It is similarly not hard to show that the velocity of energy is just c. However, when kr is small (this is the proper non-dimensional statement), other terms would seem to take over both in the kinetic energy density and the energy flow rate. For we have

$$\begin{cases} p = iAc\rho_0 kr^{-1}e^{i(kr-\omega t)}, \\ v_r = (iAkr^{-1} - Ar^{-2})e^{i(kr-\omega t)} \end{cases}$$

and $r^{-2} \gg kr^{-1}$ when $kr \ll 1$. However, there is zero average energy flow rate from the term in r^{-2} in the velocity. This is because the term

$$C\bar{D} + C\bar{D}$$

of the theorem on averages of products becomes

$$iA\bar{A} + \overline{iA}A,$$

which is exactly zero. So the rate of flow of energy through a sphere of surface $4\pi r^2$ does not vary with r. But the kinetic energy near the source is mainly that associated with the flow that would exist if the fluid were incompressible – flow out, followed by flow back, with almost no radiation. This local flow near $r = 0$, the 'near field' of the source, is dominated by this 'incompressible' motion; but the actual flow still has just enough difference from the incompressible motion to give the correct rate of radiation, even though it is not in this region equal to c times the total energy density.

(b) Example of spherically symmetric radiation

The easiest example of a spherically symmetric source is the oscillating bubble of Chapter XI. The situation is sketched in fig. XII.13. We now solve

$$c^2 \nabla^2 \phi = \partial^2 \phi / \partial t^2$$

outside $r = R(t)$, instead of solving Laplace's equation. The boundary conditions are

$$\partial \phi / \partial r = \dot{R}(t) \text{ at } r = R(t)$$

and

$$\phi = r^{-1} F(r - ct),$$

so that radiation is outward at large distances.

This requires

$$-R^{-2} F(R - ct) + R^{-1} F'(R - ct) = \dot{R}(t),$$

which is a nasty type of equation to solve. Let us assume small oscillations

Fig. XII.13. Radial motion outside a bubble of oscillating radius $R(t)$.

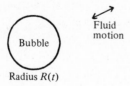

Fluid
motion

Bubble

Radius $R(t)$

in R about a mean radius a, so as to get an easier problem:

$$R(t) = a(1 + \varepsilon e^{-i\omega t}).$$

Now try the solution

$$\phi = Ar^{-1}e^{i(kr - \omega t)},$$

with the boundary condition imposed at $r = a$ rather than $r = R$, as a good approximation provided ε is small. We need to impose

$$-Aa^{-2}e^{i(ka - \omega t)} + ikAa^{-1}e^{i(ka - \omega t)} = -ia\omega\varepsilon e^{-i\omega t},$$

which determines A:

$$A = i\varepsilon e^{-ika}\omega a^3(1 + ika)/(1 + k^2 a^2).$$

We assume that a and ω have been determined sufficiently well from the incompressible solution of Chapter XI; really we should start completely afresh and re-solve the whole problem – if we do, we shall not find too much change, just because the energy in the sound is rather smaller than the energy in the incompressible solution. Using the values given previously for a and ω then, we find that

$$ka = 1.5 \times 10^{-2}$$

approximately. Let use this as a reason for approximating A. Since

$$\begin{cases} e^{-ika} \doteqdot 1 - ika - \tfrac{1}{2}k^2 a^2, \\ (1 + k^2 a^2)^{-1} \doteqdot 1 - k^2 a^2, \end{cases}$$

we obtain

$$A \doteqdot i\varepsilon\omega a^3$$

with error of size about $\tfrac{1}{2}k^2 a^2$ times the term retained. This is just the result appropriate to the incompressible solution. However ϕ is now such as to represent radiation, through terms of order ka which come into the pressure near the bubble.

The smallness of the bubble on a wavelength scale, $ka \ll 1$, is an example of a very common phenomenon: emitters of sound are usually of small dimension compared with the wavelength of sound they emit. This is because most sources of sound match frequencies with the sound wave, and the frequency of the source is usually determined by the scale of the emitter and the speed of elastic waves in it; and this speed is usually much higher than the wave speed in the transmitting medium.

It may now be easily verified that the average rate of working at the surface of the bubble is exactly equal to the rate of energy radiation at infinity. There is, of course, nowhere else for the energy to go in a steadily oscillating situation. Actually the situation is not quite steady, as the radiation reduces the energy in the oscillation (so do other effects like viscosity, but we are neglecting them). Let us calculate the efficiency of this radiation as

$$\eta = \frac{\text{energy radiated in one cycle}}{\text{total energy in near field}}$$

Since $A = i\varepsilon\omega a^3$, the radiation in one cycle is

$$2\pi\rho_0\omega k(\varepsilon\omega a^3)^2 2\pi/\omega.$$

And the total kinetic energy present (integrate from a to ∞ to get it) in the solution

$$\phi = (A/r)e^{-i\omega t}$$

is, averaged over a cycle,

$$\pi\rho_0|A|^2/a = \pi\rho_0(\varepsilon\omega a^3)^2/a.$$

Now the energy in the near field is almost all kinetic, so

$$\eta = 4\pi ka$$

approximately, which has value 0.18 approximately. This means that the radiation is efficient enough to reduce the oscillation energy by 18% in each cycle, rapidly removing it entirely.

You should notice that this has been an example where the sound has propagated through water. The only change needed is in the wave speed.

(c) The dipole solution

The oscillating source solution

$$\phi = r^{-1} \exp\{i(kr - \omega t)\}$$

provides us with further basic solutions in just the same way as previously. For example

$$-\partial\phi/\partial z$$

will also satisfy the wave equation whenever ϕ does; and this will give a dipole along the z-axis (sketched in fig. XII.14) if we operate on the given

Fig. XII.14. A dipole along the z-axis.

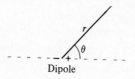

Dipole

source solution. The general dipole solution must be

$$- \boldsymbol{\mu} \cdot \nabla \{ Ar^{-1} e^{i(kr - \omega t)} \},$$

with strength and direction given by $\boldsymbol{\mu}$.

The dipole solution

$$\cos \theta (r^{-2} - ikr^{-1}) \exp \{ i(kr - \omega t) \}$$

has the near field (for $kr \ll 1$)

$$r^{-2} e^{-i\omega t} \cos \theta,$$

and we should already have met the solution

$$U(t) \cos \theta / r^2.$$

It occurs in one of the exercises to Chapter XI, and describes irrotational flow round a sphere which moves at $U(t)$ in a fluid at rest. So this dipole solution should give the sound radiation from an oscillating sphere. Let us examine this in detail.

The boundary conditions we intend to satisfy are firstly, as shown in fig. XII.15,

$$\partial \phi / \partial r = U e^{-i\omega t} \cos \theta$$

on $r = a$; it is taken at $r = a$ on the assumption that the motion is small enough for this to be reasonable. But we also need to have a solution that represents outward radiation at infinity. So we choose to try

$$\phi = Ar^{-2} (1 - ikr) e^{i(kr - \omega t)} \cos \theta.$$

Now, as was explained above, we expect to have $ka \ll 1$ in any reasonable physical set-up, so we shall simplify the mathematics by using this condition when imposing the boundary value on $r = a$. Very much as in (b) we find that

$$A = - \tfrac{1}{2} a^3 U \{ 1 + O(k^2 a^2) \},$$

so we use $A = - \tfrac{1}{2} a^3 U$ as an adequate approximation.

The radiation that we get from this solution is

$$\phi = \tfrac{1}{2} ika^3 U r^{-1} \cos \theta \, e^{i(kr - \omega t)},$$

by taking the largest term as $r \to \infty$. You will notice that this is zero for

Fig. XII.15. The boundary condition on the sphere.

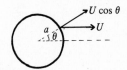

Fig. XII.16. The elementary area for integration.

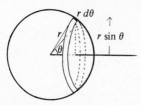

$\theta = \frac{1}{2}\pi$: there is no radiation sideways from a sphere moving backwards and forwards. Because the radiation is not the same for all θ, we must recalculate the energy flow outwards at large distances. We still need to multiply pressure and outward velocity and average over a cycle; but then we must integrate over a spherical surface by using strips (as in fig. XII.16) corresponding to small changes of angle $d\theta$ at angle θ. If the energy flow rate at angle θ is

$$F(\theta),$$

the total flow rate is

$$\int_{\theta=0}^{\pi} F(\theta) 2\pi r^2 \sin \theta \, d\theta.$$

At large distances the pressure is given by

$$p \sim -\tfrac{1}{2}\omega k \rho_0 a^3 U r^{-1} \cos \theta \, e^{ikr} e^{-i\omega t}$$

and the radial velocity is similarly

$$v_r \sim -\tfrac{1}{2}k^2 a^3 U r^{-1} \cos \theta \, e^{ikr} e^{-i\omega t}.$$

We do not need the velocity v_θ, as this gives no outward radiation of energy. The average of pv_r over a cycle is, from the theorem on averaging,

$$\tfrac{1}{8}\rho_0 \omega k^3 a^6 U^2 r^{-2} \cos^2 \theta :$$

this is $F(\theta)$. Carrying out the integration of $F(\theta)$ finally gives a total energy flow rate of

$$\tfrac{1}{6}\pi \rho_0 \omega k^3 a^6 U^2.$$

This is clearly a much smaller radiation than that from the bubble, because it has a higher power of ka in it, and ka is taken to be a small quantity. The proper measure of this is to calculate the efficiency of the dipole radiation in much the same way as before, and we do this next.

The time average total kinetic energy in the corresponding incompressible flow is

$$\tfrac{1}{4}\rho_0 \int_V (\nabla \phi)^2 dV,$$

where

$$\phi = -\tfrac{1}{2}a^3 U r^{-2} \cos\theta$$

and V is the region outside the sphere, i.e. $r > a$. This may be rather easily calculated as a surface integral, since

$$(\nabla\phi)^2 = \nabla\cdot(\phi\nabla\phi) - \phi\nabla^2\phi$$

and $\nabla^2\phi = 0$ for this solution. Hence

$$\tfrac{1}{4}\rho \int_V (\nabla\phi)^2 dV = -\tfrac{1}{4}\rho_0 \int_S \phi(\partial\phi/\partial r)dS,$$

where the $-$ sign comes because the normal to S *out* of V is in the direction of r decreasing. Now $\partial\phi/\partial r$ on S is just $U\cos\theta$ – that was the boundary condition we had to satisfy. So we need to evaluate

$$\tfrac{1}{4}\rho_0 \int_{\theta=0}^{\pi} \tfrac{1}{2}aU^2 \cos^2\theta \times 2\pi a^2 \sin\theta \, d\theta = \tfrac{1}{6}\pi\rho_0 a^3 U^2 .$$

The efficiency is therefore

$$\eta = 2\pi(ka)^3 .$$

This efficiency is really rather small if ka is small. Typically one gets oscillating bodies which produce sound in string instruments like the guitar. It might seem better to use a two-dimensional theory for the oscillation of a guitar string, but this is not entirely appropriate, as the strings are usually not much longer than the sound wavelengths generated. Anyway, in such cases the shortest wavelengths are around 10^{-1} m, whereas the radius of the wire is about 10^{-3} m. Thus $ka = 6 \times 10^{-2}$ at most, and so η is about 10^{-3}. Hence the direct radiation of energy may not be the most important cause of the decrease in amplitude of the string's oscillation, as it would take several seconds to reduce the energy of any note to a negligible size.

Exercises

1. Use $\nabla\cdot$(Euler's equation), $\partial/\partial t$ (continuity equation) and $dp'/d\rho' = c^2$ to find a wave equation for p' with a non-linear term involving density and velocity. Show that it linearises to the usual wave equation.

2. If the basic state is not one of rest, but has a velocity $U(y)\mathbf{i}$, we must take the total velocity to be

   $\mathbf{v} + U(y)\mathbf{i}$

   where $\mathbf{v}$ is 'small'. Suppose also that the scale over which U changes is much greater than the wavelength of the sound. Show that the governing equations reduce to

$(\partial/\partial t + U\partial/\partial x)^2 p = c^2 \nabla^2 p + \text{a term in } dU/dy$

approximately. This is a start on describing sound waves in a wind.

3. Acoustic tiles absorb energy, so the boundary condition on a tiled wall cannot be $\partial\phi/\partial x = 0$. Assume that it is modelled by

$\partial\phi/\partial x = -\lambda\partial\phi/\partial t,$

[to give $v = \text{constant} \times p'$], and calculate the reflected wave when the incident wave

$\phi = Ae^{i(kx - \omega t)}$

for $x < 0$ meets the wall $x = 0$.

4. Repeat Q3 for an incident wave

$\phi = A\exp\{i(kx\cos\alpha + ky\sin\alpha - \omega t)\}.$

Discuss the case $\lambda = c^{-1}\cos\alpha$.

5. Two fluids meet along $z = 0$. The density and speed of sound for $z > 0$ are ρ_1 and c_1, while for $z < 0$ they are ρ_2 and c_2. A plane sound wave in one fluid is incident on the interface at angle α. Calculate the reflected and transmitted waves, assuming that the interface remains almost plane. What can you deduce about sound transmission between air and water?

6. When a cork is taken out of an empty tube a brief note is heard. Model this situation as follows to determine the frequency of the note.

  The tube lies between $x = 0$ where $v = 0$ and $x = l$ where $p' = 0$. At $t = 0$ the air in the tube is at rest at pressure $p_0 - \alpha$. The sound wave in the tube is a plane standing wave (non-progressive, as in §4(b)). Calculate the pressure for $t > 0$ and the lowest frequency in the sound.

7. A tube of length l has one end open and at the other end a piston moves with velocity $a\omega\cos\omega t$, as in fig. XII.17, where a is small (compared with l, or what?) Calculate the air motion in the tube on the assumption that it is linear and that $p' = 0$ at $x = l$. Is this realistic for all frequencies?

8. A layer of air in the atmosphere which acts as a wave guide is modelled by

$$\begin{cases} \partial\phi/\partial z = 0 & \text{on } z = 0, \\ \phi = 0 & \text{on } z = h. \end{cases}$$

Find the relation between frequency and wavenumber component in the x-direction for waves propagating along the x-direction in this wave guide.

Fig. XII.17. A piston oscillating at one end of a tube.

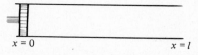

$x = 0$ $\qquad\qquad\qquad\qquad x = l$

9. A large organ pipe has square section of side a and length $20a$ between open ends. Find at what frequency the first mode of vibration in the pipe occurs whose frequency is not an integer multiple of the fundamental frequency.

10. Show that the energy flow in the transmitted wave in §5(a) together with the energy flow in the reflected wave balances the energy flow in the incident wave.

11. For the solution of §5(c), calculate the average energy per unit length in the x-direction, and also the energy flow rate in the x-direction. Hence calculate the speed of flow of energy in this direction.

12. A spherical balloon bursts at $t = 0$, giving the following initial values.
$$\begin{cases} r < a, p = p_0 + \alpha, \mathbf{v} = 0; \\ r > a, p = p_0, \mathbf{v} = 0. \end{cases}$$
The solution of this problem is not as easy as it looks. You need to take
$$\phi(r, t) = r^{-1}\{F(r - ct) + G(r + ct)\}$$
with the given initial values, and also:
(i) the radiation condition, that there is no inward radiation from $r = \infty$ – this does not mean $G = 0$ for *all* $r + ct$;
(ii) a condition at $r = 0$, that describes the absence of any source of sound there.
Show that (ii) requires
$$\lim_{r \to 0} \{r^2 \partial\phi/\partial r\} = 0,$$
and solve the problem to find F and G.

13. A simple sinusoidal source of sound is at $(0, 0, a)$ above the rigid wall $z = 0$, as in the sketch in fig. XII.18. Show that an image source may be

Fig. XII.18. Definition sketch for Q13.

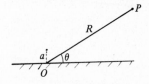

Fig. XII.19. Definition sketch for Q14.

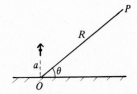

used to solve this problem, and calculate the potential at the point P given by (R, θ) in the diagram, where $R \gg a$.

14. A dipole sound source replaces the simple source of Q13, and $ka \ll 1$, where k is the wavenumber of the sound. The situation is illustrated in fig. XII.19. Show that the potential at large distances is approximately

$$AR^{-1} \sin^2 \theta \, e^{i(kr - \omega t)}$$

for some constant A. [This is a 'quadrupole' source due to almost cancelling dipoles.]

15. Find the general form of waves that can propagate along a pipe of circular cross-section. [Leave this question until after Chapter XIII if you have not yet met Bessel functions.]

References

The simple theory of sound is a large and well-worked field, which has been only started on in this chapter. The real properties of musical instruments, and the real acoustics of buildings, are complicated subjects, far beyond this text, and indeed not everything is yet understood in these areas. The references are provided for the simple theory, but usually cover much more.

(a) *Theory of Sound* (2 volumes), Lord Rayleigh; the most recent edition is Dover, 1945. The best of texts, blending mathematics and physics skilfully, but not always easy for present day students to follow since his background was so different.

(b) *Dynamical Theory of Sound* (2nd edn), W. Lamb, Dover 1960. Rather an old text, but the material of this chapter is covered thoroughly, and there is added background material.

(c) *Theoretical Acoustics*, P. M., Morse and K. U. Ingard, McGraw-Hill 1968. A very full text on much the same material as Lamb, all of the basic theory of sound and vibration is here.

(d) *Waves in Fluids*, M. J. Lighthill, C.U.P. 1978. Modern in approach, at an advanced level, with much material well beyond this text.

(e) There are several film loops on waves on strings, which exhibit many of the features of plane sound waves. But the analogy should not be pressed too far. Useful loops are S-81591–2 *Reflection of Pulses*, S-81590 *Superposition of Pulses*, S-81299 *Standing Sound Waves*, S-81303 *Rubber Membrane – Standing Waves Patterns*.

(f) The water tank simulation of sound waves in the loop FM-105 *Ripple-Tank Radiation Patterns of Source, Dipole and Quadrupole* also has some interest.

XIII

Water waves

1. Background

It is part of everyone's experience that waves and oscillations occur on the surface of water, and they indeed form a wide range of phenomena for any mathematical theory to deal with. Ocean waves from the tsunami – the 'tidal wave' which is generated by an earthquake and which may travel at high speed for huge distances – and the tides themselves, down to small ripples in a calm sea. Other types of disturbance occur on rivers, such as the tidal bore which travels up the River Severn, and the surface waves which can be caused by an irregularity on the bottom. A lake may have a 'seiche' in it, which is an oscillation to-and-fro of the surface of the fluid (very much like the same phenomenon in a bath). On a wet day you may see waves on thin sheets of water flowing down a road.

Real waves have awkward properties: they may break, or have their tops blown off by the wind; they may be suppressed by a small amount of oil on their surfaces; or be running over water of non-uniform depth. We model the waves that you see on water in rather a drastic fashion, to give approachable mathematics, and get some very interesting results; but we cannot cope with too much reality, and so will not predict every visible effect.

This chapter deals with the easy aspects of waves in oceans, lakes and smaller containers. We cannot deal here with such interesting topics as:

(i) the generation of waves by the wind;

(ii) the interactions of waves which meet;

(iii) the changes in waves as they approach the shore;

(iv) waves due to ships, or interaction of waves and ships;

(v) energy generation from waves.

Some aspects of waves on rivers will be covered in Chapter XV, so we assume in this chapter that the water is at rest apart from near its disturbed surface.

2. The linear equations

(a) *The boundary conditions and their linearisation*

As a first model we consider waves of small amplitude on the surface of deep water – both amplitude and wavelength are much less than the depth. We neglect viscosity, as there are no solid boundaries at which it could cause marked effects; and we also neglect compressibility and surface tension. The atmosphere, too, will be neglected, being replaced by a uniform pressure p_0 on the upper surface of the water. With these assumptions we may use Kelvin's theorem to show that

$$\nabla \times \mathbf{v} = 0 \quad \text{and so} \quad \mathbf{v} = \nabla \phi,$$

and then

$$\nabla \cdot \mathbf{v} = 0 \quad \text{implies} \quad \nabla^2 \phi = 0.$$

We will solve this governing equation in the region

$$-\infty < z < \zeta(x, y, t),$$

where ζ is the elevation of the surface above the undisturbed level $z = 0$. The boundary conditions are of four types:

(i) as $z \to -\infty$, $\nabla\phi \to 0$;

(ii) on the surface $p = p_0$;

(iii) the vertical motions of the surface are related to the velocities in the fluid;

(iv) on any rigid boundary $\partial\phi/\partial n = 0$.

In addition we may restrict attention to radiation proceeding in one direction. Let us examine the first three boundary conditions in turn (the fourth is now too well known to need more attention).

(i) Separate variables in $\nabla^2 \phi = 0$ by writing

$$\phi(x, y, z, t) = \phi_0(x, y, t)f(z).$$

Then we must have

$$f(z)\{\partial^2\phi_0/\partial x^2 + \partial^2\phi_0/\partial y^2\} + \phi_0 f''(z) = 0.$$

This gives

$$f''(z)/f(z) = -\phi_0^{-1}\{\partial^2\phi_0/\partial x^2 + \partial^2\phi_0/\partial y^2\},$$

and hence, as usual, each is a constant. The only choice of constant which will give

$$\nabla\phi \to 0 \quad \text{as} \quad z \to -\infty$$

is $f''(z)/f(z) = k^2$, and we must then choose k positive and

$$f(z) = e^{kz}$$

as the z-dependence in ϕ.

(ii) The pressure on the surface is derived from Bernoulli's equation for unsteady potential flow:

$$\partial\phi/\partial t + p_0/\rho + \tfrac{1}{2}\mathbf{v}^2 + gz = F(t)$$

on the surface $z = \zeta$. Now (as in sound waves) we neglect $\tfrac{1}{2}\mathbf{v}^2$ as a small term for small amplitude waves; and we may take $F(t)$ and p_0/ρ into the potential ϕ, where they do not affect any velocities. This leaves us with

$$(\partial\phi/\partial t)_{z=\zeta} + g\zeta = 0.$$

Finally, we use Taylor's theorem to approximate the first term by its value at $z = 0$, since ζ is small. The boundary condition is thus taken to be

$$(\partial\phi/\partial t)_{z=0} + g\zeta = 0.$$

(iii) It might be thought that the relation between surface motion and fluid velocity was just

$$\partial\zeta/\partial t = w = \partial\phi/\partial z,$$

but this is *not* exact, as can be seen by considering the fixed surface $\zeta(x)$ shown in fig. XIII.1. There must be an upward velocity w if there is a horizontal velocity u in the fluid, and the relation would be

$$w = u d\zeta/dx.$$

Fig. XIII.1. Velocity components near the fixed surface at height $\zeta(x)$.

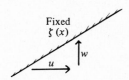

Fixed
$\zeta(x)$

u w

The full relation is found by noticing that the surface moves with the fluid, and hence that (for a particle on the surface)

$D/Dt\{z - \zeta(x, y, t)\} = 0.$

When we use $D/Dt = \partial/\partial t + \mathbf{v} \cdot \nabla$ we get

$w - \partial\zeta/\partial t - u\partial\zeta/\partial x - v\partial\zeta/\partial y = 0,$

where $\mathbf{v} = (u, v, w) = \nabla\phi$. This clearly fits with the special case above, and it may be linearised, first to

$(\partial\phi/\partial z)_{z=\zeta} = \partial\zeta/\partial t,$

and then, using Taylor's theorem again, to

$(\partial\phi/\partial z)_{z=0} = \partial\zeta/\partial t.$

This is our second boundary condition on the surface.

(b) The wave equation for the surface elevation

The boundary conditions (ii) and (iii) provide a single condition on ϕ at $z = 0$. When we eliminate ζ from them, we find

$\partial^2\phi/\partial t^2 + g\partial\phi/\partial z = 0 \quad \text{on} \quad z = 0.$

The method of solving a problem is thus to solve $\nabla^2\phi = 0$ with this boundary condition on $z = 0$ and with whatever other boundary conditions are needed. Then we derive the surface elevation (which is the wave that we see) from

$\zeta = -g^{-1}(\partial\phi/\partial t)_{z=0}.$

However, at the moment it is not at all obvious that the solutions of $\nabla^2\phi = 0$ should represent waves – previously a contrast was drawn between solutions of Laplace's equation and solutions of the wave equation. It is this upper boundary condition on the free surface that gives a wave character to the solution, and we may see this by constructing an equation for ζ.

Differentiate the linearised boundary condition in (a) (ii) to get

$$\frac{\partial^2\zeta}{\partial x^2} + \frac{\partial^2\zeta}{\partial y^2} = -\frac{1}{g}\frac{\partial}{\partial t}\left(\frac{\partial^2\phi}{\partial x^2} + \frac{\partial^2\phi}{\partial y^2}\right)_{z=0}$$

because the order of the derivatives on the right can be changed. But Laplace's equation is just

$$\partial^2\phi/\partial x^2 + \partial^2\phi/\partial y^2 = -\partial^2\phi/\partial z^2,$$

and from (a) (i)

$$\partial^2\phi/\partial z^2 = \partial^2/\partial z^2\{e^{kz}\phi_0(x, y, t)\}$$
$$= \partial/\partial z\{ke^{kz}\phi_0(x, y, t)\}$$
$$= k\partial\phi/\partial z;$$

and from the linearised boundary condition in (a) (iii), on $z = 0$,

$$\partial\phi/\partial z = \partial\zeta/\partial t.$$

Hence, combining these results, we obtain

$$\nabla_1^2\zeta = -g^{-1}\partial/\partial t(-k\partial\zeta/\partial t),$$

where $\nabla_1^2 = \partial^2/\partial x^2 + \partial^2/\partial y^2$ (the two-dimensional Laplacian operator). So the wave elevation $\zeta(x, y, t)$ satisfies the wave equation

$$(g/k)\nabla_1^2\zeta = \partial^2\zeta/\partial t^2.$$

This means that waves propagate on the surface with speed $(g/k)^{1/2}$. Consequently ϕ has a wave character in x, y, t as well as the exponential behaviour with z.

3. Plane waves on deep water

(a) *The dispersive character of plane waves*

The easiest waves in the theory of sound were plane waves, and we look at them first here too. If we take

$$\phi = e^{kz}f(x)g(t)$$

as a separation of variables, we easily find that $\nabla^2\phi = 0$ can only be satisfied by

$$f''(x) = -k^2f(x).$$

We get a wave character for the solutions by taking

$$\begin{cases} f(x) = e^{ikx}, \\ g(t) = e^{-i\omega t}, \end{cases}$$

so that

$$\phi(x, z, t) = Ae^{kz}e^{i(kx-\omega t)}.$$

This is a plane wave travelling in the x-direction with wavenumber k and wavelength $\lambda = 2\pi/k$; it penetrates in depth only as far as $z = -\lambda$, in effect, because below there the factor e^{kz} is negligibly small ($e^{-2\pi} = 1.87 \times 10^{-3}$).

The frequency ω is related to k because the speed has to be $(g/k)^{1/2}$, as we have seen above. Or we can see that the boundary condition

$$\partial^2\phi/\partial t^2 + g\partial\phi/\partial z = 0 \quad \text{on} \quad z = 0$$

requires

$$-\omega^2 + gk = 0,$$

i.e.

$$\omega = (gk)^{1/2}.$$

This relation is importantly different from the relation between ω and k for sound waves. There the speed was a constant, here it depends on frequency or wavelength

$$c = (g/k)^{1/2} = (g\lambda/2\pi)^{1/2}.$$

Hence plane waves of different length travel at different speeds; this 'dispersive' character of even the simplest water waves is rather like the behaviour of sound waves in a wave guide, as we saw in Chapter XII §5(*c*). But for water waves no wave guide is needed, long waves (with k small) naturally travel faster than short waves.

The table below shows the wave speed in relation to wavelength and period.

λ, (m)	0.1	1	10	100	1000
c, (m s^{-1})	0.4	1.25	4	12.5	40
T, (s)	0.25	0.8	2.5	8	25

The first sign of an Atlantic storm is usually provided by the arrival at the west coast of Britain of very long waves of rather small amplitude (swell), which have been generated by the storm. Typically they will have period around 30 s, and so a speed near 47 m s^{-1}. Thus they might be expected to travel about 4000 km in a day, and hence arrive well before the storm, which can take several days to cross the Atlantic. As the storm approaches, so the shorter waves will start to arrive, which have been travelling more slowly.

The longest waves, those due to earthquakes or tides, would seem on this basis to have enormously high speeds. But the theory cannot apply to them in its present form, as the water will not be deep compared to a wavelength. For the seas round Europe, the depth of water over the continental shelf is far less than 1000 m, so that a revised theory is needed; this will be given in §6 below.

(b) Reflection of a plane wave

Some of the properties of plane surface waves are very like those of plane sound waves, but it would be as well to present the appropriate theory here also.

A plane wave travelling at angle θ to the positive x-axis has the surface elevation

$$\zeta = A \exp\{i(kx \cos\theta + ky \sin\theta - \omega t)\}$$

and so the potential

$$\phi = -(iAg/\omega)e^{kz} \exp\{i(kx \cos\theta + ky \sin\theta - \omega t)\}$$

to satisfy $(\partial\phi/\partial t)_{z=0} = -g\zeta$. It is probably better to keep A for the amplitude of the elevation wave, as it, rather than the potential, is the observable.

When such a wave meets a plane barrier at $x = 0$ (which must be thought of as a cliff in this context), then there will be a reflected wave as shown in fig. XIII.2. It is easily verified that a reflected wave

$$\phi_r = -(iAg/\omega)e^{kz} \exp\{i(-kx \cos\theta + ky \sin\theta - \omega t)\}$$

is appropriate, because then

$$(\partial/\partial x)(\phi + \phi_r) = 0 \quad \text{on} \quad x = 0$$

for all y and t, which is the required boundary condition at a rigid barrier. That is, the incident wave is reflected without change of amplitude, just the expected change of direction.

There are no obvious analogues to the plane wave solutions in pipes that occupied us in the last chapter. Waves can occur in canals, but the more interesting ones are long compared to the depth, and so come later.

(c) Paths of the particles

In a sound wave the fluid particles moved backwards and forwards along the direction of propagation. For a water wave the behaviour

Fig. XIII.2. Reflection of a plane wave from a cliff.

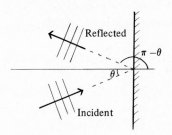

is rather different. Take the plane wave

$$\phi = Be^{kz}e^{i(kx - \omega t)}$$

where B is related to the surface amplitude A by

$$B = -iAg/\omega.$$

The velocity at (x, z), where $z < 0$, is

$$\begin{cases} u = \partial\phi/\partial x = ikBe^{kz}e^{i(kx - \omega t)} \\ w = \partial\phi/\partial z = kBe^{kz}e^{i(kx - \omega t)} \end{cases},$$

It is clear that $u = iw$, the velocities have equal magnitude but $u = we^{i\pi/2}$, so that the x-component is $\frac{1}{2}\pi$ ahead of the z-component. This just gives a circular motion. More directly, let the particle have position

$$\begin{cases} x = x_0 + X(t), \\ z = z_0 + Z(t), \end{cases}$$

where X and Z will be the small displacements of the particle from its mean position x_0, z_0. We then approximate the velocities of the particle as

$$\begin{cases} u = -kBe^{kz_0}\sin(kx_0 - \omega t) = X'(t), \\ w = kBe^{kz_0}\cos(kx_0 - \omega t) = Z'(t), \end{cases}$$

where we have taken the real parts now for convenience and assumed B to be real. These integrate to

$$\begin{cases} X = x - x_0 = -(kB/\omega)e^{kz_0}\cos(kx_0 - \omega t), \\ Z = z - z_0 = -(kB/\omega)e^{kz_0}\sin(kx_0 - \omega t). \end{cases}$$

The path is therefore the circle

$$X^2 + Z^2 = (kB/\omega)^2 e^{2kz_0}.$$

That is, the path of each particle is a circle, with radius decreasing downwards like e^{kz}. A particular case is shown in fig. XIII.3. Naturally the amplitude on the surface must be just A, and so further down it is Ae^{kz}. For a wave moving to the right, the particle at the wave crest is moving

Fig. XIII.3. Particle paths in a progressive plane wave on deep water.

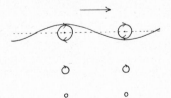

to the right at speed $A\omega$, while the particle in the trough is moving *left* at the same speed. This is easily observed when you swim in the waves of moderate amplitude and frequency near a shore – with $A = 1$ m and $\omega = 1$ radian per second, these speeds are 1 m s^{-1}, which is readily noticed.

4. Energy flow and group velocity

(a) *Energy density in a plane wave*

The amount of energy in water waves can be very large; even the moderate waves mentioned just above had fluid velocities of up to 1 m s^{-1}, and the velocity would still be noticeable many metres down: so a lot of water is moving.

Consider the wave

$$\begin{cases} \zeta = Ae^{i(kx - \omega t)} \\ \phi = Be^{kz}e^{i(kx - \omega t)}, \end{cases}$$

where

$$B = -iAg/\omega \quad \text{and} \quad \omega^2 = gk.$$

The velocities of the particles in this wave are given in §3(c) above, and the square of the speed is (when averaged over a cycle)

$$\mathbf{v}^2 = k^2|B|^2e^{2kz}.$$

Hence the kinetic energy per area of surface is

$$\tfrac{1}{2}\rho \int_{-\infty}^{0} k^2|B|^2e^{2kz}dz,$$

on taking the surface to be $z = 0$ approximately. This gives the value

$$\tfrac{1}{4}\rho k|B|^2 = \tfrac{1}{4}\rho g|A|^2$$

for the kinetic energy per area.

Similarly the potential energy of the fluid in the wave, above what it had in its mean level, is

$$\rho g \int_{0}^{\zeta} zdz = \tfrac{1}{2}\rho g\zeta^2.$$

When we take the average of this over a cycle, by the theorem in the last chapter, we get

$$\tfrac{1}{4}\rho g|A|^2.$$

The total energy per area is thus equally divided (as it was in sound waves too) between kinetic and potential energy.

(b) *Energy flow rate and energy velocity*

Surface waves carry energy from one place to another, and this energy flow rate is calculated in much the same way as for sound waves. Neglecting energy transport by the velocities in the waves, which is of third order in the amplitude, we concentrate on the rate of working across a surface $x = $ constant. This is

$$\int_{-\infty}^{0} pu\,dz$$

where again we have approximated the surface by $z = 0$. Now p, from Bernoulli's equation, is

$$-\rho\partial\phi/\partial t - \rho gz$$

and $u = \partial\phi/\partial x$. So for the average energy flow rate we must calculate the average of

$$-\rho\int_{-\infty}^{0} \{(\partial\phi/\partial t)(\partial\phi/\partial x) + gz\partial\phi/\partial x\}\,dz.$$

The second term has average zero, as z is independent of t and $\partial\phi/\partial x$ has average zero. We use the averaging theorem again to get the energy flow rate (per unit length in the y-direction) to be

$$\tfrac{1}{2}\rho|B|^2\omega k\int_{-\infty}^{0} e^{2kz}dz = \tfrac{1}{4}\rho\omega|B|^2 = \tfrac{1}{4}\rho g|A|^2\omega/k.$$

The velocity of energy transfer is obtained as before by dividing energy flow rate by energy density, and the result (rather surprisingly) is

$$\tfrac{1}{2}\omega/k,$$

which is just half the wave speed. Having checked for errors (and found none) we must accept that in these waves the energy travels less fast than the individual wave crests.

It is of interest to do a sample calculation on this energy flow formula. Take an amplitude $A = 1$ m and a wavelength $\lambda = 100$ m, corresponding to a period of 8 s. The energy flow rate per metre length of wave in the y-direction is then

$$3.2 \times 10^4 \text{ J m}^{-1}\text{ s}^{-1}$$

or

$$3.2 \times 10^4 \text{ W m}^{-1}.$$

Clearly the energy that is transported towards the British Isles even in such a gentle wave is vast. Taking a length of 10^6 m for the British Isles gives

3.2×10^4 MW. The total power in waves approaching the coast will be far larger than this. There are, however, considerable problems in using this power.

(c) *The group velocity*

Let us investigate how the wave crests can travel faster than the wave energy. We would like to set off a limited disturbance like that in fig. XIII.4 at $t = 0$, and see how it develops. Unfortunately this is mathematically rather hard, but we can set up a rather similar periodic pattern for which the mathematics is easier.

Consider a surface elevation composed of two waves of almost equal wavenumbers $k - \delta k$ and $k + \delta k$, and corresponding frequencies $\omega - \delta\omega$ and $\omega + \delta\omega$:

$$\zeta = A\cos\{(k - \delta k)x - (\omega - \delta\omega)t\} + A\cos\{(k + \delta k)x - (\omega + \delta\omega)t\}.$$

This may be rearranged as

$$\zeta = 2A\cos(kx - \omega t)\cos(\delta kx - \delta\omega t),$$

which is the product of a long wave, of wavelength $2\pi/\delta k$, and a short wave of wavelength $2\pi/k$. This gives, as fig. XIII.5 shows, a 'packet' of waves which are almost isolated by a region of calmer water from the next 'packet'. The individual waves have speeds

$$(\omega \pm \delta\omega)/(k \pm \delta k) \doteqdot \omega/k,$$

but the 'modulating' wave

$$\cos(\delta kx - \delta\omega t)$$

has speed

$$\delta\omega/\delta k,$$

which may not be equal to ω/k.

Fig. XIII.4. An initial wave packet.

Fig. XIII.5. The superposition of two waves of slightly different frequency and wavelength.

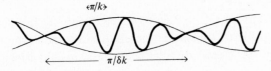

What happens is that the individual waves move relative to the packet at speed

$$\omega/k - d\omega/dk$$

approximately. They appear, apparently from nowhere (but really from the next packet) at the rear of the packet, move through it, and disappear at the front. This behaviour can be seen in the waves created by a splash; you can see new waves appear at the back of the advancing disturbance, travel through it, and disappear again.

Thus it seems that the wave packet as a whole moves at speed $d\omega/dk$, and this can indeed be proved by more advanced mathematics. For plane sound waves, where $\omega = ck$, the speed of the wave packet is just c again, agreeing with our previous finding that wave velocity and 'group velocity' (as it is usually known) are equal. For plane water waves, where we have found $\omega = (gk)^{1/2}$, we derive

$$d\omega/dk = \tfrac{1}{2}(g/k)^{1/2} = \tfrac{1}{2}c$$

and this confirms that the group velocity is the velocity of energy travel.

As a third example of the group velocity $d\omega/dk$ being the velocity of energy travel, consider the sound waves travelling down a corridor of Chapter XII §5(c); there it was shown that

$$\omega = c(k^2 + m^2\pi^2/a^2)^{1/2}$$

where k was the wavenumber along the corridor, a was the width of the corridor and m was an integer. In this case

$$\frac{d\omega}{dk} = \frac{ck}{(k^2 + m^2\pi^2/a^2)^{1/2}}$$

which is just $c \cos \delta$, in the notation of that section. And this was the energy speed deduced for that situation in Chapter XII §6(d).

As a result of these calculations we must change our estimate of how long a swell on the Atlantic takes to reach Britain after being generated by a storm. The group velocity will only be some 23 m s^{-1}, and the group cannot travel more than 2000 km in a day. However this is still fast enough to give early warning of the approach of a storm.

5. Waves at an interface

(a) The simplest case

Waves occur on the surface of water essentially because air is lighter than water, and so a disturbed surface has potential energy which can be exchanged for kinetic energy – heavy fluid (water) is raised and

Fig. XIII.6. A warmer fluid of density ρ_1 and potential ϕ_1 above a colder fluid of density ρ_2 and potential ϕ_2.

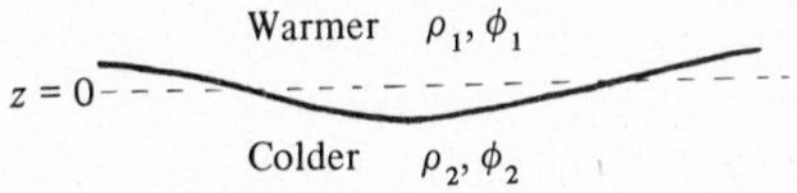

Warmer ρ_1, ϕ_1

$z = 0$

Colder ρ_2, ϕ_2

light fluid (air) is lowered. All that is needed for waves, then, is a lighter fluid on top of a heavier one, and this occurs more widely than just at the air–water interface. Frequently the upper layers of water are made warmer by sunlight and mixing than the lower layers; there is then an interface between warmer and colder water, and there ought to be the possibility of waves on this interface. A similar situation can be produced where less salty water lies on top of more salty lower layers. Fig. XIII.6 illustrates the former case.

The assumptions and analysis will be very much those of §§2–4. At the boundary we

(i) match pressures using an approximate form of Bernoulli's equation, and

(ii) match velocities with the derivative $\partial\zeta/\partial t$; we also

(iii) assume exponential decay away from the boundary in each direction, and

(iv) assume a plane wave form proportional to

$$e^{i(kx - \omega t)}.$$

So for $z > 0$ we take

$$\phi_1 = B_1 e^{-kz} e^{i(kx - \omega t)}$$

as a suitable solution of Laplace's equation, and for $z < 0$ we take

$$\phi_2 = B_2 e^{kz} e^{i(kx - \omega t)}.$$

The condition (i) is

$$\rho_1 \partial\phi_1/\partial t + \rho_1 g\zeta = \rho_2 \partial\phi_2/\partial t + \rho_2 g\zeta$$

and the condition (ii) is

$$\partial\zeta/\partial t = -kB_1 e^{i(kx - \omega t)} = kB_2 e^{i(kx - \omega t)}.$$

Thus

$$\begin{cases} B_1 = -B_2, \\ \zeta = -(ikB_1/\omega)e^{i(kx - \omega t)}, \\ \omega^2 = kg(\rho_2 - \rho_1)/(\rho_2 + \rho_1). \end{cases}$$

Now $(\rho_2 - \rho_1)/(\rho_2 + \rho_1)$ is likely to be quite small for the situations we have discussed: a 10 K temperature difference would give this ratio the value 4.5×10^{-4} approximately, and a change from a 0% to a 5% solution of salt (by weight) would give the ratio the value 1.8×10^{-2} approximately. The wave speed ω/k is therefore $\varepsilon(g/k)^{1/2}$ where ε is a small number, at most 10^{-1}: waves on this interface travel rather slowly. Similarly the group velocity $d\omega/dk$ will be correspondingly small.

The energy densities will also be changed, but not in the same way. Both fluids will now have kinetic energy, so the total kinetic energy will be apparently doubled; however the frequency is now much smaller, which reduces the kinetic energy. The potential energy will be reduced greatly, because when the lower liquid is raised, an equal amount of upper liquid of almost equal density is lowered. If you carry out the calculations, you get a kinetic energy density

$$\tfrac{1}{4}(\rho_1 + \rho_1)k|B_1|^2 = \tfrac{1}{4}(\rho_2 - \rho_1)g|A|^2$$

and an equally reduced potential energy density

$$\tfrac{1}{4}(\rho_2 - \rho_1)g|A|^2.$$

There is still an equal division between kinetic and potential energy. The total energy is now

$$\tfrac{1}{2}(\rho_2 - \rho_1)g|A|^2,$$

which has a very small factor in it, and so if the motion has been set off with a reasonable amount of energy, a quite large amplitude can result. Large waves of this sort have frequently been observed in seas and oceans. One particular case is in fjords in Norway, where fresh water from rivers lies over sea water; a ship of moderate size can reach the interface and create considerable waves there, adding greatly to the resistance to the motion of the ship. Similar waves have been found in Loch Ness in Scotland, where warmer water lies over colder; because of their large amplitude they were greeted by newspapers as 'monster waves'. In both cases the upper layer of fluid may be rather shallow, and the calculations for this case come in §6(d) below.

(b) Instability due to relative velocity

In some of the geophysical situations where less dense fluid lies over denser fluid there is also a relative velocity between the two layers. For example, the water of the Amazon tends to flow out in a large sheet over the denser waters of the South Atlantic; and in the atmosphere there is often such a relative velocity, though the interface is rarely a sharp one. In this section therefore we include a relative velocity in the calculations.

Fig. XIII.7. A wave on the interface between fluid at rest and fluid in motion.

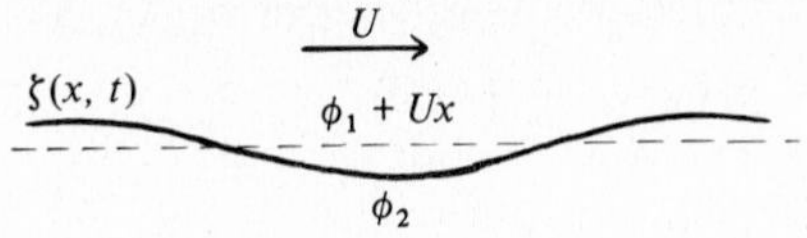

Indeed to make the working as easy as possible, we consider the case when there is *only* a relative velocity, and neglect the density difference. The more general case when there is also a density difference is left to an exercise.

Take, then, a potential $\phi_1 + Ux$ for $z > 0$, and ϕ_2 for $z < 0$, as shown in fig. XIII.7. The interface will be taken to be

$$z = \zeta(x, t)$$

and we are assuming that ζ, ϕ_1, ϕ_2 are all small so that the problem may be linearised by neglecting second order terms. We shall assume that there is an x- dependence

$$e^{ikx}$$

in all three functions, to give a wave-like shape. Thus we take

$$\begin{cases} \zeta(x, t) = A(t)e^{ikx}, \\ \phi_1(x, z, t) = B(t)e^{-kz}e^{ikx}, \\ \phi_2(x, z, t) = C(t)e^{kz}e^{ikx}, \end{cases}$$

so that ϕ_1 and ϕ_2 satisfy $\nabla^2\phi = 0$ and are zero for points well away from the interface.

The boundary conditions at the interface must be those of §2(a), on pressure and normal velocity; but they have to be modified to take into account the velocity U of the upper layer, which is not a small quantity.

Bernoulli's equation taken above and below the interface gives

$$g\zeta + (\partial\phi_1/\partial t)_{z=\zeta} + p/\rho + \tfrac{1}{2}\{(U + \partial\phi_1/\partial x)^2 + (\partial\phi_1/\partial z)^2\}_{z=\zeta}$$
$$= g\zeta + (\partial\phi_2/\partial t)_{z=\zeta} + p/\rho + \tfrac{1}{2}(\nabla\phi_2)^2_{z=\zeta} + K,$$

where the constant K allows for the different value of the Bernoulli constant above and below the vortex sheet. This is approximated as before, by taking the derivatives at $z = 0$ and by neglecting second order terms, and we derive

$$(\partial\phi_1/\partial t)_{z=0} + U(\partial\phi_1/\partial x)_{z=0} = (\partial\phi_2/\partial t)_{z=0}$$

on choosing $K = \tfrac{1}{2}U^2$ so that the equation still holds in a state of rest, or

$$\dot{B} + ikUB = \dot{C}.$$

The condition on normal velocity must be taken in its *full* form from §2(a)(iii), at least above the surface. We find, for this upper layer,

$$(\partial\phi_1/\partial z)_{z=\zeta} = \partial\zeta/\partial t + \{(U\mathbf{i} + \nabla\phi_1)\cdot\nabla\zeta\}_{z=\zeta}.$$

Linearising this as usual gives

$$-kB = \dot{A} + ikUA.$$

The condition for the lower layer is the usual one, because all the velocities are small there:

$$(\partial\phi_2/\partial z)_{z=0} = \partial\zeta/\partial t$$

to our approximations, and so

$$kC = \dot{A}.$$

We now eliminate B and C from these equations, and find

$$\ddot{A} + ikU\dot{A} - \tfrac{1}{2}k^2U^2A = 0.$$

This linear equation for $A(t)$ has solutions

$$\exp \sigma t$$

where

$$\sigma = -\tfrac{1}{2}ikU \pm \tfrac{1}{2}kU.$$

The imaginary part of σ corresponds to waves travelling along the vortex sheet at the average speed $\tfrac{1}{2}U$. The real part of σ shows that (at least for any real initial conditions) there is an exponential growth of the wave. Naturally the above linearised analysis cannot describe the large amplitudes which are soon reached in such an exponential growth; but the analysis does show that small waves are 'linearly unstable' and will grow into something else. We have seen before, in Chapter IX, an instability of a flow in circles where the instability led in fact (at moderate speeds) to a system of cells. The present flow is rather a hypothetical one, being hard to set up exactly in practice; but many such flows with vortex sheets lead directly to turbulence rather than to a steady, large-amplitude disturbance.

You should note that, in (a) above, if the heavier fluid is on top of the lighter, i.e. if $\rho_1 > \rho_2$, then you get $\omega^2 < 0$ and linear instability. This is a very obvious instability, where the upper fluid falls into the lower fluid; but how it does it is rather less obvious, and will not be attempted here.

When the upper fluid has velocity U and is also lighter, you get competing tendencies: the vortex sheet tends to make the situation unstable, but the density difference tends to make it stable. The result is (as you will see when you have done the exercise) that for this 'Kelvin–Helmholtz' instability there is stability up to a certain critical speed, instability beyond it. It would be more realistic to include viscosity in the calculation, and also to allow the transition between the two regions of fluid to take place smoothly rather than suddenly. For such further details you must read texts on hydrodynamic stability.

6. Waves on shallower water

(a) Plane waves on shallower water

Many of the waves mentioned above may well occur in water that is not particularly deep compared with the wavelength. So we now revise the theory to allow for depth h of water below the mean free surface.

The boundary condition of §2(a)(i) must now be

$$(\partial\phi/\partial z)_{z=-h} = 0,$$

i.e. no velocity through the bottom of the region. The separation of variables proceeds as before to give a z-dependence $f(z)$ where

$$f''(z) = k^2 z,$$

but we now choose the solution

$$f(z) = \cosh k(z + h),$$

which has $f'(-h) = 0$. Thus our plane wave solution of §3(a) becomes

$$\phi(x, z, t) = B \cosh k(z + h)e^{i(kx - \omega t)}$$

and the boundary condition on ϕ at $z = 0$,

$$\partial^2\phi/\partial t^2 + g\partial\phi/\partial z = 0,$$

gives the equation

$$\omega^2 = gk \tanh kh.$$

This is the new 'dispersion relation', the equation connecting ω and k. The wave speed ω/k is now

$$c = (gk^{-1} \tanh kh)^{1/2}.$$

This wave speed formula has two obvious properties:
  (i) as $h \to \infty$, $c \to (g/k)^{1/2}$, which is the formula for infinite depth that we found above;
  (ii) as $kh \to 0$, $c \to (gh)^{1/2}$, which provides the wave speed for very long waves in rather shallow water.

The wave speed is most easily plotted against wavelength, in the non-dimensional form

$$c/(gh)^{1/2} = \{(\lambda/2\pi h)\tanh(2\pi h/\lambda)\}^{1/2}.$$

Fig. XIII.8. Dependence of wave speed c on wavelength λ.

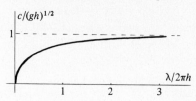

The graph is sketched in fig. XIII.8. For $\lambda < 3h$, the deep water formula is a good approximation; for $\lambda > 15h$, the shallow water formula is adequate. Between the two, the full formula must be used.

It is interesting to calculate the maximum speed of ocean waves. With a depth of (say) 2000 m, the speed of long waves of wavelength exceeding 3×10^4 m is some 140 m s^{-1}. These are extremely long waves of periods exceeding 200 s. On the continental shelf the depths will be more like 100 m, and the maximum speed about 32 m s^{-1}.

The surface elevation ζ is still found from

$$\zeta = -g^{-1}(\partial\phi/\partial t)_{z=0}.$$

In the present case of a plane wave on moderately shallow to very shallow water it is given by

$$\zeta = i\omega Bg^{-1} \cosh kh \ e^{i(kx - \omega t)}.$$

Thus the constant B in the potential of the wave is related to the amplitude A of the surface elevation by

$$B = A(g/i\omega)\operatorname{sech} kh$$
$$= -iA\{2g/(k \sinh 2kh)\}^{1/2}.$$

That is, for small kh you get large values of B corresponding to modest amplitudes A.

(b) Particle paths in shallow water waves

The reflection properties of plane waves do not depend on the depth factor $f(z)$ in the potential ϕ, and so we need not investigate any further the reflection of plane waves.

However the motion of the particles in a wave *is* affected by the change from e^{kz} to $\cosh k(z + h)$ in ϕ. The two velocity components are now

$$\begin{cases} u = \partial\phi/\partial x = ikB \cosh k(z + h)e^{i(kx - \omega t)}, \\ w = \partial\phi/\partial z = kB \sinh k(z + h)e^{i(kx - \omega t)}. \end{cases}$$

It is no longer true that $u = iw$: proceeding as in §3(c) we have

$$\begin{cases} u \doteqdot -kB \cosh k(z_0 + h)\sin(kx_0 - \omega t) = X'(t), \\ w \doteqdot kB \sinh k(z_0 + h)\cos(kx_0 - \omega t) = Z'(t). \end{cases}$$

These integrate to

$$\begin{cases} (x - x_0)/\cosh k(z_0 + h) = -(kB/\omega)\cos(kx_0 - \omega t), \\ (z - z_0)/\sinh k(z_0 + h) = -(kB/\omega)\sin(kx_0 - \omega t) \end{cases}$$

so that

$$\left\{ \frac{x - x_0}{\cosh k(z_0 + h)} \right\}^2 + \left\{ \frac{z - z_0}{\sinh k(z_0 + h)} \right\}^2 = k^2 B^2/\omega^2.$$

This is an ellipse of semi-axes

$$(kB/\omega)\cosh k(z_0 + h) \quad \text{and} \quad (kB/\omega)\sinh k(z_0 + h).$$

It is better to put these in terms of the surface amplitude A as

$$\frac{A \cosh k(z_0 + h)}{\sinh kh} \quad \text{and} \quad \frac{A \sinh k(z_0 + h)}{\sinh kh}.$$

The vertical amplitude tends to zero as $z_0 \to -h$ and it is less than the horizontal amplitude by a factor $\tanh k(z_0 + h)$. Fig. XIII.9 shows a particular case. This factor of course tends to 1 as $kh \to \infty$, i.e. for very deep water.

When the water is very shallow, or the wave is extremely long, then $\tanh k(z_0 + h)$ is always very small, and the motions are effectively horizontal. Moreover, the velocity in the horizontal direction is then independent of depth as $\cosh k(z_0 + h) \doteqdot 1$. This gives an alternative approach to long waves in shallow water, but we do not follow it here.

(c) The group velocity in waves on shallow water

The energy properties of waves in shallower water are different from those for very deep water, because of the $\cosh k(z + h)$ factor in

Fig. XIII.9. Particle paths for a progressive plane wave on water with $kh = 2.4$.

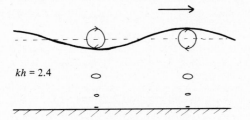

$kh = 2.4$

ϕ. For example, the average kinetic energy per unit area of surface requires an integral

$$\tfrac{1}{4}\rho k^2 B^2 \int_{-h}^{0} \{\cosh^2 k(z+h) + \sinh^2 k(z+h)\}dz.$$

However the calculations are straightforward enough to leave to an exercise, and we can calculate the energy flow rate from the group velocity formula

$$d\omega/dk.$$

Since we have here

$$\omega^2 = gk \tanh kh,$$

we calculate $d\omega/dk$ quite easily from

$$2\omega d\omega/dk = g \tanh kh + gkh \operatorname{sech}^2 kh$$

as

$$\frac{d\omega}{dk} = \tfrac{1}{2}c\left\{1 + \frac{2kh}{\sinh 2kh}\right\}.$$

This group velocity tends, at it must, to $\tfrac{1}{2}c$ when $kh \to \infty$; and to c when $kh \to 0$, as then the system is non-dispersive.

(d) A shallow layer over deep water

As a final example on waves in a shallower layer of water, we calculate the waves at an interface between two layers of different density, and *also* at the upper surface, when the lower layer is deep. Very often, after all, it is only a rather thin layer of water at the top of the sea or a lake that has been heated from above.

We now have two linked problems for the potentials shown in fig. XIII.10.

(i) For $z < 0$

$$\begin{cases} \nabla^2\phi_2 = 0, \\ \phi_2 \text{ decreases like } e^{kz}, \\ \phi_2 \text{ is a progressive wave in the } x\text{-direction.} \end{cases}$$

(ii) For $0 < z < h$

$$\begin{cases} \nabla^2\phi_1 = 0, \\ \phi_1 \text{ has exponential or hyperbolic functions in } z, \\ \phi_1 \text{ is a progressive wave in the } x\text{-direction.} \end{cases}$$

Fig. XIII.10. Densities ρ_i and potentials ϕ_i for a layer of thickness h lying above deep water.

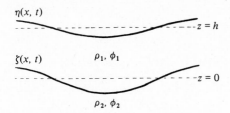

(iii) At $z = h$ the boundary conditions are

$$\partial\phi_1/\partial t + g\eta = 0,$$

and

$$\partial\eta/\partial t = \partial\phi_1/\partial z$$

as in §2(a).

(iv) At $z = 0$ the conditions on pressure and velocity are

$$\rho_1\partial\phi_1/\partial t + \rho_1 g\zeta = \rho_2\partial\phi_2/\partial t + \rho_2 g\zeta,$$

and

$$\partial\zeta/\partial t = \partial\phi_1/\partial z = \partial\phi_2/\partial z.$$

To satisfy (i) we take

$$\phi_2 = Be^{kz}e^{i(kx-\omega t)},$$

and to satisfy (ii) we take

$$\phi_1 = (Ce^{kz} + De^{-kz})e^{i(kx-\omega t)}.$$

Then the boundary conditions at $z = h$, which are equivalent to

$$\partial^2\phi_1/\partial t + g\partial\phi_1/\partial z = 0 \quad \text{at} \quad z = h,$$

give

$$\omega^2(Ce^{kh} + De^{-kh}) = gk(Ce^{kh} - De^{-kh}).$$

Those at $z = 0$ are equivalent to

$$\rho_1(\partial^2\phi_1/\partial t^2 + g\partial\phi_1/\partial z) = \rho_2(\partial^2\phi_2/\partial t^2 + g\partial\phi_2/\partial z)$$

and also $\partial\phi_1/\partial z = \partial\phi_2/\partial z$. These are, when we substitute the forms for ϕ_1 and ϕ_2,

$$C - D = B,$$

and

$$\omega^2(\rho_2 B - \rho_1 C - \rho_1 D) = (\rho_2 - \rho_1)gkB.$$

We have here equations for C and D in terms of B, and also an equation for ω^2 in terms of the wavenumber k and the physical parameters g, h, ρ_1, ρ_2. We also have equations, as yet unused, for the elevations $\eta(x, t)$ and $\zeta(x, t)$ of the surface and the interface.

It is a tedious calculation to show that, if $D \neq 0$,

$$\omega^2 = \frac{gk(\rho_2 - \rho_1)\sinh kh}{\rho_2 \cosh kh + \rho_1 \sinh kh},$$

and that

$$\begin{cases} C = \frac{1}{2}(\rho_2/\rho_1)(1 - \coth kh)B, \\ D = \{\frac{1}{2}(\rho_2/\rho_1)(1 - \coth kh) - 1\}B. \end{cases}$$

The case $D = 0$ gives $\omega^2 = gk$, and is really just a (modified) surface wave. The detail is left as an exercise. All these formulae can easily be seen to have the behaviour required by the results in §5 when $kh \to \infty$, as then $\cosh kh \sim \sinh kh$ and $\coth kh \to 1$.

Finally, the interface elevation is

$$\zeta = iBk\omega^{-1}e^{i(kx - \omega t)},$$

and the free surface elevation is

$$\eta = (1 - \rho_2/\rho_1)e^{-kh}\zeta.$$

This result shows that the interface elevation will greatly exceed the surface elevation whenever $kh \gg 1$; but that if kh is small the two are approximately related by

$$\eta = (1 - \rho_2/\rho_1)\zeta.$$

So in this case also you get a small surface disturbance provided that

$$\rho_2 \doteqdot \rho_1.$$

The wave motion described in this section is essentially an 'interface wave', with little motion away from the interface. The other mode, with $D = 0$, gives the surface wave which can also exist, mainly near $z = h$. It depends how you disturb the fluid which mode will predominate.

It would be more realistic to assume a continuous variation of density with depth, but that requires more advanced mathematical treatment than we can give here.

7. Oscillations in a container

(a) A rectangular container

Liquids have a (sometimes unfortunate) tendency to slosh back and forth in a container. The tea in a teacup on a train which is rocking may slosh out of the cup; liquid in road tankers must be prevented from sloshing, or else the tanker becomes uncontrollable. In this section we study the motion of a liquid with a free surface in a container.

The easiest container has rigid walls

$$\begin{cases} x = 0,\ x = l, \\ y = 0,\ y = b, \\ z = -h. \end{cases}$$

We have to solve $\nabla^2\phi = 0$ in this region, with $\partial\phi/\partial n = 0$ on the rigid walls, and with the condition

$$\partial^2\phi/\partial t^2 + g\partial\phi/\partial z = 0 \quad \text{on} \quad z = 0.$$

It is easy to see, from previous work on waves and on separation of variables, that

$$\phi = B\cos(m\pi x/l)\cos(n\pi y/b)\cosh k(z+h)e^{-i\omega t}$$

will satisfy all the conditions provided that

$$(m\pi/l)^2 + (n\pi/b)^2 = k^2$$

(to satisfy $\nabla^2\phi = 0$), and

$$\omega^2 = gk\ \tanh\ kh$$

(to satisfy the surface condition).

These last two equations determine the frequency ω in terms of the physical dimensions l, b, h and the mode shape given by m and n. As an example of this theory consider Loch Ness. It can be modelled by $l = 3 \times 10^4$ m, $b = 2 \times 10^3$ m, $h = 100$ m. When a wind blows from the south-west, the water of the loch is driven to its north-east end. When the wind drops, the water oscillates at small amplitude about its mean level, in the mode $m = 1, n = 0$, as is shown in fig. XIII.11. In this case we have, from the formulae above,

$$k = \pi/l,$$

and so

$$\omega = \{(\pi g/l)\tanh(\pi h/l)\}^{1/2}.$$

Fig. XIII.11. The sloshing stationary wave in a rectangular container.

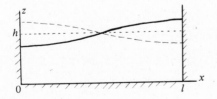

Now in this case $\pi h/l$ is very small, giving

$$\omega \doteqdot \pi(gh)^{1/2}/l$$
$$= 3.3 \times 10^{-3}\,\text{s}^{-1}.$$

The period of this sloshing mode is thus about 32 minutes. Its surface amplitude will not be large.

It is worth noticing that in this mode

$$k(z + h)$$

varies between 0 and $\pi h/l \doteqdot 10^{-2}$, so that the cosh $k(z + h)$ factor in ϕ is almost indistinguishable from 1. The shallow water theory which assumes that velocities are independent of z would clearly be quite adequate here.

The motion of the fluid particles is given by

$$\begin{cases} u = \partial\phi/\partial x, \\ w = \partial\phi/\partial z, \end{cases}$$

and for this particular solution it is easily found that the path of the particles is approximately a straight line; this is quite different from the case of travelling waves, where the paths were ellipses.

(b) Separation of variables for a circular cylinder

Waves in a circular container can be easily demonstrated – for example with a plate of soup and a spoon (though soup is not a very Newtonian fluid). This situation also illustrates some mathematics which we have been avoiding so far.

The problem is to solve

$$\nabla^2\phi = 0$$

with a rigid boundary at $r = a$, with no singularity at $r = 0$, with a bottom at $z = -h$, and with

$$\partial^2\phi/\partial t^2 + g\partial\phi/\partial z = 0$$

on the surface $z = 0$. We know quite well by now that the appropriate z-dependence is cosh $k(z + h)$, so we seek a solution

$$\phi = f(r)g(\theta)\cosh k(z + h)e^{-i\omega t}.$$

Laplace's equation is satisfied provided that

$$\{r^2 f''(r) + rf'(r) + k^2 r^2 f(r)\}/f(r) = -g''(\theta)/g(\theta),$$

in the usual separation of variables manner. Each side of the equation must be a constant, as usual, and so

$$g''(\theta) = -\text{ constant} \times g(\theta).$$

If $g(\theta)$ is to be continuous for all θ, and this seems a sensible requirement for surface waves, the constant must be the square of an integer n, or zero:

$$g''(\theta) + n^2 g(\theta) = 0, n \geq 0.$$

That is, the possibilities for $g(\theta)$ are:

    (i) a constant;

    (ii) $\cos(n\theta + \alpha)$;

    (iii) $e^{in\theta}$.

We will examine these in turn later.

Now return to the equation for $f(r)$. It is

$$f''(r) + r^{-1}f'(r) + \{k^2 - n^2/r^2\} f(r) = 0.$$

The two standard solutions of this second order equation are known as Bessel's functions of the first and second kind, with symbols

$$J_n(kr) \quad \text{and} \quad Y_n(kr).$$

The details of these functions can be found in texts on differential equations (some of the more useful properties are given in §8 below). Their general properties are *somewhat* (but not entirely) like those of $r^{-1/2} \sin nkr$ and $r^{-1/2} \cos nkr$ in that

$$(1) \quad \begin{cases} J_n(kr) \text{ is well behaved as } r \to 0, \\ Y_n(kr) \text{ is singular as } r \to 0. \end{cases}$$

    The exact behaviour is *not* identical, but the *type* of behaviour is the same.

    (2) $J_n(kr)$ and $Y_n(kr)$ both oscillate and decrease as $r \to \infty$.

    (3) General functions can be represented as series of these functions.

    (4) Different functions (different values of k) give functions which are orthogonal, in that some integral involving the product is zero.

These properties should be compared with those for the Legendre polynomials in Chapter XI §5.

We reject the solution $Y_n(kr)$, as it has a singularity at $r = 0$. This leaves three types of solutions.

    (i) $\phi = BJ_0(kr)\cosh k(z + h)e^{-i\omega t}$,

    (ii) $\phi = B\cos(n\theta + \alpha)J_n(kr)\cosh k(z + h)e^{-i\omega t}$,

    (iii) $\phi = BJ_n(kr)\cosh k(z + h)e^{i(n\theta - \omega t)}$.

(c) *The solution independent of angle*

Take the solution

$$\phi = BJ_0(kr)\cosh k(z + h)e^{-i\omega t}$$

of $\nabla^2 \phi = 0$. This will satisfy $\partial \phi / \partial r = 0$ on $r = a$ provided that

$$\left\{ \frac{d}{dr} J_0(kr) \right\}_{r=a} = 0.$$

Now $J_0'(ka) = 0$ is an equation whose roots have been tabulated: the first four roots are given by

$$ka/\pi = 0, \ 1.2197, \ 2.2330, \ 3.2383$$

and the roots eventually are given by

$$ka/\pi = m + \tfrac{1}{4},$$

where m is an integer. Hence we have values for the constant k, and since we will still have

$$\omega^2 = gk \tanh kh$$

(from the surface condition on ϕ), the values of ω are known once h and a are known. That is, the frequencies of the normal modes – the eigenvalues of the system – have been found.

The motion of the surface is given by

$$\zeta = -g^{-1}(\partial \phi / \partial t)_{z=0}$$
$$= -i\omega g^{-1}BJ_0(kr)e^{-i\omega t}.$$

From tables you may find the zeroes of $J_0(kr)$ to be at

$$kr/\pi = 0.7655, \ 1.7571, \ 2.7546, \ \ldots$$

and eventually at $m - \tfrac{1}{4}$ for integers m.

These modes of oscillation are independent of θ, and can be quite easily excited by forcing an oscillation in the centre of a dish. The shapes of two of these modes are shown in fig. XIII.12.

If you use a cylindrical bowl of radius 15 cm and depth 10 cm, for

Fig. XIII.12. Stationary wave profiles independent of angle in a circular container. Higher modes have more points of zero displacement.

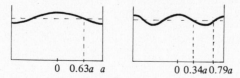

0   0.63*a*   *a*   0   0.34*a* 0.79*a*

example, then the first J_0 mode has

$$ka/\pi = 1.2197$$

and so

$$k = 25.545 \text{ m}^{-1}.$$

Hence

$$kh = 2.5545$$

and

$$\omega = 15.735 \text{ s}^{-1}.$$

So in this case the periodic time of the forcing needs to be about 0.40 s. The second J_0 mode, with

$$ka/\pi = 2.2330,$$

leads to a period of about 0.29 s. In both cases you will find that

$$\tanh kh \doteqdot 1,$$

so that a deep water theory could have been used.

(d) *Solutions depending on angle*
The solution

$$\phi = BJ_1(kr)\cos\theta\cosh k(z+h)e^{-i\omega t}$$

will do to represent the second kind of solution. Again we must satisfy

$$J_1'(ka) = 0,$$

Fig. XIII.13. The first and second modes of $J_1(kr)\cos\theta$; Views are from above and the side.

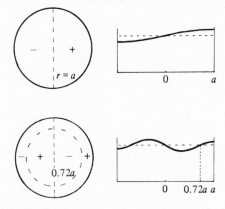

and this has tabulated solutions, starting with

$$ka/\pi = 0.586, 1.697, 2.717$$

(and becoming asymptotically $m - \frac{1}{4}$, much as before). As before the frequencies of oscillation are given by

$$\omega^2 = gk \tanh kh$$

once the value of k for the particular mode has been found.

In this case we need two diagrams (in Fig. XIII.13) for each mode, one to show r-dependence and one for θ-dependence. Note that $J_1(ka) = 0$ has the same roots as $J'_0(ka) = 0$; it may be shown that $J'_0 = J_1$.

The solution

$$\phi = B J_1(kr)\cosh k(z + h)\, e^{i(\theta - \omega t)}$$

may be taken to represent the third type of solution. As before we need

$$J'_1(ka) = 0,$$

which fixes k, and so fixes the frequencies ω of this type of oscillation. The angular dependence is now a progressive wave round the container. The elevated region moves round the surface with angular velocity ω. Fig. XIII.14 provides diagrams for this wave.

A motion of roughly this type occurs as a component of tidal oscillations in the North Sea. Naturally in this case the boundaries are not circular and the bottom is not flat, and there is a complicated forcing of the motion from the Moon (and Sun), from the Earth's rotation, and from adjoining tidal regions. However, the result is partly rotating waves; and if we put

$$a = 600 \text{ km}, \quad h = 100 \text{ m}$$

we get a period of about 9 hours, a moderate approximation for the tidal period of $12\frac{1}{2}$ hours. Note that in this case kh is very small and a shallow water theory would be good enough.

You should notice that in this mode the water does not rotate as a whole: it is the wave that travels while the fluid particles oscillate with small

Fig. XIII.14. The first mode of the rotating wave solution $J_1(kr) \exp i(\theta - \omega t)$.

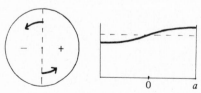

amplitudes, the velocity components being

$$\begin{cases} v_r = \partial\phi/\partial r \text{ radially,} \\ v_\theta = r^{-1}\partial\phi/\partial\theta \text{ transversely,} \\ v_z = \partial\phi/\partial z \text{ vertically.} \end{cases}$$

(e) *Waves along canals*

The last solution has propagation in one direction, with a standing wave character in the other direction. This behaviour is closely analogous to the wave guide propagation that was seen for sound waves. We may set up a similar theory here, for waves travelling along a canal.

Take a canal with sides $y = 0$, a and bottom $z = -h$, and consider a wave travelling along it in the x-direction. This suggests that we should take

$$\phi = B \cos(n\pi y/a)\cosh k(z + h)e^{i(Kx - \omega t)}.$$

Then $\nabla^2\phi = 0$ is satisfied provided that

$$k^2 - (n\pi/a)^2 - K^2 = 0;$$

and the surface condition gives

$$\omega^2 = gk \tanh kh$$

as usual. The dispersion relation between ω and K is

$$\omega^2 = g\{K^2 + (n\pi/a)^2\}^{1/2} \tanh\{K^2 h^2 + (n\pi/a)^2 h^2\}^{1/2}.$$

This looks harder than the corresponding one in the theory of sound: ordinary travelling waves going at the usual speed are being reflected repeatedly from the two walls $y = 0$ and $y = a$. If you work out the group velocity $d\omega/dK$, you will find the energy flows along the canal at a rate totally in keeping with this view of the motion.

8. Bessel functions

A summary of the main properties of the Bessel functions that are needed in elementary fluid dynamics is worth giving at this stage, to save hunting through other texts. But the other texts must be consulted for derivations, for further properties, and for more detail.

(a) *Bessel's equation*

The differential equation

$$f''(r) + r^{-1}f'(r) + (k^2 - n^2/r^2)f(r) = 0$$

arose in the solution of Laplace's equation by separation of variables, and can arise elsewhere. The two standard solutions are

$$J_n(kr) \text{ and } Y_n(kr)$$

when n is an integer. When n is *not* an integer, the two standard solutions may be taken to be

$$J_n(kr) \text{ and } J_{-n}(kr).$$

Any other solution is a linear combination of these standard solutions. Compare the equation

$$f''(r) + k^2 f(r) = 0$$

which has standard solutions

$$e^{ikr} \text{ and } e^{-ikr};$$

the solutions $\cos kr$ and $\sin kr$ are just combinations of the standard ones. The related differential equation

$$f''(r) + r^{-1} f'(r) - (k^2 + n^2/r^2) f(r) = 0$$

has solutions called $I_n(kr)$ and $K_n(kr)$. $I_n(kr)$ is related to $J_n(kr)$ in much the same way as e^{kr} is related to $\cos kr$.

Many differential equations can, with a bit of juggling, be transformed into one or other of these Bessel equations. For example,

$$f''(r) + 2r^{-1} f'(r) + \{k^2 - n(n+1)/r^2\} f(r) = 0$$

can be transformed into Bessel's equation for $g(r)$ by putting

$$f(r) = r^{-1/2} g(r).$$

(b) *Approximations for Bessel functions*
Near the origin,

$$J_n(kr) \doteqdot (\tfrac{1}{2}kr)^n/n!$$
$$Y_n(kr) \doteqdot -\pi^{-1}(n-1)!\,(\tfrac{1}{2}kr)^{-n}, \ n \neq 0,$$
$$Y_0(kr) \doteqdot 2\pi^{-1}\ln(\tfrac{1}{2}kr),$$
$$I_n(kr) \doteqdot (\tfrac{1}{2}kr)^n/(n!),$$
$$K_n(kr) \doteqdot \tfrac{1}{2}(\tfrac{1}{2}kr)^{-n}(n-1)!, \ n \neq 0,$$
$$K_0(kr) \doteqdot -\ln(\tfrac{1}{2}kr).$$

When kr is large,

$$J_n(kr) \sim (\tfrac{1}{2}kr)^{-1/2} \cos(kr - \tfrac{1}{2}n\pi - \tfrac{1}{4}\pi),$$
$$Y_n(kr) \sim (\tfrac{1}{2}kr)^{-1/2} \sin(kr - \tfrac{1}{2}n\pi - \tfrac{1}{4}\pi),$$
$$I_n(kr) \sim (2\pi kr)^{-1/2} e^{kr},$$
$$K_n(kr) \sim (2kr/\pi)^{-1/2} e^{-kr}.$$

In these formulae n need not be an integer.

(c) Radiation solutions

For radiation problems we have used solutions like e^{ikx} and e^{-ikx} in place of the more obvious solutions $\sin kx$ and $\cos kx$. The same combinations of J_n and Y_n are needed to solve radiation problems in polar coordinates.

We define

$$H_n^{(1)}(kr) = J_n(kr) + iY_n(kr)$$

to correspond to e^{ikx} above, and

$$H_n^{(2)}(kr) = J_n(kr) - iY_n(kr)$$

to correspond to e^{-ikx}. They are called Hankel functions. Both are infinite at the origin, rather in the way that the source solution in sound theory

$$\phi = r^{-1} e^{i(kr - \omega t)}$$

becomes infinite as $r \to 0$. At large distances

$$\begin{cases} H_n^{(1)}(kr) \sim (\tfrac{1}{2}kr)^{-1/2} \exp\{i(kr - \tfrac{1}{2}n\pi - \tfrac{1}{4}\pi)\}, \\ H_n^{(2)}(kr) \sim (\tfrac{1}{2}kr)^{-1/2} \exp\{-i(kr - \tfrac{1}{2}n\pi - \tfrac{1}{4}\pi)\}. \end{cases}$$

The factors $r^{-1/2}$ in these formulae give energy flows (quadratic in amplitudes) like r^{-1}, and so constant total energy flow through any circle.

(d) Integral formulae and series expansions

The two main integrals involving Bessel functions are

$$\int_0^a r J_n(kr) J_n(lr) dr = \{a/(k^2 - l^2)\}\{l J_n(ka) J_n'(la) - k J_n(la) J_n'(ka)\}$$

and

$$\int_0^a r\{J_n(kr)\}^2 dr = \tfrac{1}{2}a^2\{[J_n'(ka)]^2 + [1 - n^2/(k^2 a^2)][J_n(ka)]^2\}.$$

These may be used to expand reasonably well behaved functions as series of Bessel functions on $[0, a]$, in just the same way that Fourier series may be constructed.

Suppose that k_i are the roots of

$$J_n(ka) = 0,$$

and suppose that

$$f(r) = \sum_{i=1}^{\infty} A_i J_n(k_i r).$$

Multiply both sides by $r J_n(k_j r)$ and integrate from $r = 0$ to $r = a$. The only term that remains on the right, because $J_n(k_i a) = J_n(k_j a) = 0$, is

$$\int_0^a A_j r \{J_n(k_j r)\}^2 dr.$$

The left side is

$$\int_0^a rf(r) J_n(k_j r) dr.$$

Hence

$$A_j = \frac{2}{a^2 \{J'_n(k_j a)\}^2} \int_0^a rf(r) J_n(k_j r) dr,$$

which is (in principle) calculable

Similar calculation be done when the k_i are instead roots of

$$J'_n(ka) = 0.$$

The resulting series in both cases are called Fourier–Bessel series.

(e) An example

A number of problems in fluid motion with polar coordinates need Bessel functions. As an example we shall take up the rotation of a long circular cylinder containing fluid which is initially at rest (see Chapter IX and fig. XIII.15).

The equation we need to solve for the velocity

$$\mathbf{v} = U(r, t)\hat{\boldsymbol{\theta}}$$

is

$$\partial U/\partial t = v\{\partial^2 U/\partial r^2 + r^{-1}\partial U/\partial r - U/r^2\}.$$

The boundary conditions are

  (i) at $t = 0$, $U = 0$ for $r < a$,

  (ii) on $r = a$, $U = \Omega a$ for $t > 0$.

Now as $t \to \infty$ we expect to get rigid body rotation, and so

$$U(r, t) \to \Omega r$$

Fig. XIII.15. A rotating cylinder containing fluid.

Ωa

inside the cylinder, and it turns out to be more convenient to use

$$V(r, t) = \Omega r - U(r, t)$$

as the unknown function. It is easily shown that V satisfies the equation

$$\partial V/\partial t = v\{\partial^2 V/\partial r^2 + r^{-1}\partial V/\partial r - V/r^2\}.$$

The boundary conditions are:

   (iii) at $t = 0$, $V = \Omega r$ for $r < a$,

   (iv) on $r = a$, $V = 0$ for $t > 0$,

   (v) as $t \to \infty$, $V \to 0$ for fixed r.

We now separate variables in the equation for V, by taking $V(r, t) = f(r)g(t)$. The usual manipulations lead to

$$g'(t) = \text{constant} \times g(t)$$

and we take the constant to be negative so as to satisfy boundary condition (v). That is

$$g'(t) = - k^2 vg(t)$$

and so

$$g(t) = \exp(- k^2 vt).$$

The equation for $f(r)$ is then

$$f''(r) + r^{-1}f'(r) + (k^2 - 1/r^2)f(r) = 0$$

which has solutions

$$J_1(kr) \text{ and } Y_1(kr).$$

We reject $Y_1(kr)$ as the flow must be reasonable at $r = 0$. And we impose the boundary condition (iv) to get solutions

$$V(r, t) = J_1(k_i r)\exp(- k_i^2 vt),$$

where the k_i satisfy $J_1(k_i a) = 0$.

   The full solution may be assumed to be a Fourier–Bessel series

$$V(r, t) = \sum_{i=1}^{\infty} A_i J_1(k_i r)\exp(- k_i^2 vt),$$

where the coefficients are chosen to satisfy condition (iii):

$$\Omega r = \sum_{i=1}^{\infty} A_i J_1(k_i r).$$

The coefficients are determined by the method given in (d) above:

$$A_j = \frac{2\Omega}{a^2\{J_1'(k_j a)\}^2} \int_0^a r^2 J_1(k_j r)dr.$$

Now it is a standard result (i.e. go and look in a fuller text) that

$$\int_0^c x^n J_{n-1}(x)dx = c^n J_n(c), \quad n > 0$$

and so

$$\int_0^a r^2 J_1(k_j r)dr = k_j^{-1} a^2 J_2(k_j a), \quad k_j \neq 0.$$

So the coefficients A_j are given by

$$A_j = \frac{2\Omega J_2(k_j a)}{k_j \{J_1'(k_j a)\}^2},$$

which has in fact the value $-2\Omega a \{k_j a J_0(k_j a)\}^{-1}$, for $k_j \neq 0$.

The solution to this problem is not in a very manageable form. But it does show the dependence on the two dimensionless variables

$$vt/a^2 \text{ and } r/a.$$

The first comes in the time dependence, when we write

$$k_j a = n_j,$$

where n_j is a pure number; and the second comes in the space dependence for a similar reason. The dependence on

$$\exp\{-n_j^2(vt/a^2)\}$$

just serves to remind us of the time it takes to diffuse vorticity a distance a.

The values of n_1 and n_2 are 0 and 3.822. The corresponding values of A_1 and A_2 can be calculated, and are

$$\begin{cases} A_1 = 0, \\ A_2 = -1.30\Omega a. \end{cases}$$

So the leading term in $U(r, t)$ is

$$\Omega r - A_2 J_1(3.822 \, r/a)\exp(-14.61 \, vt/a^2),$$

showing that it takes a time about $\frac{1}{4}a^2/v$ to set up the rigid body rotation.

Exercises

1. In §2(a) $\phi(x, y, z, t) = \phi_0(x, y, t)e^{ekz}$. Show that ϕ_0 satisfies a wave equation.

2. When surface tension (energy) γ is allowed for, the pressure p in the liquid at the surface is less than atmospheric pressure p_0 by γ/R, where R is the radius of curvature of the surface. For plane waves with elevation $\zeta(x, t)$ the radius of curvature is given by

$$R^{-1} = \partial^2\zeta/\partial x^2 \{1 + (\partial\zeta/\partial x)^2\}^{-3/2}.$$

[$R > 0$ means a surface *concave* upwards.] Modify the boundary condition in §2(a)(ii) to allow for surface tension, and show that the new dispersion relation is

$$\omega^2 = gk(1 + \gamma k^2/\rho g).$$

Plot the wave velocity against k and find the minimum wave speed for a clean water–air interface.

3. One model for the generation of waves by wind (a poor one) uses a pressure distribution

$$p_0 + p_1 \cos k(x - ct)$$

on the surface for some speed c. Assume a wave with ζ proportional to $\cos k(x - ct)$ and determine its amplitude in terms of p_1.

4. The bottom of an ocean of depth h has a vertical velocity

$$a \cos k(x - ct).$$

Find the surface wave, as in Q3. If this is used as a model of an earthquake generated wave, then greatly exceeds any surface wave speed.

5. Consider water of small depth h with short waves on it, so that both water depth and surface tension have to be included. Derive the dispersion relation. Show that for water of depth 5×10^{-3} m the wave speed is constant for wave lengths exceeding about 2×10^{-2} m.

6. Calculate the energy density and energy flow rate for waves in water of depth h.

7. Calculate the group velocity for the waves in Q2 above. Plot it on your previous diagram.

8. Determine the criterion for the stability of the flow of a stream of speed U of liquid of density ρ_1 above a liquid of density $\rho_2 > \rho_1$ which is at rest. Neglect surface tension and viscosity, and take each liquid to be semi-infinite in extent.

9. Show that in §6(d) there is a special case which gives $D = 0$ and $\omega^2 = gk$. How are the surface and interface motions related in this wave?

10. Consider the special case in §7(a) when the container has equal sides. Describe the motion of the surface in the cases
   (i) $m = n = 1$,
   (ii) $m = n = 2$,
   (iii) $m = 1, n = 2$.

11. Calculate the particle paths for Q10(i).

12. Describe the motion of the water surface in a circular container in the case of standing waves and
   (i) $n = 1$, third mode,
   (ii) $n = 3$, first mode.

13. Waves travelling along a canal are discussed in §7(e). Take the case $n = 1$

and find the angle that constituent plane waves make with the x-axis. Show that the group velocity $d\omega/dK$ gives the correct speed for the propagation of energy, by discussing the speed of these waves at an angle to the x-axis.

14. Exercise 10 in Chapter IX describe a flow started suddenly at $t = 0$ in a channel. Try to do exactly the same problem for the corresponding flow in a round pipe; the equation you get to start with should be

$$\partial U/\partial t = \nu(\partial^2 U/\partial r^2 + r^{-1}\partial U/\partial r) + G/\rho,$$

and when you separate variables in the equation for $V(r, t)$, you will find Bessel's equation. The solution for $V(r, t)$ will be a 'Fourier–Bessel' series.

15. Return to Exercise 16 in Chapter X if you haven't done it yet. And to Exercise 15 in Chapter XII.

16. Waves on deep water have the potential

$$\phi_0(r, \theta)e^{kz}e^{-i\omega t}.$$

Show that there is a solution

$$\phi_0 = H_0^{(1)}(kr).$$

In what sense does this represent a source at $r = 0$?

References

This chapter has barely touched on the theory of surface waves. Even 100 years ago, when Lamb's *Hydrodynamics* was first written, the subject took up some 200 pages; and there is a renewed interest in the linear theory at present when wave power seems to be a practical source of energy. However it is inappropriate to refer to advanced texts, as the mathematics involved is usually daunting; so we refer here to some more introductory works.

(a) *Hydrodynamics*, W. Lamb, C.U.P./Dover 1945 provides a much fuller version of the material here, but is rather old in style and outlook.

(b) *Waves in Fluids*, M. J. Lighthill, C.U.P. 1978 gives a modern account of wave motions; the mathematical level is rather more advanced, but not more so than necessary. Parts could be read with considerable profit.

(c) *Theoretical Hydrodynamics*, L. M. Milne-Thomson, Macmillan 1949 has a large section on water waves. It, like Lamb, was written rather before the modern developments in water wave theory took place.

(d) *Wind Waves*, B. Kinsman, Prentice-Hall 1965. An excellent and readable text to follow on from this introduction.

(e) There are some film loops, which should be supplemented with observation of actual waves. The loops are: FM-108 on small-amplitude waves, FM-139 on small-amplitude waves in a channel, FM-91 on modes of sloshing in tanks.

(f) The properties of Bessel functions can be found in many texts on differential equations, on special functions, or on mathematical methods. The little that is required here can be obtained from any accessible and readable text.

(g) *Physical Fluid Dynamics*, D. J. Tritton, Van Nostrand Reinhold 1977 has a readable and elementary account of stability theory.

XIV
High speed flow of air

1. Subsonic and supersonic flows

(a) The speed of sound as a variable

In most of the previous chapters we have assumed that the density of the fluid is constant – except in Chapter VII where there was a hydrostatic compression of the air in the atmosphere, and in Chapter XII where there were very slight density changes in sound waves. In this chapter we investigate some of the effects of compressibility when the changes of density are due to quite large changes of pressure because of large fluid speeds.

As before, we try to deal with one thing at a time, and so we assume that effects of viscosity and heat conduction can be neglected over most of the flow. This allows us to assume that entropy is constant over most of the flow, and hence that

$$p = k\rho^{\gamma}$$

is a suitable version of the equation of state for the flows of air that we shall be studying. In Chapter XII we found that

$$(dp/d\rho)_{\rho = \rho_0}$$

or

$$\gamma p_0/\rho_0$$

was the square of the speed of sound in otherwise undisturbed air, in a linear approximation. Now it is in general true for an isentropic flow of a perfect gas that

$$dp/d\rho = \gamma p/\rho,$$

and that this has the dimensions of a speed squared; we introduce a new variable which is particularly convenient for this part of the work. Define the speed a (for a perfect gas) by

$$a^2 = \gamma p/\rho.$$

We shall call this new variable a the local speed of sound, though we cannot prove at this stage that it has this property; but it is easily seen that when the local fluid speed is small

$$a^2 \doteq a_0^2 = (\gamma p/\rho)_{\rho = \rho_0}$$
$$= c^2,$$

the speed of sound squared from Chapter XII.

What we are really doing is to set up a new variable of state a, which has the same status as all the others in Chapter VII (p, ρ, T, E, S); any pair of them (except E and T) will give an adequate description of the state of the air in the flow. In large parts of this chapter we shall be using the pair

$$a, S$$

as those with which we work; and then the others can be found in terms of them from the equations of Chapter VII and the new definition

$$a^2 = \gamma p/\rho.$$

(b) *The thermodynamic relations*

A summary of the basic thermodynamic equations is useful here. We shall assume the equation of state

$$p = R\rho T$$

for air, and that air is such that

$$E = c_v(T - T_0)$$

is an adequate equation for the internal energy E, where c_v is a constant. For such a gas we showed that

$$S = c_v \ln(p/\rho^\gamma) + \text{constant}.$$

Finally, we shall need to remember that

$$\Delta Q = 0 \Rightarrow dS \geq 0$$

where ΔQ is heat addition and there is equality for a reversible process.

For example, consider a point in a flow where a and S have the values a_1 and S_1, and take S to have value zero when $p = p_0$ and $\rho = \rho_0$. Then

$$\begin{cases} a_1^2 = \gamma p_1/\rho_1, \\ S_1 = c_v \ln(p_1 \rho_0^\gamma/p_0 \rho_1^\gamma) \end{cases}$$

can be solved to give, in sequence,

$$\begin{cases} \rho_1 = \rho_0(a_1/a_0)^{2/(\gamma-1)}\exp\{-S_1/c_v(\gamma-1)\}, \\ p_1 = a_1^2 \rho_1/\gamma, \\ T_1 = p_1/R\rho_1, \\ E_1 = c_v(T_1 - T_0). \end{cases}$$

(c) The Mach number

In a steady air flow without the molecular diffusive effects of viscosity and conductivity we have Bernoulli's equation, and when body forces are unimportant and density changes are small this has the form

$$p_0 - p = \tfrac{1}{2}\rho \mathbf{v}^2$$

where $p = p_0$ corresponds to a stagnation point. Let us write this as

$$\delta p = \tfrac{1}{2}\rho U^2.$$

Now

$$\delta \rho \doteqdot (d\rho/dp)\delta p$$

provided the changes are not excessively large, and since

$$dp/d\rho = a^2,$$

we derive

$$\delta \rho/\rho = \tfrac{1}{2}U^2/a^2.$$

That is, for changes of moderate size, the relative change of density is proportional to the square of the ratio of local speed to local sound speed. For example, a local speed of 100 m s^{-1} gives (with $a \doteqdot a_0$)

$$\delta \rho/\rho \doteqdot 0.04;$$

for this sort of speed it is clear that the approximation of incompressibility is excellent. But for a speed U approaching a_0 we cannot use such an approximation, as density changes will no longer be small. This chapter is devoted to speeds approaching or beyond a_0.

It is usual to call

$$U/a$$

the local Mach number M. This is of course a variable in a flow, firstly because U varies from place to place, but also because the speed of sound a varies. When $M < 1$, the flow is said to be subsonic, and when $M > 1$ it is supersonic.

Let us consider a steady flow when the compressibility is important but body forces are not. We expect to have Bernoulli's equation, in the form

$$\int \rho^{-1} dp + \tfrac{1}{2}\mathbf{v}^2 = \text{constant}$$

along a streamline for which diffusive effects are negligible. Since we have

$$p = k\rho^\gamma,$$

the integral can be rewritten as

$$\int k\gamma \rho^{\gamma-2} \, d\rho.$$

Using the definition of a^2 as $\gamma p/\rho$, Bernoulli's equation may be put in the form

$$a^2/(\gamma - 1) + \tfrac{1}{2}\mathbf{v}^2 = \text{constant}$$

along such a streamline. Hence the local speed of sound varies in an easy fashion with the local speed in such situations.

Take as example a flow at a high subsonic speed, as shown in fig. XIV.1. Suppose that upstream at A the conditions are

$$p_0, \rho_0, U.$$

Here the speed of sound must be just a_0, and the Mach number is

$$M_0 = U/a_0.$$

At the stagnation point B the speed is zero, and from Bernoulli's equation the local sound speed is

$$a_B^2/(\gamma - 1) + 0 = a_0^2/(\gamma - 1) + \tfrac{1}{2}U^2$$

or

$$a_B^2 = a_0^2\{1 + \tfrac{1}{2}(\gamma - 1)M_0^2\}.$$

So the speed of sound at B is higher than it is at A. But at C the local speed V exceeds U (because all the air has to get round the body) and so

$$a_C^2 = a_0^2\{1 - \tfrac{1}{2}(\gamma - 1)(V^2 - U^2)/a_0^2\}$$

Fig. XIV.1. Flow past a stagnation point.

which is less than a_0. The local Mach number at C is V/a_C, which can therefore be considerably greater than the Mach number of the initial flow at A, both because $V > U$ and $a_C < a_0$.

For example, if you take $\gamma = 7/5$ and $V = U\sqrt{3}$ – a value typical of those in earlier chapters for flow round blunt-nosed cylinders – you get

$$a_C^2 = a_0^2(1 - 2M_0^2/5)$$

and $M_C > 1$ whenever $M_0^2 > 5/17$. But it is found in practice that the transition from the subsonic upstream flow to the supersonic flow near C is not achieved in a continuous fashion, so this discussion must not be pressed too far. The difference between wholly subsonic flow and flow with some supersonic regions is in fact a considerable one, and the dividing line at $M = 1$ is a sharp one.

(d) Spreading of a disturbance for $M > 1$

The difference between $M < 1$ and $M > 1$ can be well shown by considering a small source of sound moving in a straight line and emitting pulses at regular intervals. Take the case $M < 1$ first, say $M = \frac{1}{2}$ so that the source moves at speed $\frac{1}{2}a_0$; and let the source be at $x = 0$ at $t = 0$ and emit a pulse every second.

At $t = 1$, the source is at $x = \frac{1}{2}a_0$, and the pulse emitted at the origin at $t = 0$ has spread to give a circle of radius a_0 in the (x, y) plane, as shown in fig. XIV.2 (of course it is a sphere in three dimensions, but this is more trouble to draw). At $t = 2$, the source is at $x = a_0$, the first pulse has travelled out to a radius $2a_0$, and the one emitted at $t = 1$ is now forming a circle of radius a_0 about the point $(\frac{1}{2}a_0, 0)$.

By drawing more diagrams (see fig. XIV.3) you can confirm that the source is always inside the circular wave fronts from previous pulses. If we impose a velocity $-\frac{1}{2}a_0$ on air and source we see that in this subsonic flow the sound pulses spread out to reach all upstream positions, though at a reduced speed. Similar sketches in fig. XIV.4 for a source of Mach number exceeding 1, say $\frac{3}{2}$, show that the source leaves behind the pulses spreading from previous positions, and the wave fronts overlap. The

Fig. XIV.2. The wave emitted at $t = 0$ by a source travelling at $\frac{1}{2}a_0$, shown at $t = 1$.

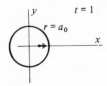

Fig. XIV.3. Waves emitted at $t = 0$ and $t = 1$ shown at $t = 2$, firstly for a moving source, secondly for a fixed source and moving air.

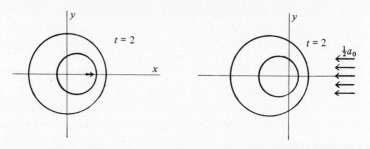

Fig. XIV.4. Formation of an envelope of wave fronts when the source speed or the air speed is $\frac{3}{2}a_0$.

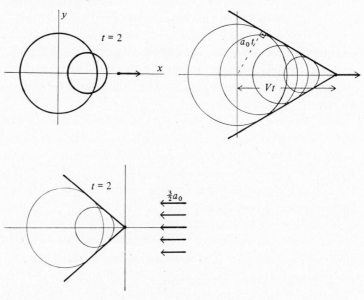

circles in this case have an 'envelope' which all of them touch. In three dimensions this envelope is a cone which has an angle

$$\sin^{-1}(a_0 t / V t)$$

at its vertex. This is known as a 'Mach cone' and the angle

$$\sin^{-1}(M^{-1})$$

is the 'Mach angle'.

When we take this latter case and impose the velocity

$$-\tfrac{3}{2}a_0$$

to reduce the source to rest, the envelope passes through the origin, and forms a boundary beyond which the wave fronts from the pulse cannot pass. There is thus a clear distinction between a subsonic flow, for which the pulses reach all of space eventually; and a supersonic flow, for which the pulses never propagate upstream, and indeed can never pass the Mach cone whose semi-angle at the vertex is

$$\sin^{-1}(M^{-1}).$$

2. The use of characteristics

(a) Unsteady, one-dimensional flow

In a new area of a mathematical subject it is always sensible to start off by trying the easy problems first. In the high speed flow of a gas it turns out that one-dimensional motion is an 'easy' problem, in that some examples can be solved completely, using some mathematics which may be unfamiliar but which is not essentially hard. So in this section we consider one-dimensional, unsteady motion of a perfect gas with constant entropy.

Take the velocity to be $u(x, t)\mathbf{i}$; the equation of mass conservation is

$$\partial\rho/\partial t + u\partial\rho/\partial x + \rho\partial u/\partial x = 0$$

because

$$\nabla\cdot(\rho u\mathbf{i}) = \partial/\partial x(\rho u).$$

As explained above, we shall use the sound speed a as a variable, and so we replace ρ here by using the equations

$$\begin{cases} a^2 = \gamma p/\rho, \\ p = k\rho^\gamma. \end{cases}$$

When p is eliminated, these equations may be written as

$$\ln\rho = \ln a^{2/(\gamma-1)} + \text{constant},$$

and so

$$d\rho/\rho = \{2/(\gamma-1)\}\,da/a.$$

This allows us to eliminate ρ from the mass conservation equation and replace it by a:

$$a\,\partial u/\partial x + \{2/(\gamma-1)\}(\partial a/\partial t + u\partial a/\partial x) = 0.$$

The only advantage so far of this manoeuvre is that the equation is entirely in terms of speeds; however we may also treat the momentum equation

$$\partial u/\partial t + u\partial u/\partial x + \rho^{-1}\partial p/\partial x = 0$$

in the same way, to obtain

$$\partial u/\partial t + u\partial u/\partial x + \{2a/(\gamma - 1)\}\partial a/\partial x = 0.$$

This still does not look helpful; but when we add our two transformed equations we get

$$\left\{\frac{\partial}{\partial t} + (u + a)\frac{\partial}{\partial x}\right\}\left\{\tfrac{1}{2}u + \frac{a}{\gamma - 1}\right\} = 0.$$

And subtracting gives

$$\left\{\frac{\partial}{\partial t} + (u - a)\frac{\partial}{\partial x}\right\}\left\{\tfrac{1}{2}u - \frac{a}{\gamma - 1}\right\} = 0.$$

Both these equations are of the form

'derivative of combination $= 0$',

and consequently can be integrated to give

'combination $=$ constant on a curve'.

We have arrived at an area known as the 'theory of characteristics', and we need to investigate it for a while before going on with the fluid dynamics.

(b) Derivatives along curves

We start this diversion on characteristics by supposing that we have a function

$$f(x, t)$$

defined in the x, t plane (or some region of it), and also a curve C in the plane which is given in terms of some parameter η by

$$x = \xi(\eta), t = \tau(\eta).$$

For example, a parabola in the x, t plane can be given by

$$x = A\eta^2, t = 2A\eta,$$

because on elimination you get

$$t^2 = 4Ax.$$

This curve, with values of η, is sketched in fig. XIV.5. The parameter η *could* be the length along the curve, but it doesn't have to be, and isn't in the example given.

Suppose we are interested in how $f(x, t)$ behaves on the curve C; substitute the forms for x and t which define C and we get

$$f(\xi(\eta), \tau(\eta)) = F(\eta) \text{ say.}$$

Fig. XIV.5. The parabola $x = A\eta^2$, $t = 2A\eta$.

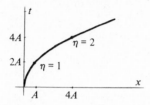

For example, take the curve to be the parabola above, and take

$$f(x, t) = e^{x+ct}$$

where c is a constant. Then

$$\begin{aligned}
f(x, t)_{\text{on } C} &= f(A\eta^2, 2A\eta) \\
&= \exp(A\eta^2 + 2Ac\eta) \\
&= F(\eta).
\end{aligned}$$

Since F depends only on η, we may ask what the derivative $dF/d\eta$ is. In the example it is just

$$(2A\eta + 2Ac)\exp(A\eta^2 + 2Ac\eta);$$

in general it is given by the 'chain rule'

$$\frac{dF}{d\eta} = \frac{\partial f}{\partial x}\frac{dx}{d\eta} + \frac{\partial f}{\partial t}\frac{dt}{d\eta}$$

(where $dx/d\eta$ really means $d\xi/d\eta$, but using $dx/d\eta$ gives a more memorable formula). In the example

$$\begin{cases}
\partial f/\partial x = e^{x+ct}, \\
\partial f/\partial t = ce^{x+ct}, \\
dx/d\eta = 2A\eta, \\
dt/d\eta = 2A,
\end{cases}$$

confirming the earlier calculation of $F'(\eta)$.

Now suppose we had been given the equation

$$t\,\partial f/\partial x + 2A\,\partial f/\partial t = 0.$$

If we rewrite this as

$$2A\eta\,\partial f/\partial x + 2A\,\partial f/\partial t = 0$$

it is just in the form

$$\frac{\partial f}{\partial x}\frac{dx}{d\eta} + \frac{\partial f}{\partial t}\frac{dt}{d\eta} = 0$$

where η is the parameter on the parabola. That is, the equation may be put as

$$dF/d\eta = 0 \text{ along the parabola.}$$

This is easily solved. The solution is just

$$F = \text{constant along the parabola,}$$

and if you know the value of f (or F) at one point of the parabola, you know it at all points.

This preliminary work and example show us how to approach an equation like

$$A(x, t)\,\partial f/\partial x + B(x, t)\partial f/\partial t = 0.$$

We seek to write it as

$$\frac{dx}{d\eta}\frac{\partial f}{\partial x} + \frac{dt}{d\eta}\frac{\partial f}{\partial t} = 0$$

by looking for x and t such that

$$\begin{cases} dx/d\eta = A(x, t), \\ dt/d\eta = B(x, t). \end{cases}$$

These equations define a family of curves C, the 'characteristics' of the equation, for which

$$dt/dx = B(x, t)/A(x, t),$$

and on these characteristic curves we must have

$$dF/d\eta = 0,$$

or

$$F = \text{constant.}$$

So the solution of

$$A(x, t)\,\partial f/\partial x + B(x, t)\partial f/\partial t = 0$$

is that

$$\begin{cases} f(x, t) = \text{constant, on the characteristics given by} \\ dt/dx = B(x, t)/A(x, t). \end{cases}$$

(c) *An example*

Let us use these ideas to solve the equation

$$e^t\,\partial f/\partial x + \partial f/\partial t = 0 \quad \text{for } t > 0$$

with

$$f(x, 0) = \cos \tfrac{1}{2}\pi x \quad \text{for } x \geqslant 0$$

being given as an initial condition on f.

First we find the characteristics, given by

$$dt/dx = 1/e^t;$$

this equation separates as

$$e^t dt = dx,$$

and so has solution

$$e^t + c = x.$$

The characteristics are the family of curves

$$x = e^t + c$$

or

$$t = \ln(x - c)$$

for different values of the constant c, which are shown in fig. XIV.6.

Now from the theory we have just done,

$$f(x, t)$$

is constant along each curve of the family. Each curve cuts the x-axis just once and 'picks up' a value of f there, given by

$$f(x, 0) = \cos \tfrac{1}{2}\pi x$$

and this is the value of f all the way along this characteristic. For example, the curve that passes through $x = 3$ and $t = 0$ is the curve of the family with

$$c = 2,$$

that is

$$t = \ln(x - 2).$$

Fig. XIV.6. The characteristics $t = \ln(x - c)$ for five values of c.

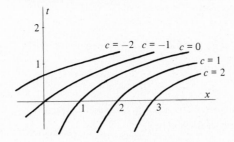

Fig. XIV.7. Constant values of *f* along the sample characteristics.

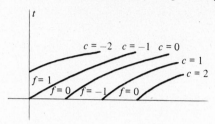

Now at $x = 3$ the boundary condition is

$$f(3, 0) = \cos \tfrac{3}{2}\pi$$
$$= 0.$$

Hence $f = 0$ all the way along the curve

$$t = \ln(x - 2),$$

as is shown in fig. XIV.7.

So we see that *f* is constant whenever we are on a characteristic curve, and in this example that means whenever

$$x - e^t$$

is constant, as the characteristics are just

$$x - e^t = c.$$

Hence the function $f(x, t)$ can only have the form

$$f(x, t) = F(x - e^t).$$

(An alternative derivation of this formula is set as an exercise.) But we know that

$$f(x, 0) = \cos \tfrac{1}{2}\pi x,$$

and so

$$F(x - 1) = \cos \tfrac{1}{2}\pi x.$$

This equation shows that

$$F(x) = \cos \tfrac{1}{2}\pi(x + 1),$$

by a simple change of variable. Finally we substitute $x - e^t$ for x to get

$$F(x - e^t) = \cos \tfrac{1}{2}\pi(x - e^t + 1)$$
$$= f(x, t)$$

as the solution of the equation. You may easily verify that it does satisfy the equation, and also the initial condition.

(d) *Initial and other conditions*

You should have noticed that there is no value of f marked in the last diagram on the characteristic with $c = -2$. This is because this characteristic does not cut the positive x-axis, where the condition on f is given. The problem as set up does not determine $f(x, t)$ in the whole quadrant $x \geqslant 0 \; t \geqslant 0$ but only in the 'range of influence' of the positive x-axis, the set of points that characteristics through the positive x-axis pass through.

What is happening here is that the initial information, say at x_0, moves on to larger values of x at later times t; say x_1 at time t_1, as in fig. XIV.8. This is exactly the same sort of behaviour as you would get with the propagation of sound, though here the speed dx/dt is not a constant as it was in Chapter XII.

Now suppose that the initial value that was given for $f(x, 0)$ had been discontinuous, say

$$\begin{cases} f(x, 0) = 0 & \text{for } 0 < x < x_0, \\ f(x, 0) = 1 & \text{for } x > x_0. \end{cases}$$

Then the points on the x-axis to the left of x_0 will give

$$f(x, t) = 0 \text{ to the left of the characteristic through } x_0$$

(but $f(x, t)$ is still undefined to the left of the characteristic through the origin) and

$$f(x, t) = 1 \text{ to the right of the characteristic through } x_0.$$

Fig. XIV.8. Information from x_0 at $t = 0$ reaches x_1 at $t = t_1$ along the characteristic.

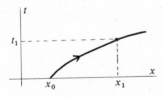

Fig. XIV.9. A solution discontinuous across a characteristic.

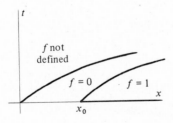

This behaviour is sketched in fig. XIV.9, where it is clear that the solution $f(x, t)$ can have discontinuities across characteristics. This property is important in the following work, where we find discontinuities occurring in the flow.

A discontinuity in a sound field can happen rather easily. When we start some sound, then this change from silence to sound spreads out at the speed of sound as a sharp charge. A person at a distance hears just the same sudden start to the sound, though at a later time, as the person who starts the sound.

A further property of the characteristics in this problem is that they do not cross. It would be most unfortunate if they did, because at the crossing point each characteristic would bring a different value of $f(x, t)$; and clearly $f(x, t)$ cannot have two different values at the same point in any real problem.

This example has been based on an initial value of $f(x, t)$, given on the positive x-axis. But we could equally well think of a value being given for $f(x, t)$ on the line

$$x = 0,$$

say for $t > 0$. Then the characteristics pick up values on the t-axis (as shown in fig. XIV.10), and as before $f(x, t)$ is constant along these characteristics. Naturally there is now a different range of influence for this different boundary on which $f(x, t)$ is given.

Fig. XIV.10. Characteristics from conditions given on $x = 0$.

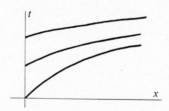

Fig. XIV.11. Characteristics from a quarter circle on which values of f are given.

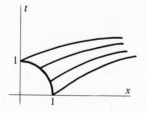

Fig. XIV.12. Characteristics from a hyperbola on which values of f are given.

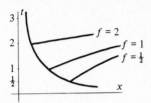

The curve on which $f(x, t)$ is given need not be one of the axes. The only requirement is that it cuts each characteristic only once – or else we are likely to get two (or more) values of $f(x, t)$ picked up by the characteristic. For example, if $f(x, t)$ has given values on the quarter circle $x^2 + t^2 = 1$ (shown in fig. XIV.11) then $f(x, t)$ is known on all the characteristics which cut this quadrant. Alternatively, if we let

$$f(x, t) = t$$

on the hyperbola

$$xt = 1,$$

then $f(x, t)$ is known at all points above and to the right of the curve, as is shown in fig. XIV.12.

3. The formation of discontinuities

(a) Characteristics of the flow equations

We return to the fluid dynamic equations in the forms involving u and a only:

$$\left\{\frac{\partial}{\partial t} + (u + a)\frac{\partial}{\partial x}\right\}\left\{\tfrac{1}{2}u + \frac{a}{\gamma - 1}\right\} = 0,$$

$$\left\{\frac{\partial}{\partial t} + (u - a)\frac{\partial}{\partial x}\right\}\left\{\tfrac{1}{2}u - \frac{a}{\gamma - 1}\right\} = 0.$$

It is convenient to define new unknown functions r and s instead of u and a, by

$$\begin{cases} r = \tfrac{1}{2}u + a/(\gamma - 1), \\ s = -\tfrac{1}{2}u + a/(\gamma - 1), \end{cases}$$

which are called the Riemann invariants. If you know r and s in the flow, then clearly you know u and a also; and from these you can calculate any of the other variables of state,

$$p, \rho, T, E$$

from the formulae in §1 in regions where the entropy S is constant.

Fig. XIV.13. A network of characteristics.

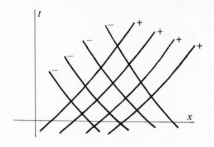

Now r and s satisfy simple first order equations of the type we have just discussed:

$$\begin{cases} \partial r/\partial t + (u + a)\partial r/\partial x = 0, \\ \partial s/\partial t + (u - a)\partial s/\partial x = 0. \end{cases}$$

The solutions are therefore

$$\begin{cases} r = \text{constant on the curves } dx/dt = u + a, \\ s = \text{constant on the curves } dx/dt = u - a. \end{cases}$$

The two sets of characteristic curves

$$\begin{cases} dx/dt = u + a \text{ (the positive characteristics)}, \\ dx/dt = u - a \text{ (the negative characteristics)} \end{cases}$$

will usually form a network in the x, t plane.

The $+$ curves in fig. XIV.13 have a positive slope, corresponding to

$$dx/dt = u + a,$$

while the $-$ curves will have a negative slope whenever a is greater than u. On each curve of one family, r is constant; and on each curve of the other, s is constant. At each intersection both r and s are given in terms of some boundary values, and hence u and a are known.

There is one drawback to this attractive solution of the equations: the characteristics are not known, and the equations

$$dx/dt = u \pm a$$

cannot be solved because we do not know $u(x, t)$ and $a(x, t)$. We can find a way round this difficulty however, by choosing to have rather simple initial (or boundary) conditions; we now do some examples of this sort, which reveal important effects quite easily.

(b) An example which leads to a discontinuity

Take, as a simple example, one dimensional flow which has the initial conditions, as shown in fig. XIV.14,

$$u(x, 0) = U_0\{1 - \tanh(x/l)\}.$$

This corresponds to air moving at a subsonic speed $2U_0$ into a region which is at rest, with the transition between the two taking a distance about $4l$ (because $\tanh \pm 2 = \pm 0.96$). Now we shall impose an initial value on $a(x, t)$; this is equivalent to giving an initial value for the pressure (or for the density, or for the temperature). We choose rather a special initial value

$$a(x, 0) = \tfrac{1}{2}(\gamma - 1)u(x, 0) + a_0,$$

because then we have

$$s(x, 0) = a_0/(\gamma - 1)$$

which is just a constant. The a_0 that we have included in $a(x, 0)$ is the speed of sound in air at rest; this means that $a(x, 0) \to a_0$ as x becomes large, where the air is at rest.

The other Riemann invariant has the initial value

$$r(x, 0) = \tfrac{1}{2}u(x, 0) + a(x, 0)/(\gamma - 1)$$
$$= U_0\{1 - \tanh(x/l)\} + a_0/(\gamma - 1).$$

Fig. XIV.14. A compressive initial velocity profile.

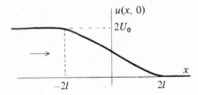

Fig. XIV.15. A rough sketch of the characteristics for §3(b).

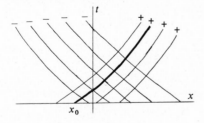

This has different constant values as $x \to \pm \infty$, with a transition region between.

Now consider the diagram of characteristics in fig. XIV.15. Since s is constant on the x-axis, and s is constant along each characteristic of the $-$ family, and since a member of this family passes through each point in the diagram (there is a solution to $dx/dt = u - a$ at every point), we must have

$$s = \text{constant} = a_0/(\gamma - 1)$$

at every point in the x, t diagram.

But we also have that r is a constant along any characteristic of the $+$ family, keeping the value $r(x_0, 0)$ that it picks up where it meets the x-axis. It follows at once from

$$u = r - s$$

that u has the value

$$r(x_0, 0) - a_0/(\gamma - 1) = u(x_0, 0)$$

all along this characteristic. And similarly, since

$$a = \tfrac{1}{2}(\gamma - 1)(r + s),$$

it follows that a has the value

$$\tfrac{1}{2}(\gamma - 1)u(x_0, 0) + a_0$$

all along this characteristic.

So along any $+$ characteristic both u and a are constants; but the characteristic curve is the solution of the equation

$$dx/dt = u + a,$$

and if $u + a$ is constant on the curve, then the curve must be a straight line. That is, the characteristics of the $+$ family have the equation

$$x = \{u(x_0, 0) + a(x_0, 0)\}t + x_0.$$

Fig. XIV.16. The $+$ characteristics for §3(*b*).

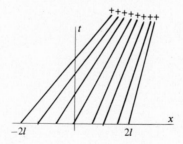

These are straight lines, whose gradients in fig. XIV.16 are

$$\{u(x_0, 0) + a(x_0, 0)\}^{-1}.$$

Since we have already found $a(x, 0)$ in terms of $u(x, 0)$ we may write the gradient as

$$\{\tfrac{1}{2}(\gamma + 1)u(x_0, 0) + a_0\}^{-1}.$$

When we use the given form for $u(x_0, 0)$, we see that the straight lines have gradient a_0^{-1} for $x > 2l$, and have gradient

$$\{(\gamma + 1)U_0 + a_0\}^{-1}$$

for $x < -2l$. Between $-2l$ and $+2l$ the slopes change smoothly from one value to the other.

Evidently these lines meet, as you see in fig. XIV.17. Those through $-2l$ and $+2l$ meet at (approximately)

$$t_s = \frac{4l}{(\gamma + 1)U_0}.$$

As we have seen in §2, characteristics must not meet, as then you get two values of each function at one point. Since there seems to be nothing wrong with the method of solution up as far as t_s, we must assume that something goes wrong after t_s. All that can happen is that the solution ceases to be continuous: the two values of each function coexist side by side along a discontinuity. For example, there are different values of $u(x, t)$

Fig. XIV.17. Characteristics from $-2l$ and $2l$ meet at t_s.

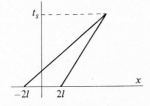

Fig. XIV.18. A shock wave forms from near t_s.

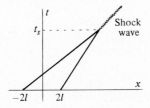

on the two sides of the discontinuity in the x, t diagram in fig. XIV.18. These discontinuities are known as 'shock waves', and we spend some time below in investigating how their two sides fit together.

(c) All compression waves steepen

Before we go on to discuss shock waves, it is worth examining the solution we have found in more detail.

We have found that both $u(x, t)$ and $a(x, t)$ are constant on the straight line characteristics

$$x - \{u(x_0, 0) + a(x_0, 0)\}t = x_0,$$

which in this example are

$$x - [\tfrac{1}{2}(\gamma + 1)U_0\{1 - \tanh(x_0/l)\} + a_0]t = x_0.$$

Call these characteristics

$$x - m_0 t = x_0$$

for simplicity. As before, since $u(x, t)$ is constant on the line

$$x - m_0 t = \text{constant},$$

we must have that

$$u(x, t) = F(x - m_0 t)$$

for some function F; and since $u(x, 0)$ is known, it can only be that

$$u(x, t) = u(x - m_0 t, 0)$$

so as to get the correct initial value. Hence in this example

$$u(x, t) = U_0\{1 - \tanh(x - m_0 t)/l\}$$

where

$$m_0 = \tfrac{1}{2}(\gamma + 1)U_0\{1 - \tanh(x_0/l)\} + a_0.$$

Note that $m_0 = u(x_0, 0) + a(x_0, 0) = u(x, t) + a(x, t)$ in terms of the co-ordinates of a point in the x, t diagram. Similarly

$$a(x, t) = \tfrac{1}{2}(\gamma - 1)U_0\{1 - \tanh(x - m_0 t)/l\} + a_0.$$

Both of these functions have the general form

$$f(x - m_0 t),$$

which is the form we have seen in Chapter XII for a wave travelling (to the right) at speed m_0. The speed of this wave is just

$$m_0 = u(x, t) + a(x, t),$$

in other words a speed $a(x, t)$ relative to the fluid which is moving at $u(x, t)$.

This provides a (partial) justification for calling $a(x, t)$ the speed of sound: at least in this example it provides the speed at which a disturbance is moving relative to the fluid.

Following up this point, it is easily seen that when U_0 is very small

$$a(x, t) \doteqdot a_0$$

everywhere in the flow, because tanh is a function bounded by ± 1. So the example degenerates into an example on sound waves travelling at speed a_0 when the disturbance is very small, and this is what we ought to expect, as then the linearisation done in Chapter XII must be valid. However, even when U_0 is very small, the characteristics eventually meet, though the time

$$t_s = 4l/\{(\gamma + 1)U_0\}$$

will be very large. The effect of non-linearities in sound waves must eventually be important, unless the sound waves have already been destroyed by viscous dissipation, heat conduction, or imperfect reflection. If we take typical values for U_0 and l in a sound wave from Chapter XII, we get the time scale t_s at which non-linear effects become important.

	l	$U_0(\text{m s}^{-1})$	t_s (s)	Frequency (Hz)
Weak sounds	$\begin{cases} 1.65 \text{ m} \\ 16.5 \text{ cm} \end{cases}$	3×10^{-6} 3×10^{-6}	9.2×10^5 9.2×10^4	200 2000
Loud sounds	$\begin{cases} 1.65 \text{ m} \\ 16.5 \text{ cm} \end{cases}$	3×10^{-4} 3×10^{-4}	9.2×10^3 9.2×10^2	200 2000

You can see that, even for a loud sound of short wavelength, a travel time of over 100 s is needed before non-linear effects become important. This justifies the neglect of them in the chapter on sound waves. It is only when U_0 reaches a noticeable fraction of the speed of sound that t_s reduces enough to give a discontinuity within a reasonable range of x. For example,

Fig. XIV.19. The width of the transition region reduces as t increases.

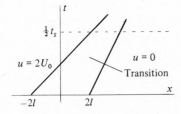

with $l = 16.5$ cm and $U_0 = 3$ m s^{-1}, the distance x_s at which the discontinuity forms is about 34 m.

The length of this 'compression wave' (it is a compression because faster air is meeting slower air) changes as t increases. At $t = 0$ the transition region occupies a length $2l$, but we can see from the diagram of characteristics in fig. XIV.19 that at later times the transition region is narrower. This result may also be disentangled from the formula for $u(x, t)$

$$U_0\{1 - \tanh(x - m_0 t)/l\}$$

but with more difficulty.

But it is clear from the characteristics diagram that at later times the profile for $u(x, t)$ still lies between

$$u = 2U_0$$

and

$$u = 0,$$

with the transition spreading over shorter and shorter regions. The gradient

$$\partial u/\partial x$$

thus gets steeper (as you see in fig. XIV.20) at the transition until finally a discontinuity as in fig. XIV.21 seems inevitable. In fact you do not get

Fig. XIV.20. Steepening of the gradient $\partial u/\partial x$ as t increases; $x_1 = \frac{1}{2}(U_0 + a_0)t_s$.

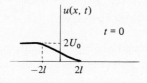

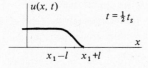

Fig. XIV.21. The expected discontinuity when $t = t_s$: $x_2 = (U_0 + a_0)t_s$.

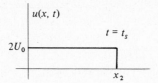

Fig. XIV.22. Initial velocity and characteristics for an expansion.

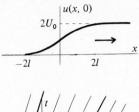

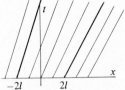

an actual discontinuity, because the model we have used of an inviscid, non-heat-conducting gas breaks down when the scale of the transition region approaches the mean free path of the molecules in the gas.

The work on this example has really been independent of the exact shape of the initial velocity distribution $u(x, 0)$. We have used a tanh profile as a representative function which changes from

$$u = 2U_0 \quad \text{for } x < -2l$$

to

$$u = 0 \quad \text{for } x > 2l.$$

However the conclusions about the steepening of the velocity gradient $\partial u/\partial x$ only require characteristics which are converging towards a point; and this just needs a compression wave.

If instead of having a compression wave you have an 'expansion wave', such as the one in fig. XIV.22, which could be modelled as

$$u(x, 0) = U_0\{1 + \tanh(x/l)\},$$

then similar arguments show that the characteristics diverge from each other, as shown in fig. XIV.22. This means that an expansion wave flattens rather than steepening, and no shock wave is formed from it.

(d) Motion of a piston in a tube

Our next example concerns the motion of a piston in a tube, and the consequent air motion. Assume that the piston and the air in the tube start at rest, and then at $t = 0$ the piston starts to accelerate, reaching a (subsonic) speed U_1 which is then maintained. A suitable graph for $U(t)$ is sketched in fig. XIV.23. For example, you could take the piston's speed

Fig. XIV.23. The velocity of the piston in the tube.

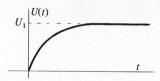

to be

$$U(t) = U_1 \tanh \alpha t.$$

The displacement of the piston for $t > 0$ is just

$$X(t) = \int_0^t U(\tau)d\tau,$$

and with the given $U(t)$ this is

$$X(t) = (U_1/\alpha)\ln(\cosh \alpha t).$$

When we draw the x, t diagram (as in fig. XIV.24), we have initial conditions along $t = 0$, that $u(x, 0) = 0$ and $a(x, 0) = a_0$; we also have the boundary condition that $u = U(t)$ on $x = X(t)$. There will clearly be a compression wave between the piston and the $+$ characteristic through the origin; and to the right of this characteristic the gas will be at rest, as the wave has not reached this region.

Along the x-axis we have

$$s = -\tfrac{1}{2}u + a/(\gamma - 1)$$
$$= a_0/(\gamma - 1),$$

a constant. As before, this means that s must have this constant value everywhere, because a $-$ characteristic from the x-axis passes through all points of the region we are considering. On the piston we know that

$$\begin{cases} u = U, \\ s = -\tfrac{1}{2}u + a/(\gamma - 1) = a_0/(\gamma - 1). \end{cases}$$

Fig. XIV.24. The piston path $x = X(t)$ and the $+$ characteristic through the origin.

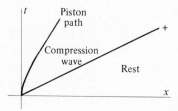

Fig. XIV.25. Converging characteristics from the piston give a shock.

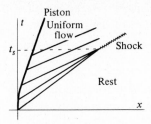

Hence on the piston

$$a = a_0 + \tfrac{1}{2}(\gamma - 1)U.$$

This exceeds a_0, as it must to give

$$p > p_0$$

at the piston, and so a force opposing the motion of the piston. A knowledge of both u and a at the piston enables us to find r there:

$$r = \tfrac{1}{2}u + a/(\gamma - 1),$$
$$= U + a_0/(\gamma - 1).$$

As before, since r is constant on the $+$ family of characteristics and s is constant everywhere, we must have both u and a constant on the $+$ characteristics, which must be straight lines in the x, t diagram drawn in fig. XIV.25. The slope of the characteristics is

$$(a + u)^{-1},$$

and on the piston this is

$$\{a_0 + \tfrac{1}{2}(\gamma + 1)U\}^{-1}.$$

These characteristics are bound to converge, and a shock wave must again form at some time t_s which could be calculated from the form of $U(t)$.

It should be noticed in the last x, t diagram that conditions between the piston and the shock are uniform. The characteristics are parallel straight lines because the piston stops accelerating, and then both U and a (and so r) are constants on the piston. This suggests that the conditions in the

Fig. XIV.26. Two uniform regions separated by a shock.

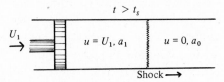

tube at this stage are just two uniform regions separated by a discontinuity which travels at some as yet unknown speed into the region of rest. The speed U_1 in fig. XIV.26 is the final speed of the piston, and a_1 is given by

$$a_1 = a_0 + \tfrac{1}{2}(\gamma - 1)U_1.$$

We must next investigate what the speed of the shock is, and we do this by examining what is conserved across such a discontinuity.

4. Plane shock waves

(a) Discontinuity at a shock wave

The mathematical solutions in the last section have suggested that surfaces of discontinuity will be expected to occur in some circumstances. Such shock waves (not a very good name, but generally accepted) are indeed found to occur widely in high speed flows of gases. They are normally very thin regions of rapid change of the thermodynamic variables and the velocity rather than actual discontinuities, but we shall model them as having no thickness. In reality the thickness is some three or four mean free paths, say 5×10^{-7} m; this is so small that the continuum model does not obviously apply (it can be used because, as in sound theory, a reasonable area of shock wave contains a large number of molecules), and we shall certainly not investigate in detail the structure of a shock wave. But it is well worth having a few thoughts on what can and what cannot be discontinuous across such a 'discontinuity' before we tackle the mathematics.

A shock wave is not a source of mass; no molecules are destroyed in it. Hence there is continuity of mass flow at a shock wave. And because the discontinuity is an isolated region of fluid, there can be no loss of momentum in the shock wave, from Newton's third law of motion; the rate of flow of apparent fluid momentum through the shock need not be continuous if the pressure changes across it, but the sum of pressure and momentum flow rate must be constant. Finally, there is no addition of energy from external sources at a shock – a shock is adiabatic.

The only thing that is left to be discontinuous is temperature, or its equivalent, internal energy. The shock wave is a small region where organised (average) velocity of the fluid is partially converted into random molecular motion, or heat. Thus a shock wave is a region where energy is converted from one form to another, with no external interference. This is a process which can only go one way, from organised motion to random motion, which we will see below is in accordance with the form of the second law of thermodynamics which says that entropy in an isolated system can only increase.

This is a situation in which the words 'adiabatic' and 'isentropic' must be used properly. The change is an irreversible one, so we have

$$dS > \Delta Q/T,$$

and hence $dS > 0$ even though $\Delta Q = 0$. The situation is adiabatic and irreversible, and not isentropic.

(b) *Changes in the variables across a shock*

Let us, as usual, do the easy problem. Consider a shock wave which is at rest, and take one-dimensional steady flow. The flow incident on the shock has variables u_1, p_1, ρ_1, a_1 and so on; beyond the shock the variables are u_2, p_2, ρ_2, a_2 (as in fig. XIV.27).

Our first equation represents the constancy of mass flow rate:

$$\rho_1 u_1 = \rho_2 u_2 \ldots (A)$$

The second equation reflects the fact that the shock is not a source of momentum, and so the net force (per area) balances the change of momentum flow rate:

$$p_1 - p_2 = \rho_2 u_2^2 - \rho_1 u_1^2 \ldots (B)$$

Now we only need three equations to give downstream conditions in terms of upstream ones, because we only need to know u_2 and any two of the thermodynamic variables. The third equation expresses the constancy of total energy through the shock, because none is added from outside. That is, total energy is constant on a streamline, which is just a way of saying that Bernoulli's equation for steady flow still applies, despite the fact that viscosity and heat conduction are important in the shock itself. This may be proved by taking a full form of the equation of motion from Chapter VIII §3 and rederiving Bernoulli's integral; we shall be content to use the result. So conservation of total energy along a streamline means that

$$\tfrac{1}{2}u_1^2 + a_1^2/(\gamma - 1) = \tfrac{1}{2}u_2^2 + a_2^2/(\gamma - 1) \ldots (C)$$
$$= H, \text{ say.}$$

Fig. XIV.27. Definition diagram for §4(b).

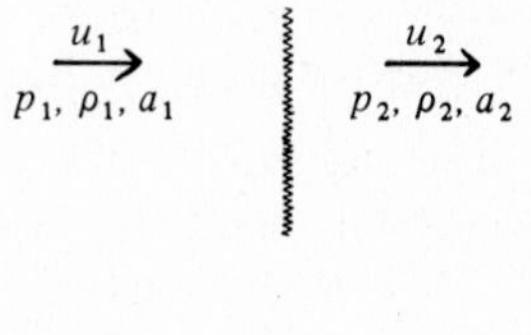

These equations are known as the Rankine–Hugoniot equations, and we now solve them to get downstream conditions in terms of upstream. It is convenient to put most of the results in terms of

$$M_1 = u_1/a_1,$$

the upstream Mach number.

We start with equation (B), and divide each term by either $\rho_1 u_1$ or $\rho_2 u_2$ – they are equal by (A), so it does not matter which we use. This gives

$$u_2 - u_1 = p_1/(\rho_1 u_1) - p_2/(\rho_2 u_2).$$

Now $\gamma p/\rho = a^2$, and so we may rewrite this as

$$u_2 - u_1 = a_1^2/(\gamma u_1) - a_2^2/(\gamma u_2).$$

Next we take (C) in the form

$$a_1^2/(\gamma u_1) = (\gamma - 1)(H - \tfrac{1}{2}u_1^2)/(\gamma u_1),$$
$$a_2^2/(\gamma u_2) = (\gamma - 1)(H - \tfrac{1}{2}u_2^2)/(\gamma u_2).$$

This leaves us with

$$u_2 - u_1 = (\gamma - 1)\{H(u_2 - u_1)/u_1 u_2 + \tfrac{1}{2}(u_2 - u_1)\}/\gamma.$$

Since were are assuming that there *is* a discontinuity of some sort, we take $u_2 \neq u_1$, and so we have derived

$$\gamma/(\gamma - 1) = H/(u_1 u_2) + \tfrac{1}{2}$$

or

$$u_1 u_2 = 2H(\gamma - 1)/(\gamma + 1).$$

This intermediate result enables us to calculate u_2/u_1 quite easily. Begin with

$$u_2/u_1 = u_1 u_2/u_1^2$$
$$= \frac{2(\gamma - 1)H}{(\gamma + 1)u_1^2}.$$

Now use (C) in the form

$$\tfrac{1}{2} + M_1^{-2}/(\gamma - 1) = H/u_1^2$$

to finish with the formula for u_2/u_1 in terms of M_1:

$$\frac{u_2}{u_1} = \frac{(\gamma - 1)M_1^2 + 2}{(\gamma + 1)M_1^2}.$$

The value of γ for air is almost exactly 7/5, and the graph in fig. XIV.28 plots the function

$$\frac{u_2}{u_1} = \frac{M_1^2 + 5}{6M_1^2}.$$

Fig. XIV.28. Variation of u_2/u_1 with M_1 across a shock.

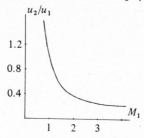

From our discussion, we expect u_1 to be greater than u_2, because upstream kinetic energy is converted into heat in the shock. So only the part of the graph with $u_2/u_1 < 1$ is relevant, and this is the part with $M_1 > 1$. The upstream flow leading into a stationary shock must be supersonic, and the flow out of the shock must have

$$u_2/u_1 > 1/6,$$

since this is the value reached as $M_1 \to \infty$.

Return next to equation (A): this may be written as

$$\rho_2/\rho_1 = u_1/u_2$$
$$= \frac{(\gamma + 1)M_1^2}{(\gamma - 1)M_1^2 + 2}.$$

which means that $\rho_2 > \rho_1$ since we have just decided that $u_1 > u_2$ or $M_1 > 1$. That is, a shock compresses the air that passes through it. Notice that there is a rise of density

$$\rho_2/\rho_1 \to 6$$

as $M_1 \to \infty$.

The pressure must rise through a shock to balance the decrease in momentum flow rate. This is shown in equation (B), by writing it as

$$p_2 - p_1 = \rho_1 u_1^2 - \rho_2 u_2^2$$
$$= m(u_1 - u_2),$$

where m is the mass flow rate $\rho_1 u_1$ or $\rho_2 u_2$. It is straightforward to show that

$$p_2/p_1 = 1 + \frac{2\gamma}{\gamma + 1}(M_1^2 - 1).$$

The pressure rise through a shock has no bound; as M_1 increases, so does p_2/p_1. It is for this reason that

$$(p_2 - p_1)/p_1 = 2\gamma(M_1^2 - 1)/(\gamma + 1)$$

is sometimes called the 'shock strength'.

Equation (C) may be written as

$$a_2^2 = a_1^2 + \tfrac{1}{2}(\gamma - 1)(u_1^2 - u_2^2),$$

which immediately shows that $a_2 > a_1$. The relation between the Mach numbers before and after the shock follows by dividing by u_2^2:

$$M_2^{-2} = M_1^{-2}(u_1^2/u_2^2) + \tfrac{1}{2}(\gamma - 1)\{(u_1^2/u_2^2) - 1\}.$$

From the properties of u_2/u_1 which we have derived above it is not hard to show that

  (i) $M_2 = 1$ when $M_1 = 1$,

  (ii) M_2 decreases as M_1 increases,

  (iii) $M_2^{-2} \to \tfrac{1}{2}(\gamma - 1)\left\{\left(\dfrac{\gamma+1}{\gamma-1}\right)^2 - 1\right\}$ as $M_1 \to \infty$.

Thus the flow downstream of a stationary shock is always subsonic, and there is a limit to the reduction of the Mach number.

The temperature rises across a shock, because mean flow energy has been converted to random molecular energy or heat. The ratio of the temperatures is easily calculated from the equation of state

$$p = R\rho T.$$

Similarly the entropy change across the shock can be calculated from

$$S = c_v \ln(p/\rho^\gamma)$$

for a thermally perfect gas (see Chapter VII). The formula is

$$S_2 - S_1 = c_v \ln\{1 + 2\gamma(M_1^2 - 1)/(\gamma + 1)\}$$
$$- \gamma c_v \ln\{(\gamma + 1)M_1^2/[(\gamma - 1)M_1^2 + 2]\}.$$

The analysis of this formula is tedious; the results are that

  (i) $S_2 = S_1$ when $M_1 = 1$,

  (ii) $S_2 > S_1$ when $M_1 > 1$,

  (iii) $S_2 - S_1 \to \infty$ as $M_1 \to \infty$,

  (iv) $S_2 < S_1$ if $M_1 < 1$.

The last of these provides the mathematical justification for our assumption on physical grounds that $u_2 < u_1$; if $M_1 < 1$ and there is a shock,

Fig. XIV.29. Type of change across a stationary shock.

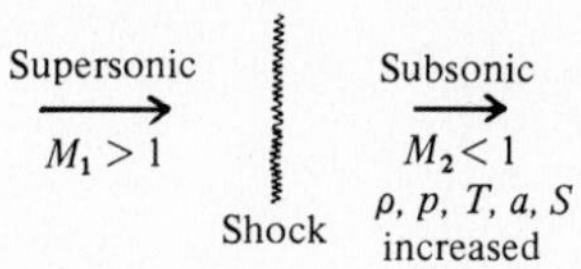

Supersonic	Shock	Subsonic
$\longrightarrow$		$\longrightarrow$
$M_1 > 1$		$M_2 < 1$
		ρ, p, T, a, S increased

then entropy decreases, which contradicts the second law of thermodynamics.

Fig. XIV.29 summarises the results we have found for a stationary shock wave in one-dimensional steady flow.

(c) *Speed of a moving shock*

In §3 above we traced the development of a shock from a compression wave, and the resulting shock which separated two regions of uniform flow was not stationary, but moved into the region of rest, as shown in fig. XIV.30. There is no great difficulty in deriving the shock speed V in this case. We take a frame of reference fixed in the shock; with respect to this the conditions are just

$$V, a_0$$

upstream of the shock and

$$V - U_1, a_1$$

downstream of it, as in fig. XIV.31.

Now we know from §3(*d*) that

$$a_1 = a_0 + \tfrac{1}{2}(\gamma - 1)U_1$$

for the region between the piston and the shock when it has become uniform. So we need to have a shock which produces a transition between

$$V, a_0 \quad \text{and} \quad V - U_1, a_0 + \tfrac{1}{2}(\gamma - 1)U_1.$$

But equation (*C*) can now be used to relate the two velocities and sound speeds:

$$\{a_0 + \tfrac{1}{2}(\gamma - 1)U_1\}^2 = a_0^2 + \tfrac{1}{2}(\gamma - 1)\{V^2 - (V - U_1)^2\}.$$

Fig. XIV.30. A shock advancing at speed V into still air.

$$U_1 \xrightarrow{} \qquad \qquad u = 0$$
$$a_1 \qquad \qquad \qquad a_0$$
$$V \longrightarrow$$

Fig. XIV.31. The equivalent stationary shock.

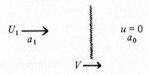

$$\xleftarrow{} V - U_1 \qquad \xleftarrow{} V$$
$$a_1 \qquad \qquad a_0$$

Shock
at rest

This equation simplifies to

$$V = a_0 + \tfrac{1}{4}(\gamma + 1)U_1.$$

Clearly with this value the upstream flow is supersonic with respect to the shock, since the relative speed is V; and the downstream flow is subsonic with respect to the shock since

$$\frac{V - U_1}{a_1} = \frac{a_0 - \tfrac{1}{4}(3 - \gamma)U_1}{a_0 + \tfrac{1}{2}(\gamma - 1)U_1} < 1.$$

It is interesting to notice that the shock speed is just the average of the disturbance speeds on each side:

$$\begin{aligned} V &= \tfrac{1}{2}a_0 + \tfrac{1}{2}(U_1 + a_1) \\ &= \tfrac{1}{2}a_0 + \tfrac{1}{2}\{U_1 + a_0 + \tfrac{1}{2}(\gamma - 1)U_1\} \\ &= a_0 + \tfrac{1}{4}(\gamma + 1)U_1. \end{aligned}$$

(d) Flow in a pipe with a contraction

A shock wave provides a transition from supersonic to subsonic flow. It is interesting to see how we may generate a supersonic air flow in the first place, other than by moving a body at high speed through mainly stationary air.

We shall consider air flowing steadily along a pipe whose cross-sectional area A changes slowly with x (as in fig. XIV.32), so that the flow can be taken to be one-dimensional. Assuming that the flow is isentropic away from shocks, we have the set of equations

$$\begin{cases} \rho u A = \text{constant}, \\ dp/d\rho = a^2, \\ \rho u\, du/dx = -\, dp/dx. \end{cases}$$

The first equation represents the conservation of mass, and the third is Euler's equation in the absence of body forces. Notice that in the second equation we must have $a(x)$. What we shall do is eliminate p and ρ from these three equations to finish with an equation relating u, a, and A.

The first equation may be differentiated and then written as

$$\rho^{-1}dp/dx + u^{-1}du/dx + A^{-1}dA/dx = 0.$$

Fig. XIV.32. Flow along a pipe of slowly changing area.

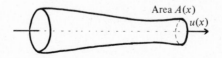

Area $A(x)$

$u(x)$

And replacing dp/dx first by $a^2 d\rho/dx$ and then by

$$- a^2\rho(u^{-1}du/dx + A^{-1}dA/dx)$$

gives a revised form of Euler's equation:

$$\rho u\, du/dx = \rho a^2(u^{-1}du/dx + A^{-1}dA/dx),$$

or

$$A^{-1}dA/dx = u^{-1}du/dx\{(u^2/a^2) - 1\}.$$

This formula shows first of all that the local velocity can only equal the speed of sound when $dA/dx = 0$: sonic velocity of the fluid requires what is known as a 'throat', where A reaches a minimum value – the proof that it is a minimum is omitted. But it is not true that

$$dA/dx = 0$$

implies

$$u = a.$$

It is quite possible for the fluid to accelerate into the throat, and then reach a subsonic maximum speed, and decelerate out of the throat. To avoid this you need to have a large pressure difference, so that the fluid accelerates up to sonic speed at the throat; then it is possible for the fluid to go on accelerating to supersonic speed. We need never have $du/dx = 0$ if $u = a$ when $dA/dx = 0$. This would seem to be such a special case that it would never arise in practice, but in fact it does, and so gives quite an easy way of generating supersonic flow. We end this section by describing, without too many calculations, how this flow of air from subsonic to supersonic speed depends on the pressure difference between the two ends.

Let us look at flow from a reservoir at a pressure p_1 through a slow contraction and out into the atmosphere which is at pressure p_e; the situation is sketched in fig. XIV.33. We start with the case when p_1 is not much larger than the exit pressure p_e. Naturally when $p_1 = p_e$ there is no flow, and when p_1 exceeds p_e only slightly there is certainly a smooth

Fig. XIV.33. Flow from a reservoir through a throat and into the atmosphere.

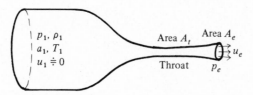

flow out of this 'Laval nozzle'. We may calculate the exit density from the equation

$$p_1 \rho_1^{-\gamma} = p_e \rho_e^{-\gamma}$$

because the flow is effectively isentropic and the pressure p_e is almost continuous through the exit jet. Then we may use Bernoulli's equation for the flow, which is almost steady if A_t is small compared to the reservoir cross-section, to work out the exit velocity u_e:

$$a_1^2/(\gamma - 1) = \tfrac{1}{2}u_e^2 + a_e^2/(\gamma - 1)$$

where

$$a_e^2 = \gamma p_e/\rho_e.$$

Conditions at the throat may now be deduced from the conservation of mass

$$\rho_t u_t A_t = \rho_e u_e A_e$$

and Bernoulli's equation

$$\tfrac{1}{2}u_t^2 + a_t^2/(\gamma - 1) = a_1^2/(\gamma - 1);$$

it looks as though there are not enough equations to deduce all of

$$\rho_t, u_t, a_t,$$

but $a_t^2 = \gamma p_t/\rho_t$ and $p_t \rho_t^{-\gamma} = p_1 \rho_1^{-\gamma}$ relate ρ_t and a_t. It is assumed that the areas A_t and A_e are known from the construction of the nozzle. The calculations needed to find u_t and so forth are lengthy, and it is sufficient to say that when p_1/p_e is not large enough, u_t is subsonic and the air accelerates into the throat, reaches a maximum speed, and then decelerates out of it. The pressure in the air similarly has a minimum at the throat, as in fig. XIV.34.

As p_1 increases, so the speed u_t increases until finally the flow is almost sonic at the throat,

$$u_t \doteqdot a_t.$$

This occurs at initial pressure p_s, say, which we could calculate from the

Fig. XIV.34. Variation of p with x when the reservoir pressure slightly exceeds the exit pressure.

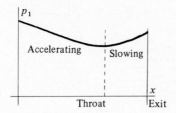

Fig. XIV.35. Pressure changes in two isentropic flows. The upper curve is sonic at the throat, the lower is just subsonic there.

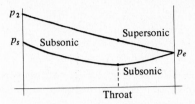

Fig. XIV.36. For intermediate reservoir pressures there is no isentropic flow and a shock forms.

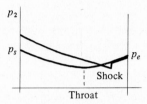

equations above; at this pressure we still get an acceleration into the throat and a retardation out of it. It would also seem possible, at a slightly greater pressure, to have a continued acceleration all along the nozzle; but isentropic flow (which we have assumed) would need to start at a considerably larger pressure (say p_2 as shown in fig. XIV.35) so as to reach p_e at the exit.

Now suppose we increase the pressure to a *little* above p_s; there can be no isentropic flow with this pressure difference. The flow cannot be sonic anywhere except at the throat, as we have seen above, and isentropic flow with sonic conditions at the throat can only start at p_s or p_2. The adjustment is achieved by means of a shock wave (as in fig. XIV.36), which occurs inside the nozzle if p_1 is slightly more than p_s, and at the exit if p_1 is nearer to (but still below) p_2.

When $p_1 = p_2$ there is no need for any shock to make an adjustment, and the flow reaches the exit travelling supersonically. And finally when $p_1 > p_2$, the one-dimensionality of the flow breaks down at the exit. But we shall not proceed to investigate such two-dimensional flows; the extension to two dimensions would fill another chapter.

Exercises

1. Show that in an isentropic flow

$$a^2 = \gamma R T.$$

Fig. XIV.37. Sketch for Q1.

A jet of air (sketched in fig. XIV.37) of speed $0.9\,a_0$ at atmospheric pressure p_0 has a stagnation point at A where the temperature is 400 K. Use Bernoulli's equation to determine the value of the variables

$a, T, \rho,$

at the jet exit. What is the local Mach number at the exit?

2. In isentropic flow show that

$\rho^{-1}\partial p/\partial x = 2a(\gamma - 1)^{-1}\partial a/\partial x,$

and hence that in one-dimensional flow

$\partial u/\partial t + u\partial u/\partial x + \{2a/(\gamma - 1)\}\partial a/\partial x = 0$

approximately. If the flow is irrotational with potential $\phi(x, t)$, find an equation for a in terms of ϕ.

3. Bernoulli's equation shows that there is a maximum possible speed q_{max} on a streamline in isentropic flow, at a point where $a = p = 0$. Calculate this speed for Q1 above, and also the speed where the local Mach number has value one.

4. The (specific) enthalpy I is defined by

$I = E + p/\rho$

where E is the internal energy of the gas. Show that

$dI = TdS + dp/\rho.$

Hence show that in steady flow

$\mathbf{v} \times \boldsymbol{\omega} = \nabla H - T\nabla S$

where

$H = I + \tfrac{1}{2}\mathbf{v}^2$

This is known as Crocco's vortex theorem.

5. Simple steady flows other than $\mathbf{v} = v(x)\mathbf{i}$ can be rediscussed with compressibility taken into account. For example, the Rankine vortex with

$\mathbf{v} = v(r)\hat{\boldsymbol{\theta}}$

(in cylindrical coordinates), where

$\begin{cases} \nabla \times \mathbf{v} = \Omega\mathbf{k} \text{ and } T = T_1 \text{ for } r < a, \\ \nabla \times \mathbf{v} = 0 \text{ and isentropic for } r > a. \end{cases}$

Discuss whether these could be reasonable modelling assumptions. For this model, show that, for $r > a$,

$$dp/dr = \rho K^2/r^3$$

for some constant K. Show also for $r < a$ that

$$RT_1 \, d\rho/dr = \tfrac{1}{4}\Omega^2 r\rho.$$

Find the central density in terms of Ω, T_1 and conditions at a large distance from the vortex.

6. The equation

$$A(x, y)\partial f/\partial x + B(x, y)\partial f/\partial y = 0$$

may be written as

$$\mathbf{A} \cdot \nabla f = 0.$$

Use vector methods to show that f is constant on curves which everywhere have the direction of the vector $\mathbf{A}$.

7. Make the change of variables

$$\begin{cases} \xi = x - e^t, \\ \eta = x + e^t, \end{cases}$$

in the equation

$$e^t \partial f/\partial x + \partial f/\partial t = 0$$

to show that the solution is

$$f(x, t) = F(x - e^t).$$

8. Derive the equations of the characteristics of the differential equation

$$t\partial f/\partial x + x^3 \partial f/\partial t = 0$$

for the quadrant $x > 0$, $t > 0$. Hence sketch the characteristics and solve the initial value problem

$$f(x, 0) = e^{-x}.$$

What is the range of influence in this problem?

9. Consider the initial value problem for one-dimensional subsonic gas flow in which

$$u(x, 0) = \begin{cases} U(1 + x/l) & \text{for } -l < x < 0, \\ U(1 - x/l) & \text{for } 0 < x < l, \\ 0 & \text{for } |x| > l, \end{cases}$$

and in which $a(x, 0)$ is such that

$$s(x, 0) = a_0/(\gamma - 1)$$

for all x. Show that the characteristics are all straight, and sketch them. Draw graphs for $u(x, t)$ for

$$t = 0, \tfrac{1}{3}t_s, \tfrac{2}{3}t_s, t_s,$$

where t_s is the time at which characteristics meet.

10. In a 'shock tube' a diaphragm at $x = 0$ initially separates two regions of gas at rest. For $x > 0$ there is low pressure and low sound speed a_0; for $x < 0$ there is high pressure and high sound speed a_1. At $t = 0$ the diaphragm is destroyed. Draw characteristics to show the various regions of flow produced. [You will find that you need an expansion region moving into $x < 0$.]

11. Derive formulae for p_2/p_1 and a_2/a_1 across a stationary normal shock, in terms of the upstream Mach number M_1.

12. A weak shock has shock strength

    $(p_2 - p_1)/p_1 = \varepsilon,$

    where ε is small and positive. Find approximations for M_1^2, u_2/u_1, and $S_2 - S_1$ in terms of ε.

13. Air flows isentropically through the Laval nozzel of §4(d), and at $x = 0$ there is a minimum of $A(x)$ with

    $(d^2 A/dx^2)_0 > 0.$

    Show that either
    (i) $u = a$ at $x = 0$ with both du/dx and dM/dx non-zero there, or
    (ii) $u \neq a$ at $x = 0$ with u having a maximum value there.

14. For isentropic flow down a Laval nozzle show that

    $dp/dx = -\rho a^2 M^2 (M^2 - 1)^{-1} A^{-1} dA/dx.$

    Deduce that $dp/dx < 0$ when $M = 1$ at the throat.

References

High speed flows of gases have been intensively investigated since the jet engine made high speed flight possible. Many engineering texts on aerodynamics naturally have a section on this area; and also some texts which are more concerned with the mathematics.

(a) A text that is at the right level for alternative reading is *Textbook of Fluid Dynamics*, F. Chorlton, Van Nostrand 1967. This has a long chapter on gas dynamics.

(b) Typical of the engineering texts at a reasonable level is *Mechanics of Fluids* (3rd edn), B. S. Massey, Van Nostrand Reinhold 1975.

(c) A fuller text, though still in one volume is, *Elements of Gas Dynamics*, H. W. Liepmann and A. Roshko, Wiley 1957.

(d) There is a full length film on *Channel Flow of a Compressible Fluid*, which illustrates some of the material of this chapter, and there is a film loop FM-129 *Compressible flow through a convergent–divergent nozzle*, which illustrates §4(d).

XV
Steady surface waves in channels

1. One-dimensional approximation

(a) General discussion

In Chapter XIII we examined surface waves of small amplitude, mainly on deep water. There are other changes in water surface level that can occur which are markedly influenced by the non-linear terms which were discarded in that chapter, and which also tend to be associated with rather shallow water, in the sense that the length scale of the surface disturbance is not small compared to the depth.

The disturbances in question are not always 'waves', as they may well be stationary phenomena on the surface of a moving stream or river; for example the water flow over an irregularity in the bed of a stream often produces a stationary disturbance on the surface and there is no obvious progress to be seen. And secondly there may be no periodicity – in this respect (and in others which we shall see later) some of these disturbances are very like the shock waves of the last chapter.

The sort of examples we have in mind in this chapter are rivers, streams, and the channels associated with civil engineering works like dams, weirs and drainage channels. We cannot at this level take into account too much of the reality of these situations; the aim is, as always, to get an understanding at a basic level of some of the major effects. For example, we shall not discuss the very important frictional force between a river and its bed, which is largely associated with the turbulent eddying motion in a

Fig. XV.1. Definition sketch for §1(*a*).

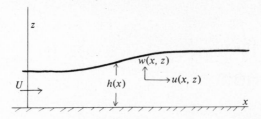

real river. We shall assume that the flows are smooth, and that viscosity can be neglected almost everywhere – the interesting and realistic predictions of the simplified theory are a good justification for doing this.

It is easier to consider steady flows and fixed wave patterns, and enough can be done in this way to make this a sensible plan. So we assume in most of what follows that we have a steady uniform flow U from the left; the flow elsewhere will be taken to be two-dimensional, with velocity

$$\mathbf{v} = u(x, z)\mathbf{i} + w(x, z)\mathbf{k},$$

under the surface shown in fig. XV.1, whose equation is

$$z = h(x).$$

Usually the bed will be flat and at $z = 0$; the slope in real rivers produces an accelerating force due to gravity which is balanced by the retarding effects of frictional forces.

Once we have completed the analysis of a steady flow pattern, we can always change the frame of reference to one moving at some speed V so that the surface shape moves at speed V and the incident stream has speed $U - V$. This technique was used in the last chapter to convert a treatment of stationary shock waves to give results for moving shocks.

In much of this chapter we shall neglect the vertical velocity $w(x, z)$. There are two reasons why this can be done without much loss of accuracy; firstly, we may choose to compare uniform upstream conditions (where $w = 0$) with uniform but different downstream conditions, where $w = 0$ also; and secondly, the surface slopes may be so small that w/u must be small everywhere. This second reason is related to what we have seen in Chapter XIII §6(*b*) on linear surface waves – where $\lambda \gg h$, the motions are very nearly horizontal, so it seems reasonable as a first try to neglect vertical velocities and accelerations. However, we shall see at the end of this chapter that for some disturbances we must keep in an estimate of the vertical velocities.

The assumption of a uniform flow $U\mathbf{i}$ upstream does not look very realistic. Even the average flow (averaging out the turbulent motions) will have to vary somewhat with z; it turns out that this variation is not too large, and anyway we can largely overcome what variation there is by using integrals over z from 0 to $h(x)$. In this chapter we shall do the easiest thing, which is to treat a uniform flow, and appeal to the reality of the results as justification.

(b) Flow over a smooth hump

As a first example we consider the flow of a stream of depth H over an obstruction on the bed. The obstruction is taken to be of smooth shape so that we can assume that the flow does not separate, and we shall also assume that there is no train of waves set up on the surface downstream of the obstacle. This is a reasonable assumption for a well-submerged smooth obstacle, and the situation is sketched in fig. XV.2. Assume that a stream of speed U passes over an obstacle whose height above the bed is given by

$$z = Z(x),$$

where $Z \ll H$ and $Z \ll L$, the length of the obstacle. The stream is assumed to resume the values U and H downstream, and over the obstacle we have speed $u(x)$ and surface height $H - d(x)$ where the depression of the surface $d \ll H$.

The equation of mass conservation is taken in the form of a constant flow rate, so that

$$UH = u(x)[H - d(x) - Z(x)].$$

This assumes that the width of the stream is a constant, independent of both x and z; this will do as a first model. The other equation for the system may be taken to be Bernoulli's equation on the surface streamline

$$\tfrac{1}{2}U^2 + gH + p_0/\rho = \tfrac{1}{2}u^2 + g(H - d) + p_0/\rho,$$

Fig. XV.2. Definition sketch for smooth flow over a hump. The surface is depressed by $d(x)$.

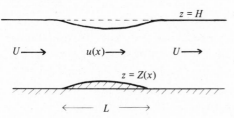

or

$$u^2 = U^2 + 2gd.$$

This is clearly just an energy equation for a fluid particle moving along the surface. Now eliminate $u(x)$:

$$U^2 H^2/(H - d - Z)^2 = U^2 + 2gd.$$

This equation for $d(x)$ may be approximated, since d and Z are both small compared to H, as

$$1 + 2(d + Z)/H = 1 + 2gd/U^2$$

or

$$d = Z/(gH/U^2 - 1).$$

This rather simple result determines the depression $d(x)$ in terms of the obstacle shape $Z(x)$. The important parameter in the result is

$$gH/U^2,$$

because this determines whether d has the same sign as Z or the opposite sign. It is conventional to call

$$U/(gH)^{1/2}$$

the Froude number F (or sometimes F is defined as U^2/gH). Thus we have

$$d(x) = Z(x)/(F^{-2} - 1),$$

so that for a slow stream ($F < 1$) there is a depression of the surface over an obstacle on the bottom; while for a fast stream ($F > 1$) the surface rises over an obstacle.

The Froude number F plays rather the same role in the present theory as the Mach number M did in the last chapter. The reason for this is that

$$gH$$

is the square of the speed of small disturbances of long wavelength on shallow water, so that

$$F = U/c$$

just as

$$M = U/a,$$

but with a different wave speed in the two cases. It is usual to call a stream with $F > 1$ 'supercritical' and one with $F < 1$ 'subcritical'; this is just like supersonic and subsonic.

It is worth noting that it is not nearly so hard to achieve $F = 1$ as it is to reach $M = 1$.

			U (m s^{-1})		
		0.1	0.5	1.0	5.0	10.0
	0.5	$F = 0.04$	0.2	0.5	2.2	4.5
H (m)	1.0	0.03	0.2	0.3	1.6	3.2
	2.0	0.02	0.1	0.2	1.1	2.3
	4.0	0.02	0.08	0.2	0.8	1.6

The critical speed for a depth of 1 m is

$$U = 3.1 \, \text{m s}^{-1},$$

and the flow due to a boat travelling faster than this speed will differ greatly from that for a slower boat. In fact the resistance is generally much less for U just greater than the critical speed than for U just less, but this is only useful on rather shallow water such as that in canals.

For the problem that we started off with, there are no downstream waves generated when $F > 1$, but for $F < 1$ there may be such waves. The calculation is complicated, and our present results are reasonably good for shallow obstacles in deep streams. This kind of result is easily observed in a swimming pool: a swimmer well below the surface creates no surface waves (check that he is travelling at a speed such that $F < 1$), but if he is near the surface the waves are obvious.

(c) *Flow of a stream through a constriction*

As our next example we take a stream flowing through a narrow region; when a river passes under a road, the piers of the bridge cause a considerable narrowing of the channel, and it is useful to know if any rise in height of the water surface can be expected locally, when the river is near to flooding.

We model the situation by the simplest possible velocity again,

$$\mathbf{v} = u(x)\mathbf{i},$$

neglecting both vertical and cross-stream velocities. For this to be reasonable we must take a gentle narrowing of the river channel as in fig. XV.3, which may not be too realistic for some bridges. So we model the geo-

Fig. XV.3. Plan view of flow through a narrow region.

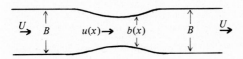

metry of the river by vertical side walls whose distance apart is $b(x)$, where

$$\begin{cases} b(x) \to B \text{ as } x \to \pm \infty, \\ db/dx \text{ is small.} \end{cases}$$

It is again convenient to use the conservation of mass equation in the form

$$u(x)b(x)h(x) = \text{constant},$$

where $h(x)$ is the local depth of water; and Bernoulli's equation for the surface streamline in the form

$$\tfrac{1}{2}u^2 + gh = \text{constant}.$$

But we now proceed much as we did in the last chapter when discussing the high speed flow of a gas through a contraction, by taking differentials:

$$\begin{cases} bhdu + uhdb + ubdh = 0, \\ udu + gdh = 0. \end{cases}$$

When we eliminate du and re-arrange, we find

$$dh(1 - gh/u^2) = - (h/b)db.$$

If we define u^2/gh to be the *local* Froude number squared, say f^2, we have

$$dh(1 - f^{-2}) = - (h/b)db.$$

This shows that in a flow which is everywhere *sub*critical (i.e. $f < 1$ throughout), then dh and db have the same sign, and a narrowing leads to a shallowing of the river. And since

$$du = - (g/u)dh,$$

this leads to an increase of the velocity and an increase of the Froude number f.

This suggests that, if we choose the initial conditions correctly, then we can send a subcritical flow into the contraction, reach $f = 1$ exactly when the width is least, and come out with a supercritical flow. This would be a rather good analogy with the flow of a gas from subsonic to supersonic through a contraction, which we saw in the last chapter.

Let us calculate the contraction that will be needed when the upstream Froude number is F. Suppose the contraction needed is from B to $(1 - \alpha)B$, and that the depth and speed are h_1 and u_1 at this contraction. Then the previous equations give

$$\begin{cases} u_1 h_1(1 - \alpha) = UH & \text{(mass)}, \\ u_1^2 + 2gh_1 = U^2 + 2gH & \text{(Bernoulli)}; \end{cases}$$

but we are now also asking to have $f = 1$ at the contraction, and so

$$u_1^2/(gh_1) = 1.$$

Elimination of u_1 from Bernoulli's equation gives

$$h_1 = \tfrac{1}{3}H(F^2 + 2);$$

and substitution in the mass equation gives

$$\tfrac{1}{3}u_1(1 - \alpha)(F^2 + 2) = U.$$

The upstream Froude number therefore satisfies

$$F^2 = U^2/(gH)$$
$$= \{u_1^2/(gh_1)\} \times \{(1 - \alpha)^2(F^2 + 2)^3/27\}.$$

So, since $u_1^2/(gh_1) = 1$ has been assumed, the relation between the contraction α and the upstream Froude number is

$$(1 - \alpha)^2 = 27F^2/(F^2 + 2)^3.$$

It is not hard to see that when F is small, then α is near $1 - a$ very large contraction is needed to accelerate a well subcritical stream. But when F is just less than 1, then α is almost zero – an almost critical stream does not need much contraction to accelerate it to a critical value.

It is worth checking that the downstream Froude number may indeed exceed 1 when the flow is subcritical upstream and the contraction has been arranged to give critical conditions at the minimum width. Let the upstream and downstream Froude numbers be F_0 and F_1 respectively, with

$$0 < F_0 < 1.$$

The calculation above shows that

$$(1 - \alpha)^2 = 27F_0^2/(F_0^2 + 2)^3$$

and that

$$0 < (1 - \alpha)^2 < 1.$$

But a similar calculation using downstream conditions must give

$$(1 - \alpha)^2 = 27F_1^2/(F_1^2 + 2)^3.$$

We must show that there is a solution of this which has $F_1 > 1$. This is not hard to do graphically: put $(1 - \alpha)^2 = 1/k$ and $F_1^2 = \Phi$ for convenience, which reduces the equation to

$$(\Phi + 2)^3 = 27k\Phi,$$

where $k > 1$. The graph of the cubic polynomial $(\Phi + 2)^3$ in fig. XV.4 cuts

Fig. XV.4. Graphical solution of the equation for F_1.

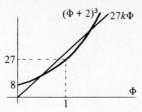

Fig. XV.5. Top and side views of a contraction which leads to supercritical flow. Return along the dashed curve to subcritical flow does not occur.

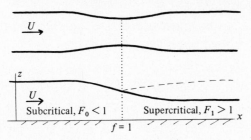

the linear graph of $27k\Phi$ at a point $\Phi > 1$ because

$$(\Phi + 2)^3 \to \infty$$

faster than $27k\Phi$ as Φ increases, and at $\Phi = 1$ the linear graph is above the cubic graph.

So we have shown that a smooth transition from subcritical to super-critical as in fig. XV.5 is possible by means of a contraction. In the last chapter we found that a similar transition from subsonic to supersonic was possible, but that unless downstream conditions were very carefully adjusted, there needed to be a shock wave to make a match possible. We must now see what plays the role of a shock wave in channel flows.

2. Hydraulic jumps or bores

(a) Equations for unsteady, one-dimensional flow

We shall consider more fully what the main effects of the non-linear terms on a surface disturbance on a shallow stream are by approximating the governing equation. The equation of mass conservation can be conveniently taken in an integrated form. From fig. XV.6 the volume in the fixed section between x and $x + dx$ is (per unit width)

$$h\,dx$$

Fig. XV.6. Definition sketch for §2(a).

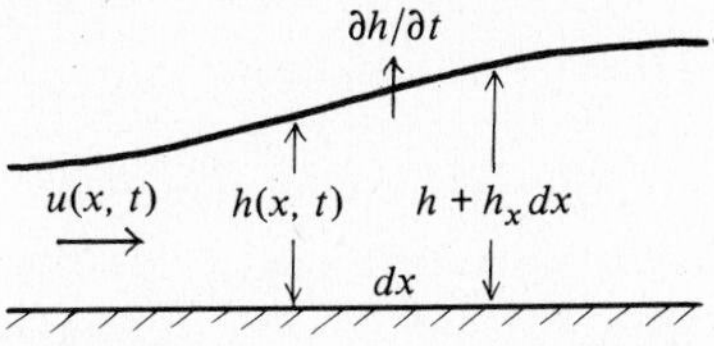

to first order in dx. This increases at rate

$(\partial h/\partial t)dx$.

But the net rate of inflow through the ends is

$uh - (u + u_x dx)(h + h_x dx)$.

Since volume is conserved (because water is incompressible, or almost so), these two terms must balance, to give (in the limit as $dx \to 0$)

$$\frac{\partial h}{\partial t} + u\frac{\partial h}{\partial x} + h\frac{\partial u}{\partial x} = 0.$$

This equation for $h(x, t)$ may equally well be derived by integrating $\nabla \cdot \mathbf{v} = 0$ from the bottom to the surface $z = h(x, t)$ and using the full surface condition from Chapter XIII §2(a) together with $u(x, t)$ being independent of z: this is left as an exercise.

The vertical component of the equation of motion is

$$\frac{\partial w}{\partial t} + u\frac{\partial w}{\partial x} + w\frac{\partial w}{\partial z} = -\frac{1}{\rho}\frac{\partial p}{\partial z} - g.$$

This will be approximated by omitting the vertical accelerations, to give

$\partial p/\partial z + \rho g = 0$.

Hence we take, on integrating from the bottom to the surface,

$p = p_0 + \rho g[h(x, t) - z]$

as the pressure at height z in the water, which is just the hydrostatic pressure.

The horizontal equation of motion is

$$\frac{\partial u}{\partial t} + u\frac{\partial u}{\partial x} + w\frac{\partial u}{\partial z} = -\frac{1}{\rho}\frac{\partial p}{\partial x}.$$

The term in $\partial u/\partial z$ is taken to be zero as u has been assumed to be independent of z (and anyway w is small). Now $\partial p/\partial x$ can be calculated from our result for p:

$\partial p/\partial x = \rho g \partial h/\partial x$.

Consequently we take as our second equation for $h(x, t)$ and $u(x, t)$

$$\frac{\partial u}{\partial t} + u \frac{\partial u}{\partial x} + g \frac{\partial h}{\partial x} = 0.$$

The approximations we have made in deriving these two equations have been two. Vertical accelerations have been neglected in deriving the pressure to be approximately hydrostatic, and the velocity u has been taken to be independent of z. The errors that have been made can be estimated (we shall not do so), and the formulation is sufficiently accurate when

$$(h/l)^2 \ll 1,$$

where l is the length scale over which h and u change appreciably. A proper derivation of these equations as the first terms in expansions in powers of h/l is beyond the present work.

(b) An example using characteristics

The equations we have found,

$$\begin{cases} \dfrac{\partial h}{\partial t} + u \dfrac{\partial h}{\partial x} + h \dfrac{\partial u}{\partial x} = 0, \\[2mm] \dfrac{\partial u}{\partial t} + u \dfrac{\partial u}{\partial x} + g \dfrac{\partial h}{\partial x} = 0, \end{cases}$$

are rather like those that we found in the last chapter for non-linear disturbances in a gas. As in that chapter we may put them into characteristic form by taking suitable combinations of them; the combinations again work best when we take the local wave speed for long waves on shallow water

$$c = (gh)^{1/2}$$

as one of the variables, instead of h.

In terms of u and c the equations are

$$\begin{cases} 2\dfrac{\partial c}{\partial t} + 2u \dfrac{\partial c}{\partial x} + c \dfrac{\partial u}{\partial x} = 0, \\[2mm] \dfrac{\partial u}{\partial t} + u \dfrac{\partial u}{\partial x} + 2c \dfrac{\partial c}{\partial x} = 0. \end{cases}$$

Simple combinations now give

$$\left\{ \frac{\partial}{\partial t} + (u + c) \frac{\partial}{\partial x} \right\} (u + 2c) = 0$$

(addition), and

$$\left\{\frac{\partial}{\partial t} + (u - c)\frac{\partial}{\partial x}\right\}(u - 2c) = 0$$

(subtraction). The differences from the equations for high speed gas flow are slight; the two speeds of propagation are $u \pm c$ as before, but the new Riemann invariants are

$$\begin{cases} r = u + 2c, \\ s = u - 2c. \end{cases}$$

The change from a factor $\pm 2/(\gamma - 1)$ to ± 2 in the Riemann invariants will make only slight changes to the algebra, so we expect to get similar results in this theory to those we derived for the high speed flow of gases. Now in the theory of the last chapter we found that shock waves were inevitably set up from any smooth compressive initial disturbance. So we must suppose that a corresponding discontinuity will be predicted in the present theory. Let us follow an example through in detail.

Take the initial profile and velocity in fig. XV.7, where $H \ll L$ and where

$$U = \tfrac{1}{3}(gH)^{1/2} = \tfrac{1}{3}c_0$$

for convenience. Then for $x < 0$ we have

$$\begin{cases} r = U + 2c = 7c_0/3, \\ s = U - 2c = -5c_0/3. \end{cases}$$

and for $x > L$ we have

$$\begin{cases} r = 2(4gH/9)^{1/2} = 4c_0/3, \\ s = -2(4gH/9)^{1/2} = -4c_0/3. \end{cases}$$

The numbers chosen here have no particular significance; they just illustrate the algebra with an easy example.

Now the slopes of the characteristic curves are

$$u \pm c,$$

Fig. XV.7. Definition sketch for the example in §2(b).

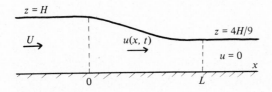

Fig. XV.8. Characteristics for the example in §2(*b*).

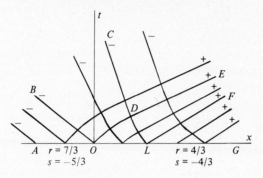

and it is easily seen that

$$\begin{cases} u + c = \tfrac{3}{4}(u + 2c) + \tfrac{1}{4}(u - 2c) \\ \qquad = \tfrac{3}{4}r + \tfrac{1}{4}s \\ u - c = \tfrac{1}{4}r + \tfrac{3}{4}s. \end{cases}$$

We use the given values of r and s on the initial line, and assume a smooth transition between $x = 0$ and $x = L$; this allows us to calculate slopes for characteristic curves and get the diagram in fig. XV.8, where c_0 is taken to have value 1 for convenience.

Start with the $+$ characteristic through L; on this r has the constant value 4/3, and each $-$ characteristic that meets it comes from LG, bringing the value $s = -4/3$. Hence on this $+$ characteristic LF we have

$$u + c = \tfrac{3}{4}r + \tfrac{1}{4}s = 2/3.$$

Now LF is defined to be the curve with

$$dx/dt = u + c$$

and so it is the straight line

$$dx/dt = 2/3$$

which has gradient 3/2 in the diagram. Hence LF, and similarly for all $+$ characteristics to its right, is straight and of gradient 3/2.

Next consider the $+$ characteristic *ODE*. At O we have

$$r = 7/3 \quad \text{and} \quad s = -5/3$$

and so

$$u + c = 4/3.$$

That means that the gradient at O of this characteristic is 3/4. Now all along *ODE* we have $r = 7/3$, but the value of s brought up by the − characteristics changes; for example at D we have

$$\dot{s} = -4/3$$

brought up from L. Hence at D

$$r = 7/3 \quad \text{and} \quad s = -4/3$$

so that

$$u + c = 17/12$$

and the slope of the characteristic is $12/17$. Along DE the value of s remains at $-4/3$, because $s = -4/3$ along LG. So the slope of the characteristic DE is constant at $12/17$. That is, DE is straight.

All the other gradients in fig. XV.8 can be calculated in a similar fashion from the values of r and s along the x-axis. This is left as an exercise. What we need here is that the gradient of the straight characteristic DE is $12/17$, which is less than the gradient of the straight characteristic LF, which is $3/2$. Hence these characteristics must meet as in fig. XV.9; in the last chapter the problem of having two values for r at this point of meeting P was resolved by having a discontinuity between two uniform regions of flow there. In the present context this discontinuity is no longer called a shock wave, but a 'hydraulic jump' or 'bore'. Often the term 'bore' is used

Fig. XV.9. Intersecting characteristics lead to the formation of a hydraulic jump.

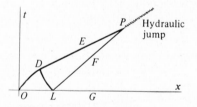

Fig. XV.10. Energy loss at a bore goes into turbulence or waves.

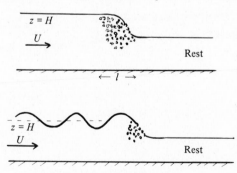

for a moving discontinuity, and 'hydraulic jump' for a stationary one. The hydraulic jump or bore has exactly the same dynamic function as a shock wave; it converts kinetic energy in the incident flow into some other form. In the shock wave the energy went into random motion of the molecules, in other words heat. In the bore the energy can be transformed into a local random motion of the water on a small scale, in other words turbulence; but in some circumstances part or all of the energy goes into generating surface waves, giving an 'undular bore'. Fig. XV.10 shows these possibilities.

(c) *Formation of a tidal bore*

Before we go on to discuss the conditions across a hydraulic jump in much the same way as we did for conditions across a shock, we should ask whether the equations we started from are indeed a good model of reality.

For some initial conditions it is true that an initial wave of elevation will steepen and eventually form a hydraulic jump or bore; this will happen when the non-linearity is quite strong. But if the initial wave of elevation is not very marked, then other terms in the full equations have an effect and a 'moving equilibrium' state may be reached which propagates unchanged in shape. We shall return to discuss this in more detail below. Another reason why a bore may not form is that friction (due, for example, to turbulence or obstructions) may be relatively important; this can remove energy at such a rate that no discontinuity needs to be formed. Friction is important in, for example, flood waves travelling down rivers and in the formation of bores due to tides at river estuaries; in these two cases the change of width of the river with depth, and the change of width with distance along the river can also be important. It is clear that a full treatment of waves in rivers will be rather too hard to do here, but it is interesting to see, in terms of our equations, why a bore can be expected to form on a river which has a large tidal range at its mouth.

A crude model of such a river is a two-dimensional channel of depth H far upstream, with the water at rest there. At the river mouth the tide forces the depth to change from H_0 to H_1 over a time T; let the mouth be at $x = 0$ and the depth there be

$$h(0, t) = f(t).$$

Then on $x = 0$ in an (x, t) diagram we know that

$$c = (gh)^{1/2}$$
$$= \{gf(t)\}^{1/2}.$$

Fig. XV.11. Variation of depth at a river mouth leads to converging characteristics and a bore.

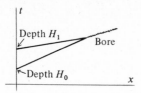

Now characteristics of the $-$ family all set off from far upstream with

$$s = 0 - 2(gH)^{1/2},$$

and bring this value to $x = 0$. But since

$$r = s + 4c,$$

we must have

$$r = 4\{gf(t)\}^{1/2} - 2(gH)^{1/2}$$

on $x = 0$. This value of r is propagated along the characteristics of the $+$ family. Take, for example, the characteristic setting off from $x = 0$ at $t = t_1$; this has

$$\begin{cases} r = 4\{gf(t_1)\}^{1/2} - 2(gH)^{1/2}, \\ s = -2(gH)^{1/2} \end{cases}$$

all along it, and so it has slope

$$\begin{aligned} u + c &= \tfrac{1}{4}(3r + s) \\ &= 3\{gf(t_1)\}^{1/2} - 2(gH)^{1/2}. \end{aligned}$$

The slope varies according to what value of t_1 you choose, and in particular the slope on the x, t diagram shown in fig. XV.11,

$$(u + c)^{-1},$$

decreases since $f(t)$ increases from H_0 to H_1.

This therefore gives converging characteristics and a bore from near their point of intersection.

This crude model predicts a bore on every river on every tide. This does not happen in practice, for the reasons discussed above. For the River Severn where a bore does occur when the tidal range is high, both friction and river narrowing have to be included to get an adequate model.

3. Changes across a hydraulic jump

(a) Transition due to loss of energy

In this section we consider abrupt changes in depth of a stream, over a length scale comparable to the depth, for which the differential

Fig. XV.12. Definition diagram for flow through a hydraulic jump.

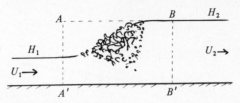

equations of the last section are not good approximations. And to make the analysis as simple as possible, we choose to consider steady flows again.

The flow in fig. XV.12 represents a uniform stream whose depth suddenly increases, through a hydraulic jump, from H_1 to H_2. The flow is uniform again after a short distance downstream of the jump, with speed U_2. Conservation of mass gives

$$U_1 H_1 = U_2 H_2,$$

and if $A'B'$ is short enough for frictional forces to be small we may balance forces with momentum flow rates for AA' and BB'. The force acting on AA' is just due to the pressure, which is hydrostatic because the flow is uniform. Hence the force on AA' is (per width of channel)

$$p_0(H_2 - H_1) + \int_{z=0}^{H_1} \{p_0 + \rho g(H_1 - z)\} dz$$
$$= p_0 H_2 + \tfrac{1}{2}\rho g H_1^2.$$

Similarly the force on BB' is

$$p_0 H_2 + \tfrac{1}{2}\rho g H_2^2.$$

The momentum flow rates are, since the flow is independent of z,

$$\rho U_1^2 H_1 \text{ through } AA',$$

and

$$\rho U_2^2 H_2 \text{ through } BB'.$$

So, equating net force to change of momentum flow rate, we find

$$\tfrac{1}{2}g H_1^2 + U_1^2 H_1 = \tfrac{1}{2}g H_2^2 + U_2^2 H_2.$$

If we eliminate U_2 between this equation and the mass flow equation, we obtain

$$U_1^2 = \tfrac{1}{2}g(H_1 + H_2)H_2/H_1,$$

or

$$F_1^2 = \tfrac{1}{2}(H_1 + H_2)H_2/H_1^2.$$

This shows immediately that if $H_2 > H_1$, as we have assumed in fig. XV.12, then

$$F_1^2 > 1.$$

A similar elimination of U_1 leads to

$$F_2^2 = \tfrac{1}{2}(H_1 + H_2)H_1/H_2^2$$

and so in the present case

$$F_2^2 < 1.$$

The transition through a hydraulic jump is from supercritical (and shallow) to subcritical (and deep). Compare this with the transition from supersonic to subsonic through a shock wave in a high speed gas flow.

As has been said above, a hydraulic jump dissipates energy into turbulence, from which it is eventually converted into heat. The rate of loss of energy in this example is not hard to calculate. The rate of working at AA' is (per unit width)

$$\int_0^{H_1} pU_1 dz,$$

and the rate of convection of (kinetic plus potential) energy through AA' is

$$\int_0^{H_1} U_1(\tfrac{1}{2}\rho U_1^2 + \rho gz)dz.$$

These integrals are easily evaluated for a uniform stream, with p being hydrostatic. They give a total contribution to energy flow

$$p_0 U_1 H_1 + \tfrac{1}{2}\rho H_1 U_1^3 + \rho g H_1^2 U_1.$$

Hence (using a similar expression for the line BB') the rate of loss of energy between AA' and BB' is

$$U_1 H_1 \{p_0 + \tfrac{1}{2}\rho U_1^2 + \rho g H_1\} - U_2 H_2 \{p_0 + \tfrac{1}{2}\rho U_2^2 + \rho g H_2\}.$$

But we know that

$$U_1 H_1 = U_2 H_2$$

from mass conservation, so the rate of (mechanical) energy loss is

$$\rho U_1 H_1 \{\tfrac{1}{2}U_1^2 + g H_1 - \tfrac{1}{2}U_2^2 - g H_2\}.$$

We may now use the previous equations to put this mainly in terms of H_1 and H_2; the energy loss rate is

$$\tfrac{1}{4}\rho U_1 g(H_2 - H_1)^3/H_2.$$

This shows conclusively that this transition must be from lower H_1 to

Fig. XV.13. Flow down a slope into a hydraulic jump.

higher H_2, as the energy loss can only be positive. This relation is rather like the change of entropy through a shock wave, which can only be positive, and which for a *weak* shock is proportional to the cube of the shock strength. But the analogy is not a complete one.

The above analysis agrees well with experiments on strong hydraulic jumps for which H_2 is considerably greater than H_1. For weaker jumps, energy losses into waves or against frictional resistance may be considerable.

The typical ways in which such a transition may occur in practice are:
  (i) water may accelerate (as in fig. XV.13) down a slope to a super-critical speed into a region which is deeper and almost at rest;
  (ii) as in §2 above an initial wave may develop into a bore, and *relative* to the bore there is a steady supercritical inflow.

(b) Transition due to loss of momentum

For the hydraulic jump in (*a*) above there was a loss of energy, while the mass flow and the momentum flow were conserved. It is possible, by inserting an obstacle in the stream, to reduce the momentum flow, while preserving energy flow and mass flow. In fig. XV.14 a force T is needed to hold the sluice gate in position, and this force must come into the equation for force and momentum flux. There is no reason for any loss of energy (other than a small amount to friction on the walls, bottom and sluice gate) so the equations we use are

$$U_1 H_1 = U_2 H_2$$

Fig. XV.14. A force T holding a sluice gate in a stream.

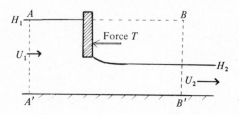

for mass conservation, and

$$\tfrac{1}{2}U_1^2 + gH_1 = \tfrac{1}{2}U_2^2 + gH_2$$

for energy conservation. This is just equivalent to Bernoulli's equation along the surface streamline.

These two equations may be put in terms of the two Froude numbers by a little manipulation. The result is (much as in §1(c))

$$(F_1^2 + 2)^3/F_1^2 = (F_2^2 + 2)^3/F_2^2.$$

This equation may be solved for F_2 in terms of F_1, but it is easier to sketch a graph, and this time we choose the graph of $(\phi + 2)^3/\phi$, sketched in fig. XV.15. Clearly the graph tends to infinity as $\phi \to 0$ and as $\phi \to \infty$, and there is a local minimum when $\phi = 1$. It follows that there is a solution for F_2 that is greater than 1 when F_1 is less than 1. That is, a subcritical flow can be converted to a supercritical flow by this sluice gate arrangement.

The value of the force T is derived by writing down a force and rate of flow of momentum equation. We obtain

$$\tfrac{1}{2}\rho g H_1^2 + \rho H_1 U_1^2 - \tfrac{1}{2}\rho g H_2^2 - \rho H_2 U_2^2 = T.$$

This may be put in terms of the upstream conditions by using the mass conservation equation and Bernoulli's equation. The result is

$$T = \tfrac{1}{2}\rho g H_1^2 \{1 + 5F_1^2/2 - \tfrac{1}{8}F_1^4 - \tfrac{1}{8}F_1(F_1^2 + 8)^{3/2}\}.$$

Clearly $T = 0$ when $F_1 = 1$, because then $F_2 = 1$ also and there is no transition. And when $F_1 \to 0$ the sluice gate cuts off all flow, leaving the hydrostatic result

$$T = \tfrac{1}{2}\rho g H_1^2.$$

This attractively simple theory unfortunately does not tell the whole story. As in the transition from supercritical to subcritical, it is possible to get a wave downstream instead of a clean jump to a supercritical stream (as in fig. XV.16). But the theory described above can be a reasonably good

Fig. XV.15. Graphical solution for F_2 in terms of F_1.

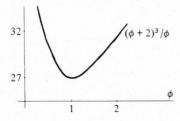

Fig. XV.16. A wave downstream of a sluice gate.

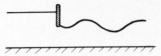

Fig. XV.17. Possible and probable results from an obstacle.

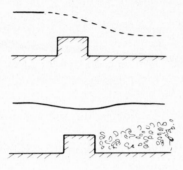

picture for strong transitions, i.e. from F_1 considerably less than 1 to F_2 clearly above 1.

It might seem that *any* obstacle in a uniform stream would lead to a force, and hence to a loss of momentum in the stream, and so a transition from subcritical to supercritical. But though such an obstacle as that shown in fig. XV.17 will certainly give a force, it will also create a turbulent wake which destroys energy of the incident stream. So the end result is quite likely to be decreases in both energy and momentum, and no transition. A sluice gate does not create much turbulence, because the separation takes place at the free surface.

4. Solitary waves

(a) An important parameter

In practice it is not always true that a wave of elevation moving into still water will automatically steepen into a bore. It is found that the behaviour of waves on shallow water is determined by the dimensionless number

$$a\lambda^2/h^3$$

where a is the wave amplitude, λ is a length scale for the disturbance (a 'wavelength') and h is the depth. This parameter can be shown to be the correct one by a careful study of the full equations, but we shall not go that far. The various behaviours are as follows.

 (i) Suppose $a/h \ll 1$ and $\lambda/h < 1$. Then we have small-amplitude surface waves, treated in Chapter XIII §§2–4. This is the case $a\lambda^2/h^3 \ll 1$.

 (ii) Suppose $a/h \ll 1$ and λ/h is moderately large, say 10 or 20. This gives linear wave theory on shallow water, mentioned in Chapter XIII §6(*b*); it is also derivable by linearising the equations in §2(*a*) above. These waves can still have $a\lambda^2/h^3$ not large, say 1 or less.

 (iii) We can get values of $a\lambda^2/h^3$ between 1 and 15 by increasing a/h and having λ/h between 1 and 10. Because a/h is not tiny, non-linear steepening will start to occur; this causes the wave form to have shorter length scale. But with such a water depth, long waves travel faster than short ones, giving a tendency for the shorter wave forms to spread. These competing tendencies of non-linear steepening and wavenumber dispersion can and do lead to a *steady* wave form, which we shall discuss in an approximate fashion below.

 (iv) For $a\lambda^2/h^3 >$ about 16 no such steady solution is possible; the non-linearity always wins, and a bore forms. These are cases for which the theory in §2 above is appropriate, where a/h may be as high as $\frac{1}{3}$ and $\lambda/h > 10$. The smooth solutions that are possible are best studied by an expansion of the full equations in terms of the parameters a/h and h^2/λ^2, but this is beyond the scope of this text. However, the results are too interesting to be abandoned entirely, so we shall carry through an approximate theory that is good enough for moderately small amplitudes a/h.

(b) The stream function approximated

We shall continue to neglect the effects of friction at the bottom, even though in real circumstances this can be an important modifying factor. So we may assume that we have an irrotational flow above a smooth bed; the flow is no longer uniform, but varies with x over a length

Fig. XV.18. Definition sketch for §4(*b*).

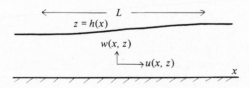

L, as shown in fig. XV.18. This length L is such that we shall neglect $(h/L)^3$.

The water is incompressible and the flow is two-dimensional, so we describe the flow by a stream function $\psi(x, z)$, with the bottom being the streamline

$$\psi(x, 0) = 0.$$

Notice that the flow is once again assumed to be steady. Now the surface $z = h(x)$ is also a streamline, the constant value of ψ on the surface being the total volume flow rate Q (see Chapter IV §3); that is

$$\psi(x, h(x)) = Q.$$

This equation is of no use as it stands, so we set about approximating to get a better version of it.

Expand $\psi(x, z)$ as a Taylor series in powers of z:

$$\psi(x, z) = \psi(x, 0) + z\psi_z(x, 0) + \tfrac{1}{2}z^2\psi_{zz}(x, 0) + \tfrac{1}{6}z^3\psi_{zzz}(x, 0) + \cdots$$

We have just seen that $\psi(x, 0) = 0$, and we know that

$$\psi_z(x, 0) = u(x, 0),$$

the speed along the bottom. The next term is rearranged by using

$$\nabla^2\psi = 0$$

in irrotational, two-dimensional flow; that is

$$\psi_{zz}(x, 0) = -\psi_{xx}(x, 0)$$
$$= \partial w/\partial x$$

on the bottom. But $w = 0$ all along the bottom so $\partial w/\partial x = 0$ on the bottom. Finally, using

$$\nabla^2\psi = 0$$

again, we replace ψ_{zzz} by $-\psi_{xxz}$, which is equal to $-\partial^2 u/\partial x^2$. So the series for $\psi(x, z)$ has been rewritten as

$$\psi(x, z) \doteqdot zu(x, 0) - \tfrac{1}{6}z^3 u_{xx}(x, 0).$$

The second term here has size at most approximately

$$\tfrac{1}{6}h^3 U/L^2$$

whereas the first has size at most

$$hU.$$

So we are retaining for the moment a term of order h^2/L^2 compared to the main term. Higher terms will be neglected. And for convenience of writing we finally put this as

$$\psi(x, z) = zs(x) - \tfrac{1}{6}z^3 s''(x)$$

where $s(x) = u(x, 0)$, the slip velocity along the bottom.

(c) *The equation for the depth*

The flow has been assumed to be smooth, steady and irrotational, so we may use Bernoulli's equation in the form

$$p + \tfrac{1}{2}\rho v^2 + \rho gz = E,$$

for some constant E. And we may also use that the force-rate-of-momentum-flow integral is a constant, because we are neglecting friction at the bottom. So

$$\int_{z=0}^{h} (p + \rho u^2)dz = M,$$

for some constant M.

The pressure is no longer hydrostatic, because the velocity v is now not a uniform stream, but varies with x and has a vertical component. But we can eliminate p between these last two equations to get

$$\int_{0}^{h} (E - \rho gz + \tfrac{1}{2}\rho u^2 - \tfrac{1}{2}\rho w^2)dz = M.$$

In this expression we must use

$$u = \psi_z, \, w = -\psi_x;$$

and these are approximately

$$\begin{cases} u = s(x) - \tfrac{1}{2}z^2 s''(x), \\ w = -zs'(x) + \tfrac{1}{6}z^3 s'''(x), \end{cases}$$

which allows us to perform the integration. It gives a rather long expression of which we only want the leading terms – that is we neglect terms of excessively high order in h/L. The approximate expression is

$$Eh - \tfrac{1}{2}\rho gh^2 + \tfrac{1}{2}\rho(s^2 h - \tfrac{1}{3}h^3 ss'') - \tfrac{1}{6}\rho h^3 s'^2 = M.$$

This equation still involves both $s(x)$ and $h(x)$. These are however connecting by the condition at the surface that

$$\psi = Q \text{ when } z = h(x).$$

So we have, approximately,

$$Q = h(x)s(x) - \tfrac{1}{6}h^3 s''.$$

The second term here is again of order $(h/L)^2$ compared to the first, so when substituting into the smaller terms it is enough to use

$$s(x) = Q/h(x).$$

This gives

$$Eh - \tfrac{1}{2}\rho gh^2 + \tfrac{1}{2}\rho Q^2/h - \rho Q^2 h'^2/(6h) = M$$

where we have needed to use

$$s(x) = Q/h + \tfrac{1}{6}h^2 s''$$

in the term $\tfrac{1}{2}\rho s^2 h$ (which is not small).

After all this intricate calculation we have a differential equation for $h(x)$ which includes correction terms of order $(h/L)^2$ and which depends on the values of the constants

$$Q, E, M$$

which determine the flow. The equation is of the form

$$(dh/dx)^2 = \text{cubic in } h(x),$$

which is of a standard form; however, the solutions are in terms of elliptic integrals which are not well known, so an elementary treatment of the equation will be given.

(d) *A solution with a uniform region: the solitary wave*
Take, then, the equation

$$\rho Q^2 (dh/dx)^2 = 3\rho Q^2 - 6Mh + 6Eh^2 - 3\rho gh^3,$$

and assume that at some stage the flow is uniform, so that

$$Q = UH,$$
$$E = \tfrac{1}{2}\rho U^2 + \rho gH = \rho gH(1 + \tfrac{1}{2}F^2)$$
$$M = \rho U^2 H + \tfrac{1}{2}\rho gH^2 = \rho gH^2(\tfrac{1}{2} + F^2),$$

where E has been evaluated on the surface and the atmospheric pressure p_0 has been omitted as it has no dynamical significance. The equation can be rewritten as

$$\left(\frac{dh}{dx}\right)^2 = \frac{3}{F^2}\left\{ F^2 - 2F^2\left(\frac{h}{H}\right) - \left(\frac{h}{H}\right) + F^2\left(\frac{h}{H}\right)^2 + 2\left(\frac{h}{H}\right)^2 - \right.$$
$$\left. - \left(\frac{h}{H}\right)^3 \right\}.$$

This factorises to

$$(dh/dx)^2 = 3(F^2 - h/H)(1 - h/H)^2/F^2,$$

which has an 'elementary' integral by separation of variables. The detail of the integration is made easier by the changes of variables

$$x = HFX/\sqrt{3} \quad \text{and} \quad h = H(1 + y),$$

which give

$$(dY/dX)^2 = Y^2(a - Y),$$

where

$$a = F^2 - 1.$$

The solution is now reasonably easy to find, and is

$$Y = a \operatorname{sech}^2(\tfrac{1}{2}a^{1/2}X)$$

or

$$h(x) = H + H(F^2 - 1)\operatorname{sech}^2\{(3F^2 - 3)^{1/2}x/(2FH)\}.$$

It is evident that this solution requires

$$F^2 > 1;$$

it can only exist on a supercritical stream. Or, to put it more realistically, if it exists on a channel of water which is at rest at large distances with depth H there, then this wave moves at a speed greater than the long wave speed $(gH)^{1/2}$.

The profile of the wave is drawn in fig. XV.19 for

$$F = 1.1, a = 0.21,$$

which is the largest value for which the theory is a reasonably good fit to experiments. (Note the exaggerated vertical scale.) The maximum excess height is at $X = 0$ and is just $Y = a = F^2 - 1$; in dimensional form the maximum depth is

$$F^2H = U^2/g.$$

Consequently the wave speed on still water is $F(gH)^{1/2}$ or $(gh_{max})^{1/2}$, which is just the long wave speed for the maximum depth. The 'length' of this wave is not clearly defined. It has fallen to about 0.1 of its maximum height

Fig. XV.19. Profile of a solitary wave with $F = 1.1$, in dimensionless form.

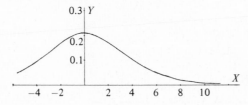

when $a^{1/2}X = 3.6$, so its length is about 16 in the non-dimensional form. Returning to the dimensional variable x, this is a length of about

$$2HF/(F^2 - 1)^{1/2}.$$

You will notice that there are only two parameters in this solution. If F is known, or equivalently if a is known, then for a given depth H in the undisturbed state, all of

 (i) the speed, $U = F(gH)^{1/2}$,
 (ii) the length, $2HF/(F^2 - 1)^{1/2}$,
 (iii) the maximum elevation, $(F^2 - 1)H$,

are known. This is quite unlike the linear theory of Chapter XIII, where amplitude and wavelength were independent.

This wave form with a sech^2 profile is known as a 'solitary wave', because it can exist in isolation with no disturbance upstream or downstream and no irregularity on the bed of the channel. It is the result of a balance between non-linearity (tending to steepen the wave) and dispersion (tending to broaden it). However its occurrence in nature is almost always as a progressive wave on water which is at rest at large distances upstream or downstream. Let us consider such a wave, sketched in fig. XV.20. The speed U is given, as above, in terms of the parameter F by

$$U = F(gH)^{1/2} = (gh_{max})^{1/2}.$$

The extra volume in the wave, above the level $z = H$, is

$$V = \int_{-\infty}^{\infty} \{h(x) - H\}dx$$

for unit width of channel. This integrates quite easily to

$$V = (4/\sqrt{3})FH^2(F^2 - 1)^{1/2}.$$

The momentum contained in the wave can also be calculated quite readily. The water velocity component is of course not just U, but is

$$U - u(x, z),$$

where $u(x, z)$ is the component of velocity in the stationary solitary wave.

Fig. XV.20. Definition sketch for a progressive solitary wave.

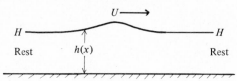

This gives a momentum

$$\rho \int_{-\infty}^{\infty} \left\{ \int_{0}^{h} (U - u)dz \right\} dx.$$

Calculation of this momentum is simplified by noticing that

$$\int_{0}^{h} u(x, z)dz$$

is just UH from the continuity equation for the stationary wave. Hence the momentum content is just

$$\rho \int_{-\infty}^{\infty} U(h - H)dx,$$

or ρU times the volume in the wave, i.e.

$$\rho U V.$$

We can now see roughly how to set up a solitary wave in a channel. We must add a volume V of water which has momentum $\rho U V$ in such a fashion that

$$a\lambda^2 / h^3$$

has a value around 10. This can be done by moving a piston forward in a channel, as in fig. XV.21. The distance X that the piston moves provides an 'extra' volume

$$V = XH$$

of water; and the extra force on the piston multiplied by its time of action gives the momentum imput. The speed of the forward wave is determined roughly by V, because V and H fix F. If the movement is done too rapidly the resulting length of the disturbance is short and the amplitude is high, which leads to the formation of a bore rather than a solitary wave. But with a carefully chosen piston motion a solitary wave will be found, after an initial period while it 'collects itself together'.

Fig. XV.21. Starting a solitary wave by moving a piston.

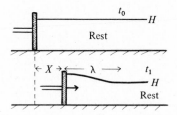

This laboratory experiment makes it sound as though a solitary wave requires careful contrivance. In fact it can occur quite often, and one recent example is both spectacular and instructive.

An earthquake in Alaska in 1958 loosened an unstable rockface above the head of Lituya Bay (which is a fjord in the coastal mountains), and an estimated 9×10^7 tonnes of rock fell into the water. Most of the rock motion was vertical, and most of the energy went into an enormous splash, which reached some 550 m up the opposite mountain. The displacement of water gave a huge solitary wave down the bay, which is roughly a channel with uniform depth. Eye-witnesses estimated the wave height to be about 15–30 m, and the speed to be around 45 m s^{-1}. The wave cleared the shore of trees to a height of over 30 m at a large distance from the original rock-fall.

Solitary wave types of solutions also occur in other areas of physics, where they are called 'solitons'. They are again caused by a balance between non-linear steepening and dispersion, and are associated with conservation equations.

(e) The general solution: cnoidal waves

The solitary wave solution occurred in (d) for very special values of the constants Q, E and M. This gave

$$\left(\frac{dh}{dx}\right)^2 = \frac{3}{F^2}\left(F^2 - \frac{h}{H}\right)\left(1 - \frac{h}{H}\right)^2,$$

where the cubic on the right, say $C_0(h)$, has the graph in fig. XV.22. There is a *double* root of $C_0(h) = 0$ at $h = H$.

Now consider more general values of Q, E and M, no longer corresponding to a uniform stream. If the changes in the constants are not large we shall get a neighbouring curve, say $C(h)$ in fig. XV.23. One possibility is that it has three (distinct) real roots, say h_1, h_2, h_3. The differential

Fig. XV.22. The graph of the cubic in h/H for a solitary wave solution.

Fig. XV.23. The cubic in h/H for a cnoidal wave.

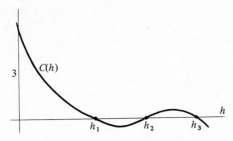

Fig. XV.24. The square root of the cubic in fig. XV.23.

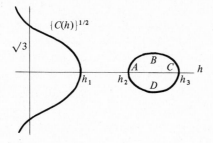

equation must now be

$$(dh/dx)^2 = -3(h - h_1)(h - h_2)(h - h_3)/h_1 h_2 h_3$$

because $(dh/dx)^2$ still has the value 3 when $h = 0$.

The graph of dh/dx against h can also be drawn quite easily be taking square roots of the values on the graph of $C(h)$, and noting that a straight line segment through h_1, for example, is converted into a parabolic arc also through h_1. Naturally, when $C(h) < 0$ there is no real graph for dh/dx. This graph for dh/dx against h in fig. XV.24 can be converted roughly into a graph for h against x, because for each value of h (between h_2 and h_3) we may associate a slope dh/dx. Given a starting value of x, and a decision as to which branch of the dh/dx graph to start on and which way to go round it, we may sketch $h(x)$.

If, for example, we start at $x = 0$ with $h = h_2$ and proceed clockwise, we derive the graph shown in fig. XV.25. The result is a periodic solution for $h(x)$. A numerical solution of the differential equation, either done directly or by integrating the equation in terms of elliptic functions and then looking up tables of values, shows that the crests are steeper and the troughs flatter than in sinusoidal waves. When the amplitude is very small, the waves are effectively sinusoidal, as they must be when linear

Fig. XV.25. The cnoidal wave for $h(x)$.

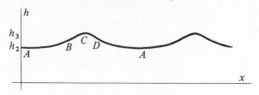

Fig. XV.26. A cubic leading to no wave solution.

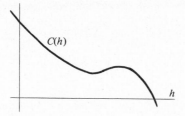

theory applies. As the amplitude increases, the wavelength also increases until a solitary wave results. If the amplitude is increased even further, then the graph for $C(h)$ has only one root, as in fig. XV.26, and there are no steady, bounded solutions of the differential equation.

These non-linear steady waves are known, for historical reasons, as 'cnoidal' waves. When waves occur at an undular bore, downstream of a sluice gate, or downstream of an obstacle on the bed, then they will be, to a good approximation, cnoidal waves unless the amplitudes are too large for the present theory to be justified.

Exercises

1. When a ship travels through a narrow, shallow canal it is known that there is a danger that the ship will ground if it goes too fast. The ship has a draught d and width b, as shown in fig. XV.27, in a canal whose width is B and which has water of depth H. If B, H do not greatly exceed b and d show

Fig. XV.27. Definition sketch for Q1, with the ship at rest.

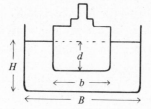

that the ship grounds if its speed U is given by

$$U^2\{B^2H^2/[d^2(B^2 - b^2)] - 1\} = 2g(H - d)$$

approximately.

[Use a simple model in which the flow past the ship is uniform all along the ship when there is a flow U in the canal upstream and the ship is at rest.]

Use the values

$B = 10$ m, $\quad b = 5$ m,

$H = 4$ m, $\quad d = 3$ m,

to calculate an appropriate value of U.

2. A stream of speed U and depth H flows along a channel of breadth B. It comes to a section where the breadth reduces slowly and smoothly and then increases again to B. The flow is everywhere smooth and steady and the upstream Froude number is $3^{-1/2}$. Show that the downstream depth is either H or $\frac{1}{2}H$, and in the latter case determine
   (i) the downstream Froude number,
   (ii) the breadth at the narrowest point,
   (iii) the depth there.

3. In many real channels the width of the water surface depends on the depth of the water. For example, a stream in flood above its usual channel could have the cross-section shown in fig. XV.28.
   Repeat the work of §1(b) for an obstacle on the bed of such a stream. Assume that $h > a$ at all points and that the flow is smooth. Do you expect your solution to be a good one if α is small?

4. Integrate $\nabla \cdot \mathbf{v} = 0$ to derive

$$\partial h/\partial t + u\,\partial h/\partial x + h\,\partial u/\partial x = 0$$

for one-dimensional flow of varying depth in a channel.

5. In §2(b) find the gradients of the characteristics OB and DC. Find three regions in which all the characteristics are straight.

6. Linearise the equations of §2(a) to derive the standard wave equation for $h(x, t)$.

7. In the tidal bore model of §2(c) find out how far upstream the bore might be expected to form.

Fig. XV.28. Definition diagram for Q3.

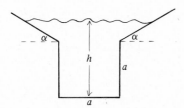

8. A bore of depth H_2 advances into stationary water of depth H_1. Show from the results of §3(a) that the speed of advance is

$$\{\tfrac{1}{2}gH_2(H_1 + H_2)/H_1\}^{1/2}.$$

If a bore of excess height $H_2 - H_1 = 0.5$ m advances at speed 2.4 m s^{-1} up a river flowing at 1.3 m s^{-1}, what is the depth of the river?

9. For the flow under a sluice gate in §3(b) show that

$$(H_2/H_1)^3 = (F_1/F_2)^2$$

and hence that

$$(F_1^2 + 2)^3/F_1^2 = (F_2^2 + 2)^3/F_2^2.$$

Solve this equation for F_2^2 in terms of F_1^2 by noting that $F_2 = F_1$ must be one root; show that $F_2 > 1$ in your solution when $F_1 < 1$.

10. Calculate the force on a sluice gate in terms of F_1 and H_1 when there is a transition to a uniform supercritical stream.

11. A subcritical stream U_1, H_1 passes over a small obstacle on the bed which experiences a *small* force T (per unit width across the stream). The stream is again uniform and subcritical after the obstacle, as shown in fig. XV.29. Calculate
   (i) the downstream parameters U_2, H_2,
   (ii) the energy loss rate due to the obstacle, in terms of U_1, H_1 and the small parameter $\alpha = T/(\rho H_1 U_1^2)$.

12. Complete the derivation of the approximate equation for dh/dx in §4(c).

13. Lituya Bay can be reasonably approximated by a channel with straight vertical sides and depth about 500 feet. What are the speeds of solitary waves of excess heights 50 and 100 feet on such a channel? What is the height of a solitary wave travelling at 100 miles per hour? What is the value of F for a wave of excess height 100 feet? Calculate the excess volume for such a wave with width 2 miles.

14. For the solitary wave solution of §4(d), determine $s(x)$ approximately (retaining terms of size h^2/L^2 in a rough fashion). Hence determine approximately the vertical velocity at the surface.

15. A uniform stream has

$$\begin{cases} Q^2 = gH_0^3, \\ F^2 = 1.4. \end{cases}$$

Fig. XV.29. An obstacle on the bed of a stream causing a change of level. A force T acts on the obstacle.

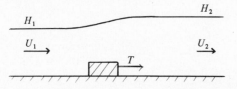

Calculate the speed U and depth H of this stream in terms of H_0; calculate also the constants M, E.

An obstacle reduces M and E to the new values

$$\begin{cases} M_1 = 217\,\rho g H_0^2/144 \\ E_1 = 217\,\rho g H_0/144. \end{cases}$$

Show that the equation for the downstream surface is

$$(dh/dx)^2 = -(3/H_0^3)(h - 8H_0/9)(h - H_0)(h - 9H_0/8),$$

and discuss the shape of the downstream surface and its average height.

References

(a) The film *Waves in Fluids* has useful sections on the hydraulic jump and the bore (or hydraulic surge wave, as it is called in the film). The relevant material is in the four film loops: FM-140 *Flattening and steepening of large-amplitude gravity waves*; FM-141 *The hydraulic surge wave*, FM-142 *The hydraulic jump*, FM-143 *Free surface flows over a towed obstacle.*

(b) Texts on fluid mechanics for engineers usually have a section on open channel hydraulics, as such flows have considerable practical importance. See, for example, *Principles of Fluid Mechanics*, W. H. Li and S. H. Lam, Addison-Wesley 1964, Chapter 10.

XVI
The complex potential

1. Simple complex potentials

(*a*) *The model*

We have seen in Chapters XI–XIII how much the existence of a potential simplifies the calculation of fluid motions. There is a still simpler version of potential theory available when the motion is two-dimensional also. This brings in the powerful methods of complex function theory, at the cost of having to learn something of those methods. It is inappropriate to insert a course on complex functions into this text; but no great depth of knowledge will be assumed: any first course in complex function theory will contain quite enough material to more depth than we shall need.

The model of fluid motion that we are using in this chapter and the next is a very restricted one. The effects of viscosity and diffusion have been neglected; the fluid is taken to be incompressible; the flow is assumed to be irrotational and two-dimensional. This would seem to be such a list of demands that very few flows will satisfy it; yet there are enough flows for which these conditions are sufficiently nearly met for it to be worth taking the trouble. Much work in this area was motivated by the study of the flow of air round an aeroplane wing; though aeroplane wings are now of a rather different shape, the exact solutions from this method still give useful understanding of what is happening in the fluid motion. And the methods can be used in other areas of fluid dynamics.

(b) *Basic potentials*

We noted in Chapter XI §3(a) the potential ϕ and stream functions ψ for several simple flows in two dimensions.

(i) The uniform stream U along the x-axis has

$$\begin{cases} \phi = Ux = Ur\cos\theta, \\ \psi = Uy = Ur\sin\theta. \end{cases}$$

(ii) The simple source of volume flux m has

$$\begin{cases} \phi = (m/2\pi)\ln(r/a), \\ \psi = (m/2\pi)\theta. \end{cases}$$

(iii) The dipole of strength A along the x-axis has

$$\begin{cases} \phi = -Ar^{-1}\cos\theta, \\ \psi = Ar^{-1}\sin\theta. \end{cases}$$

(iv) The line vortex of circulation κ has

$$\begin{cases} \phi = (\kappa/2\pi)\theta, \\ \psi = -(\kappa/2\pi)\ln(r/a). \end{cases}$$

In all these cases it is easily seen that

$$w = \phi + i\psi$$

has a rather simple form in terms of

$$z = x + iy.$$

The expressions are as follows

(i) Uniform stream along the x-axis

$$w = Uz.$$

(ii) Source of flux m at the origin

$$w = (m/2\pi)\ln(z/a).$$

(iii) Dipole of strength A at the origin, pointing along the x-axis

$$w = -A/z.$$

(iv) Vortex singularity at the origin, circulation κ

$$w = -(i\kappa/2\pi)\ln(z/a).$$

We shall henceforward drop the a in the logarithm terms (i.e. choose $a = 1$ in whatever units are used) to have source and vortex potentials proportional to $\ln z$.

Notice the notation here. The coordinates in the plane are x and y (or r and θ), and z is no longer the third cartesian direction, but is given by

$$z = x + iy = re^{i\theta}.$$

And the velocity components will be called u and v; there is no velocity component perpendicular to the plane, and w is the 'complex potential'

$$w(z) = \phi + i\psi.$$

(c) *The complex potential and the velocity*
The velocity components (u, v) are given by

$$\begin{cases} u = \partial\phi/\partial x = \partial\psi/\partial y, \\ v = \partial\phi/\partial y = -\partial\psi/\partial x. \end{cases}$$

When ϕ and ψ are taken as the real and imaginary parts of a complex function $w(z)$, these are called the Cauchy–Riemann equations, and they follow from the differentiability $w(z)$. So for *any* differentiable function $w(z)$, the real part $\phi(x, y)$ may be taken as a potential for a fluid motion (of a restricted type) and the imaginary part may be taken as the stream function for the same motion. For example

$$w(z) = Az^2$$

has

$$\begin{cases} \phi(x, y) = \mathscr{R}e\, w(z) = A(x^2 - y^2), \\ \psi(x, y) = \mathscr{I}m\, w(z) = 2Axy, \end{cases}$$

which were the potential and stream functions we used in Chapter XI §4(a) for flow against a wall. And similarly

$$w(z) = A \sin z$$

must give some flow, though we have not met this one previously.

Conversely, if we have any flow of the type described in (a) above with continuous velocity components (u, v), then a theorem (which we shall not prove) assures us that there exists some differentiable complex potential

$$w(z) = \phi(x, y) + i\psi(x, y)$$

such that

$$\begin{cases} u = \partial\phi/\partial x = \partial\psi/\partial y, \\ v = \partial\phi/\partial y = -\partial\psi/\partial x. \end{cases}$$

The derivative dw/dz must clearly be related to the derivatives of ϕ and ψ, and hence to the velocity components (u, v). Now it is an early result in complex function theory that, for a differentiable function $w(z)$,

$$dw/dz = \partial\phi/\partial x + i\partial\psi/\partial x,$$

and consequently

$$dw/dz = u - iv.$$

Hence

$$\begin{cases} u = \mathscr{R}e\,(dw/dz) = \mathscr{R}e\,w'(z), \\ v = -\mathscr{I}m(dw/dz) = -\mathscr{I}m\,w'(z). \end{cases}$$

This may also be expressed as

$$u + iv = \overline{w'(z)},$$

where an overbar denotes a complex conjugate as before. Moreover, if we put this in terms of the speed q of the flow and the angle θ that the streamlines make locally, then

$$u + iv = qe^{i\theta} = \overline{w'(z)},$$

so that

$$\begin{cases} q = \left|\overline{w'(z)}\right| = |w'(z)| \\ \theta = \arg\overline{w'(z)} = -\arg w'(z). \end{cases}$$

A knowledge of $w(z)$ therefore leads very quickly to all possible information about the velocity, just from the derivative dw/dz.

(d) Shift of origin and rotation of axes

The singularities in the simple example in (b) have all been at the origin, and the directions have all been chosen to be the x-axis. It is not hard to vary these two constraints, and we shall do so now as an illustration of the ease of the complex function formulation.

Consider a source at the point (x_0, y_0), that is, at the point z_0 in fig. XVI.1. The potential is

$$\phi = (m/2\pi)\ln r_1,$$

and the stream function is

$$\psi = (m/2\pi)\theta_1.$$

These are both relatively complicated in terms of r and θ; but the transformation has only been a shift of origin, and so we may expect to have

$$w(z) = (m/2\pi)\ln(z - z_0).$$

Fig. XVI.1. Distance r_1 and angle θ_1 at the point z_0.

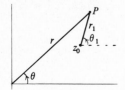

Fig. XVI.2. A stream U at angle α to the x-axis.

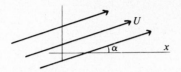

The velocity associated with this complex potential is

$$\overline{w'(z)} = (m/2\pi)\overline{(z - z_0)^{-1}},$$

and if we put $z - z_0 = r_1 e^{i\theta_1}$ we find that the velocity is

$$(m/2\pi)r_1^{-1} e^{i\theta_1}$$

which has speed $m/2\pi r_1$ for any angle θ_1, the correct value for a simple source of volume flux m at this point.

Thus we expect, and we verify in the next chapter, that a singularity at the point z_0 is derived from the corresponding singularity at the origin by the simple shift

$$z \mapsto z - z_0.$$

For a change in direction, consider a stream U at angle α to the x-axis, as in fig. XVI.2. The velocity components are

$$\begin{cases} u = U \cos \alpha, \\ v = U \sin \alpha. \end{cases}$$

Now $u - iv = dw/dz$, and so we may find w:

$$dw/dz = U \cos \alpha - iU \sin \alpha$$
$$= Ue^{-i\alpha}.$$

So for this stream

$$w = Uze^{-i\alpha}.$$

This suggests (and again we demonstrate it in the next chapter) that a rotation through α from the x-axis is achieved by the transformation

$$z \mapsto ze^{-i\alpha}.$$

As a final example here, we combine these two transformations: a dipole pointing along the direction at α to the x-axis and situated at the point z_0 has the complex potential

$$w(z) = -Ae^{i\alpha}/(z - z_0).$$

(e) The cut plane

The function $\ln z$ that has arisen for sources or vortices presents some problems in the theory of complex functions, of a sort that we have

met before in Chapter XI. When we write

$$\ln z = \ln|z| + i \arg z$$
$$= \ln r + i\theta,$$

we can only have a single valued function $\ln z$ provided the value of θ is given uniquely at each point in the plane. It is common to restrict θ, for example by

$$-\pi < \theta \leqslant \pi.$$

Then θ is single valued, but no longer continuous (or differentiable) across the line $\theta = \pi$. In complex function theory this corresponds to using a 'cut plane', with the cut in this case being from the origin to infinity along the line $\theta = \pi$. The function $\ln z$ defined on this cut plane is then differentiable in the domain (a 'domain' is a connected open set in the theory of complex functions)

$$\begin{cases} |z| > 0, \\ -\pi < \arg z < \pi. \end{cases}$$

This cut from 0 to ∞ along $\theta = -\pi$ is the customary one, but it is not by any means essential to use it. We may equally well choose to have $\ln z$ defined for

$$\begin{cases} |z| > 0, \\ 0 < \arg z < 2\pi. \end{cases}$$

if it suits us. *Any* cut from 0 to ∞ (even a curved one) will do to make $\ln z$ into a differentiable function.

As an example let us take a source at $z = ia$ and a sink at $z = -ia$. The complex potential is

$$w(z) = (m/2\pi)\{\ln(z - ia) - \ln(z + ia)\}$$

$$= \frac{m}{2\pi} \ln \frac{z - ia}{z + ia},$$

where we must take some cut in the plane to ensure that $w(z)$ is properly defined at each point. It might look as though we would need the two cuts sketched in fig. XVI.3, one from ia to ∞ and one from $-ia$ to ∞, both reaching ∞ along the negative real axis direction. But this is un-

Fig. XVI.3. The plane cut from each of $\pm ia$ to ∞.

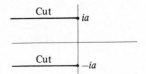

Fig. XVI.4. The equivalent single cut between $\pm ia$.

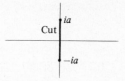

necessarily complicated. We can combine the cuts into a (topologically equivalent) single cut from ia to $-ia$, as in fig. XVI.4.

If we now take the limit as $a \to 0$ and put $ma/\pi = \mu$, we get the complex potential for a dipole along the imaginary axis:

$$\lim_{a \to 0} \frac{\mu}{2a} \ln \frac{z - ia}{z + ia} = \lim_{a \to 0} \frac{\mu}{2a} \left\{ \ln\left(1 - \frac{ia}{z}\right) - \ln\left(1 + \frac{ia}{z}\right) \right\}$$

$$= \lim_{a \to 0} \tfrac{1}{2}(\mu/a)\{ -2ia/z + \mathrm{O}(a^2) \}$$

(using the normal expansion for $\ln(1 + \zeta)$)

$$= -\mu i/z = -\mu e^{i\pi/2}/z.$$

For a dipole potential we do not need a cut plane, as z^{-1} is defined for $|z| > 0$. And we see that as $a \to 0$ the cut between ia and $-ia$ contracts to nothing.

A cut plane has to be used in connection with other functions too. For example

$$w(z) = z^{1/2}$$

is a differentiable function on the cut plane

$$|z| > 0, \; -\pi < \arg z < \pi$$

with value

$$r^{1/2} e^{i\theta/2}$$

when $z = re^{i\theta}$. In this case as well the choice of cut is arbitrary; and notice that

$$r^{1/2} e^{i(\theta + 2\pi)/2}$$

is also a differentiable function on the cut plane: it is just $-z^{1/2}$.

2. More complicated potentials

We have just seen that the complex potential $w(z) = Az^2$ generates an interesting flow. In this section we shall look at a few more functions $w(z)$ to see what flows correspond to them.

(a) *Powers of z*

Let us start with $w(z) = Az^k$ as an obvious example. Take first k to be a positive integer, and put $z = re^{i\theta}$. Then

$$\begin{cases} \phi = \mathscr{R}e(Ar^ke^{ik\theta}) = Ar^k\cos k\theta, \\ \psi = \mathscr{I}m(Ar^ke^{ik\theta}) = Ar^k\sin k\theta, \end{cases}$$

if we choose to have A real. Now the curves given by $\psi = 0$ are streamlines, and so possible boundaries for the flow; that is,

$$k\theta = n\pi$$

gives possible boundaries. Hence one such example is provided by $k = 5$ and $n = 0, 1$ or 2; the lines $\theta = 0$ and $\theta = 72°$ may be taken as boundaries, and ψ will also be zero on $\theta = 36°$. The flow is sketched in fig. XVI.5 for $A < 0$: it is a flow from the direction $\theta = 36°$ into a corner.

Other integer values of k give similar flows into corners of other angles. In particular the case $k = 2$ gives flow into an 'angle' of π, i.e. flow against a wall.

Let us next take $w(z) = Az^{1/2}$, as typical of non-integer k. This potential (with the meaning given above for $z^{1/2}$) gives

$$\begin{cases} \phi = Ar^{1/2}\cos\tfrac{1}{2}\theta \\ \psi = Ar^{1/2}\sin\tfrac{1}{2}\theta; \end{cases}$$

the streamlines are $r^{1/2}\sin\tfrac{1}{2}\theta = c$, which may be put into the form

$$2c^2/r = 1 - \cos\theta.$$

Fig. XVI.5. A flow into a corner given by $w = Az^5$.

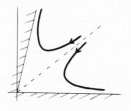

Fig. XVI.6. Potential flow round a semi-infinite plate.

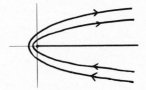

This is the equation of a parabola with focus at the origin and 'size' (the value of r when $\theta = \frac{1}{2}\pi$) $2c^2$. One of these parabolas is obtained by putting $c = 0$: it is the positive real axis. So the flow sketched in fig. XVI.6 round the positive real axis as a barrier is obtained by using $w(z) = Az^{1/2}$. This is not a very real solution, as a flow round a corner will tend to separate, as we have seen in earlier chapters.

(b) Flow past a plate

The complex potential

$$w(z) = (z^2 + a^2)^{1/2}$$

is clearly related to the $z^{1/2}$ potential in (a), because near $z = \pm ia$ the potential has this form. For example, writing $z_1 = z - ia$ gives

$$w = z_1^{1/2}(z_1 + 2ia)^{1/2},$$

which is approximated by

$$kz_1^{1/2}$$

when z_1 is small (with $k = (2ia)^{1/2}$).

We shall need a cut plane again, but as in §1(e) we can get away with having only a cut between ia and $-ia$, as in fig. XVI.7. We choose, as before, to have the square root defined so that when z is real and positive, we get the positive square root of $z^2 + a^2$. We should notice at this stage

Fig. XVI.7. The cut plane for $(z^2 + a^2)^{1/2}$.

Fig. XVI.8. Variation of $z^2 + a^2$ as z varies round a curve.

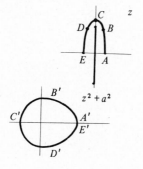

that this does *not* necessarily mean that when z is real and negative, then we take the positive square root of $z^2 + a^2$. In fact we do not, we take the negative square root. This can be conveniently seen from the diagrams in figs. XVI.8 and 9, which show a continuous variation of z from real positive to real negative, and the consequent variations in $z^2 + a^2$ and $(z^2 + a^2)^{1/2}$. At E'', $(z^2 + a^2)^{1/2}$ is real and negative. However the resulting discontinuity in $(z^2 + a^2)^{1/2}$ is only across the cut, where it is quite acceptable, as passage through the cut is not allowed.

Some aspects of the flow represented by the potential $(z^2 + a^2)^{1/2}$ should quickly be clear. The flow has to be like the flow round a half plane near $z = \pm ia$. For large $|z|$ we have

$$w(z) \sim z$$

and so the flow is a uniform stream along the x-direction. And for

$$z = iy, \ -a < y < a,$$

we have *real* values for $w(z)$ and so $\psi = 0$: that is, the cut is the streamline $\psi = 0$. So the potential must represent the flow of a uniform stream past a plate set at right angles to it, with width $2a$, as sketched in fig. XVI.10.

The velocity in this flow is given by

$$\overline{dw/dz} = z(z^2 + a^2)^{-1/2}.$$

This again shows that the velocity tends to a constant when $|z|$ is large, but it also shows that the velocity is infinite near $z = \pm ia$.

Despite the infinite velocity and the fact that a steady flow would have to separate at the rear of the plate, this flow has some use in the modelling

Fig. XVI.9. Variation of $(z^2 + a^2)^{1\,2}$ as z varies round the curve.

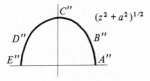

Fig. XVI.10. The flow represented by $w(z) = (z^2 + a^2)^{1/2}$.

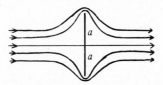

of oscillatory flows past flat plates. The infinite velocity will have to be reduced by having a plate of non-zero thickness, or by appealing to a boundary layer; and any separation will take a short while to be established, which could be longer than the period of the oscillation.

(c) *The potential sin z*

Another interesting complex potential to try is

$$w(z) = \sin z.$$

This is best taken as

$$\sin(x + iy)$$

which has expansion

$$\sin x \cos iy + \cos x \sin iy = \sin x \cosh y + i \cos x \sinh y$$

by standard results. Thus for this potential

$$\begin{cases} \phi(x, y) = \sin x \cosh y, \\ \psi(x, y) = \cos x \sinh y. \end{cases}$$

The streamlines are therefore

$$\cos x \sinh y = \text{constant},$$

and in particular (choosing the constant to be zero) we have streamlines some of which are sketched in fig. XVI.11)

$$\begin{cases} y = 0, \\ x = \tfrac{1}{2}(2n + 1)\pi. \end{cases}$$

The flow inside the region

$$\begin{cases} y > 0, \\ -\tfrac{1}{2}\pi < x < \tfrac{1}{2}\pi \end{cases}$$

will therefore be given by this function; it is easily seen that when x is small

$$\phi(x, y) \doteqdot x \cosh y,$$

Fig. XVI.11. Some streamlines for $w(z) = \sin z$.

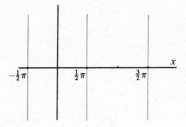

Fig. XVI.12. Streamlines in part of the plane for $w(z) = \sin z$.

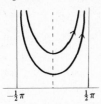

$-\tfrac{1}{2}\pi$ $\tfrac{1}{2}\pi$

which gives a velocity in the x-direction, and so a flow like that shown in fig. XVI.12, with larger velocities as y increases.

(d) *Flow into or out of a pipe*

This procedure, of taking any function of z and determining what flow it corresponds to, can be continued indefinitely. But it is not very profitable, as most of the flows are somewhat unreal. So we finish this section with one last, rather complicated example which gives a moderately real flow.

Let $w(z)$ be given as the solution (which is presumed to exist) of

$$z = w + e^{w}.$$

This is

$$x + iy = \phi + i\psi + e^{\phi}e^{i\psi},$$

or

$$\begin{cases} x = \phi + e^{\phi} \cos \psi, \\ y = \psi + e^{\phi} \sin \psi. \end{cases}$$

These equations are not easy to interpret; trial and error gives the following results.

(i) The streamline $\psi = 0$ is $y = 0$ for all x, because $\phi + e^{\phi}$ takes all values.

(ii) The streamline $\psi = \pi$ is

$$\begin{cases} y = \pi, \\ x = \phi - e^{\phi}, \end{cases}$$

and so is $y = \pi$, $-\infty < x < -1$, since $\phi - e^{\phi}$ only gives values between $-\infty$ and -1.

(iii) Similarly the streamline $\psi = -\pi$ is

$$y = -\pi, \ -\infty < x < -1.$$

(iv) If we take ϕ large and negative, and also $-\pi < \psi < \pi$, we get

$$x \sim \phi, \ y \sim \psi,$$

or

$$w(z) = \phi + i\psi \sim z.$$

That is, we have a uniform stream along the x-axis when x is large and negative and $-\pi < y < \pi$.

(v) To get x large and positive requires ϕ large and positive, and also $-\frac{1}{2}\pi < \psi < \frac{1}{2}\pi$. In this case $e^\phi \cos \psi$ greatly exceeds ϕ, and $e^\phi \sin \psi$ greatly exceeds ψ (except of course for $\psi = 0$). That is,

$$\begin{cases} x \sim e^\phi \cos \psi \\ y \sim e^\phi \sin \psi, \end{cases}$$

or

$$z \sim e^{\phi + i\psi};$$

this has solution

$$w(z) = \phi + i\psi \sim \ln z,$$

which is the potential for a source (of volume flow rate 2π) at the origin.

Let us collect all this information. It is consistent with a flow out of a channel of width 2π which ends at $x = -1$. The flow spreads out so as to seem, when viewed from large distances, to have come from a source near the origin. Fig. XVI.13 gives a sketch of the flow. Naturally, a real flow of this type would separate at the end of the channel, and emerge as a jet. But the flow with *reverse* direction, into the channel, would be more realistic except near the sharp corners at $x = -1$, $y = \pm\pi$.

For such an inflow consider the flow along the streamline $\psi = \pi - \varepsilon$ (where ε is small). It must come along the outside of the channel, round the

Fig. XVI.13. Potential flow out of a channel.

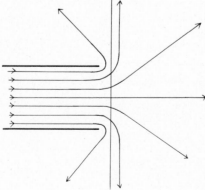

sharp edge, and then go along the inside of the channel. It is interesting to notice that this flow, like the earlier flows near sharp edges (in (*a*) and (*b*) above) has a 'square root' behaviour near the edge. We may verify this by putting

$$x = -1 + \xi, \quad y = \pi + \eta$$

where ξ and η are small as we wish to be near the edge; and we also put

$$\psi = \pi - \varepsilon, \quad \phi = -\delta,$$

where ε and δ are small as the edge corresponds to $\psi = \pi$ and $\phi = 0$. When we substitute in the equations relating x, y, ϕ, ψ and neglect third order small quantities, we derive

$$\begin{cases} \xi = \tfrac{1}{2}\varepsilon^2 - \tfrac{1}{2}\delta^2, \\ \eta = -\varepsilon\delta. \end{cases}$$

That is,

$$\delta + i\varepsilon = i2^{1/2}(\xi + i\eta)^{1/2},$$

and the potential is locally like $k(z - z_0)^{1/2}$, where $z_0 = -1 + i\pi$.

A real flow through a pipe out of a large reservoir (as in fig. XVI.14) would separate at the sharp edges, and travel as a jet inside the pipe. The corresponding jet flow for the two-dimensional case can be calculated by complex variable methods. The shape of the 'free streamlines' which bound the resulting jet can be found without undue difficulty, but it will not be done here. The theory of free streamlines is a well-developed subject which has some relevance to the separated flow behind a cylinder in a stream, as well as to the boundary between two fluids like air and water.

Fig. XVI.14. Formation of a jet bounded by free streamlines in flow out of a reservoir.

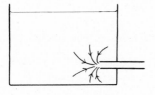

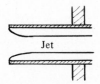

Jet

3. Potentials for systems of vortices

The potential for a vortex

$$w(z) = -(i\kappa/2\pi)\ln z,$$

where κ is the circulation, is rather an easy function; and because addition of logarithms gives multiplication of functions, sets of vortices can be easily dealt with in the complex potential theory.

Take first the example from Chapter V §5(c). The complex potential for two vortices and a uniform stream is

$$w(z) = Uz + (i\kappa/2\pi)\ln(z - ia) - (i\kappa/2\pi)\ln(z + ia)$$

when the two vortices are at $\pm ia$ on the imaginary axis, as shown in fig. XVI.15. The speed of the upper vortex is

$$\overline{dw/dz} \text{ at } z = ia$$

when we omit the contribution of the upper vortex to the potential $w(z)$ (because a vortex does not move itself). This is

$$U + i\kappa/(-4\pi ia) = U - \kappa/4\pi a$$

as before. And the points of zero velocity in the flow come from putting $dw/dz = 0$; that is, they satisfy

$$2\pi U + i\kappa/(z - ia) - i\kappa(z + ia) = 0$$

or

$$z^2 = \kappa a/\pi U - a^2.$$

From these formulae we may rederive the results of Chapter V §5(c) on the shape of the streamlines and the motion of the pair of vortices.

A more interesting example is provided by the flow of a river past a projection from the bank: a line of eddies in the direction shown in fig. XVI.16 can be shed at regular intervals, and we model this first by an infinite row of vortices. The potential for this is

$$w(z) = \sum_{n=-\infty}^{\infty} (i\kappa/2\pi)\ln(z - na).$$

Fig. XVI.15. Definition diagram for two vortices in a stream.

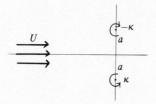

Fig. XVI.16. Vortices shed by a projection modelled by an infinite row of equal vortices.

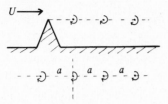

This potential can in fact be put in rather easy form, because it is a result in complex function theory that

$$\sin \pi z = \pi z(1 - z^2)(1 - \tfrac{1}{4}z^2) \ldots$$

Hence

$$\ln \sin \pi z = \ln \pi + \ln z + \ln(1 - z) + \ln(1 + z) + \cdots$$

That is,

$$w(z) = (i\kappa/2\pi)\ln \sin(\pi z/a),$$

except for a constant (which does not affect the velocity).

It is clear that the vortex at the origin (and so the others, because the origin is arbitrary) is at rest: each vortex on one side is just balanced by a vortex on the other side. However, suppose that the vortex at the origin is displaced slightly to A or to B in fig. XVI.17. At *A*, the vortices on the right have more effect than those on the left, and the vortex moves off parallel to the imaginary axis. At *B*, the nearest pair of vortices (and also each other pair) produces a resultant velocity parallel to the real axis. Thus in either case (and so in general) the displaced vortex leaves the vicinity of the origin – the row of vortices is unstable to this disturbance.

A second model of the row of vortices in the river is provided by including a set of image vortices with the directions shown in fig. XVI.18 to

Fig. XVI.17. The vortex at *O* is not stable there.

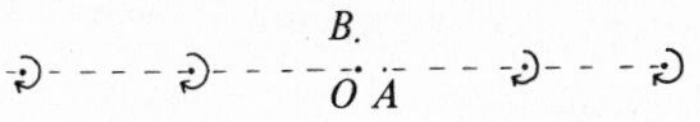

$$\text{-}\mathcal{D}\text{- - - - }\mathcal{D}\text{- - - -}\underset{O \ A}{\overset{B.}{\cdots}}\text{ - - - }\text{-}\mathcal{D}\text{- - - }\mathcal{D}$$

Fig. XVI.18. The vortex row and its image.

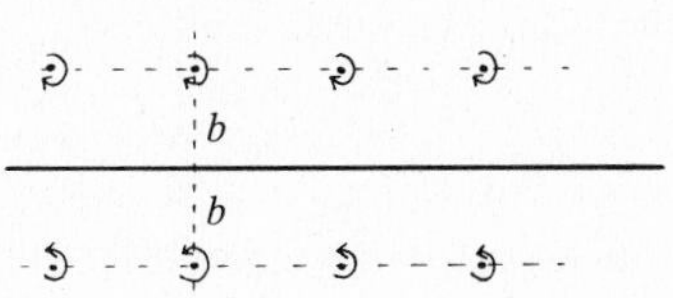

represent the effect of the river bank. By a simple shift of origin, the two rows have complex potential

$$w(z) = (i\kappa/2\pi) \ln \sin\{\pi(z - ib)/a\} - (i\kappa/2\pi) \ln \sin\{\pi(z + ib)/a\}.$$

In this case, by analogy with the single vortex and its image we expect the whole set of vortices to move to the left. We may calculate the speed of the vortex at $z = ib$ quite easily. It has no speed due to the row that it is in, as before. If we write

$$w_1(z) = -(i\kappa/2\pi) \ln \sin\{\pi(z + ib)/a\}$$

as the potential for all the images, then the velocity of the vortex at ib due to all the images is

$$\overline{w_1'(ib)}.$$

Now $dw_1/dz = -(i\kappa/2a)\cot\{\pi(z + ib)/a\}$, and so the velocity is

$$-(i\kappa/2a)\cot(2i\pi b/a) = -(\tfrac{1}{2}\kappa/a)\coth(2\pi b/a).$$

In the case of the river flowing past the projection the vortex row will be convected downstream also, so that the total speed is

$$U - (\tfrac{1}{2}\kappa/a)\coth(2\pi b/a).$$

The size of κ is of course determined by the dynamics near the projection that caused the vortices to form, and it is always such that this speed is positive.

It may be shown that this arrangement of vortices is also unstable: to do this take the one at $z = ib$ and move it to $z = ib + \zeta$, and show that its velocity relative to the row causes it to move away from $z = ib$.

Another interesting pattern of vortices forms in the wake of a cylinder at Reynolds numbers around 100. The vortices are shed alternately from the two sides of the cylinder, and form into a regular pattern, roughly sketched in fig. XVI.19. This may be modelled by a 'Kármán vortex street', which is a double row of vortices, of infinite extent and positioned as shown in fig. XVI.20. By making another easy shift of origin we get the complex potential

$$w(z) = (i\kappa/2\pi) \ln \sin\{\pi(z - ib)/a\}$$
$$- (i\kappa/2\pi) \ln \sin\{\pi(z + ib - \tfrac{1}{2}a)/a\}.$$

Fig. XVI.19. Vortices shed into the wake of a cylinder.

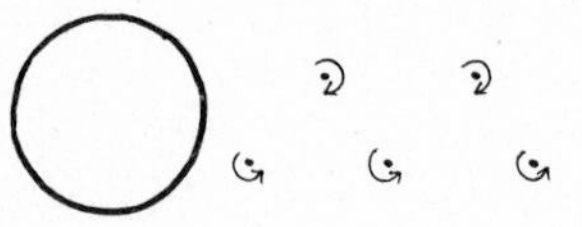

Fig. XVI.20. The Kármán vortex street.

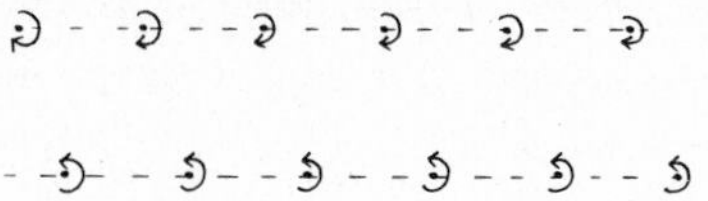

The motion of the street may be calculated as before to be a velocity to the left of amount

$$(\tfrac{1}{2}\kappa/a)\tanh(2\pi b/a).$$

The stability may be discussed, but not by the previous simple means, as we only showed instability to one easy type of displacement. A calculation beyond the scope of this text shows that the street is stable only in the case

$$2b/a = 0.281.$$

In that case the speed is $\tfrac{1}{2}\kappa/(2^{1/2}a)$.

The Kármán vortex street model clearly is not the last word on the flow behind a cylinder at this speed. Viscosity causes the vortices to weaken and spread out, as we saw in Chapter IX. The region near the cylinder cannot be described by this model. But it has considerable reality. So we ask what speed it predicts along the real axis, remembering that we must finally add in a velocity U to the right. Take, as typical point, $z = 0$. Then

$$w'(0) = (\tfrac{1}{2}i\kappa/a)\cot(-i\pi b/a) - (\tfrac{1}{2}i\kappa/a)\cot(i\pi b/a - \tfrac{1}{2}\pi)$$
$$= -(\tfrac{1}{2}\kappa/a)\{\coth(\pi b/a) + \tanh(\pi b/a)\}.$$

Now taking $2b/a = 0.281$, we get a speed

$$-1.4\kappa/a.$$

In fact if we calculate the mean velocity along the centre line of the vortex street we find

$$-\kappa/a,$$

for any value of b/a. Since the velocity in a wake is usually small, this means that we must have

$$U \doteqdot \kappa/a$$

as a (not very precise) approximation for the strength of the vortices.

4. Image theorems
(a) *Images in the wall $y = 0$*

We have seen in earlier chapters how certain problems could be solved by using images. This is particularly easily done in complex potential theory when the boundaries are planes or circular cylinders.

The easiest examples we had earlier were of images in a plane. For example, a source at $z = ia$ has as its image in the plane $y = 0$ an equal source at $z = -ia$. That is, the potential

$$A \ln(z - ia)$$

has image in $y = 0$ given by

$$A \ln(z + ia)$$

to give a total potential

$$w(z) = A \ln(z - ia) + A \ln(z + ia);$$

this combined potential has stream function

$$\psi = \mathscr{I}m \, w(z)$$

and on $y = 0$ we have

$$\begin{aligned}
\psi &= A \, \mathscr{I}m\{\ln(x - ia) + \ln(x + ia)\} \\
&= A \, \mathscr{I}m\{\ln(x^2 + a^2)\} \\
&= 0,
\end{aligned}$$

because $x^2 + a^2$ (and so its logarithm) is real. This confirms that the boundary $y = 0$ is acting as a rigid wall, with no flow across it.

It is not hard to repeat the same kind of calculation for a vortex at ia, or a dipole, or a set of sources, vortices and dipoles all lying in the region $y > 0$. In all cases we find that adding on extra terms with i replaced by $-i$ is enough to give a combined potential which has $\psi = 0$ on $y = 0$, and so represents a flow with $y = 0$ as a rigid boundary.

We can do this in general for such a simple image system. Let $f(z)$ be the potential for a flow which has all its singularities (sources, vortices, dipoles and so on) in the region $y > 0$. Then define

$$\bar{f}(z)$$

to mean the function which has i replaced by $-i$ *except in z itself*. Thus if

$$f(z) = (1 + i)\ln(z - ia),$$

our definition means that

$$\bar{f}(z) = (1 - i)\ln(z + ia),$$

where a is taken to be real.

Now consider

$$w(z) = f(z) + \bar{f}(z)$$

on the dividing plane $y = 0$. It has imaginary part on this plane

$$\mathscr{I}m\{f(x) + \bar{f}(x)\} = 0,$$

because a complex number and its conjugate add to give a real quantity, and

$$\bar{f}(x) = \overline{f(x)}$$

since x is real.

Hence

$$w(z) = f(z) + \bar{f}(z)$$

represents some flow with $y = 0$ as a streamline, and so a (possible) rigid boundary. Is it the flow we want, with appropriate singularities (sources, vortices and so on) in $y > 0$, or has $\bar{f}(z)$ brought in extra singularities? This is most easily answered by looking at examples.

    (i) If $f(z) = A/(z - a)$, a dipole at $z = a$, then

$$\bar{f}(z) = \bar{A}/(z - \bar{a}),$$

        a dipole with a different direction at $z = \bar{a}$; and if $z = a$ is in $y > 0$, then $z = \bar{a}$ is in $y < 0$, i.e. outside the flow.

    (ii) If $f(z) = \{(z - i)(z - 2i)\}^{1/2}$, which has singularities at $z = i$ and $z = 2i$ and a cut between them then

$$\bar{f}(z) = \{(z + i)(z + 2i)\}^{1/2},$$

        which has its singularities at $z = -i$ and $z = -2i$ and a cut between them, i.e. all in the region $y < 0$.

Similarly, anything which happens for $f(z)$ at $z = z_0$ will happen for $\bar{f}(z)$ at $z = \bar{z}_0$, and if z_0 is in $y > 0$ (as all the singularities of f are required to be), then $\bar{z}_0$ will be in $y < 0$ and so not in the flow region.

We may write this simple result as a 'first image theorem'. Let $f(z)$ have all its singularities in $y > 0$; then the flow with these singularities and with $y = 0$ as a rigid boundary has the complex potential

$$w(z) = f(z) + \bar{f}(z).$$

This theorem formalises some of the results we found before: images in a plane are at the corresponding point behind the plane, and with very simply related strengths. For example, a dipole at z_0 making angle α with the x-axis was shown in §2(d) to have potential

$$f(z) = -Ae^{i\alpha}/(z - z_0).$$

The image will be given by

$$\bar{f}(z) = -Ae^{-i\alpha}/(z - \bar{z}_0)$$

which is at $\bar{z}_0$ and makes an angle $-\alpha$ with the x-axis, as shown in fig. XVI.21.

Fig. XVI.21. The image of a dipole in a wall.

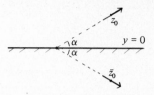

(b) Images in other plane walls

The first image theorem, on images in $y = 0$, has several very closely related theorems. For example there must be a theorem on images in $x = 0$. It is clear from fig. XVI.22 that a singularity at $x + iy$ will give rise to an image singularity at

$$-x + iy = -(x - iy) = -\bar{z}_0.$$

Hence the expected theorem is that a flow given by a function $f(z)$ whose singularities are in $x > 0$, with the rigid boundary $x = 0$, will have complex potential

$$w(z) = f(z) + \bar{f}(-z).$$

This theorem can be proved by showing that on $x = 0$, the imaginary part of $w(z)$ is zero.

Similarly, there must be a theorem on images in the rigid boundary shown in fig. XVI.23 which has equation

$$z = Re^{i\alpha};$$

Fig. XVI.22. An image in the wall $x = 0$.

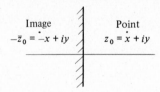

Fig. XVI.23. An image in the wall $z = Re^{i\alpha}$.

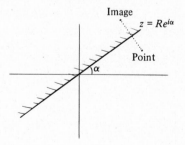

Fig. XVI.24. The image points for a corner.

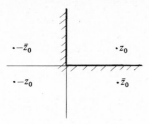

and yet another theorem on images in the boundary

$$z = z_0 + Re^{i\alpha}.$$

But these theorems are rather easy (and not much used) and may be left as exercises.

The next image theorem that we shall take some time on is the 'corner theorem'. Suppose $f(z)$ has all its singularities in the quadrant $x > 0$, $y > 0$, and we want the flow in this quadrant which has these singularities and both $x = 0$ and $y = 0$ as rigid boundaries. Clearly we shall need images in both planes, and it turns out that we need the *three* images (shown in fig. XVI.24), at $\bar{z}_0$, $-z_0$ and $-\bar{z}_0$. The appropriate potential for this flow is

$$w(z) = f(z) + \bar{f}(z) + \bar{f}(-z) + f(-z).$$

For we may show, as before, that $\mathscr{I}m\, w(z)$ is zero on $x = 0$ and on $y = 0$, and that no extra singularities are introduced by the extra terms in $w(z)$ into the quadrant where the flow is required.

In Chapter V Exercise 10 the problem was set of the motion of a vortex in a corner. Let us now solve it by using the corner theorem. A vortex at z_0 has potential

$$f(z) = -(i\kappa/2\pi)\ln(z - z_0),$$

with circulation κ coming from the real constant κ. Then the potential for the vortex together with the rigid boundaries is

$$\begin{aligned}w(z) = {} & -(i\kappa/2\pi)\{\ln(z - z_0) - \ln(z - \bar{z}_0) - \ln(-z - \bar{z}_0) \\ & + \ln(-z - z_0)\},\end{aligned}$$

which shows that we need the images (shown in fig. XVI.25)

(i) a negative vortex at $z = \bar{z}_0$,

(ii) a negative vortex at $z = -\bar{z}_0$,

(iii) a positive vortex at $z = -z_0$.

The velocity of the (real) vortex is derived from the 'image' potential

Fig. XVI.25. The images in a corner of a line vortex.

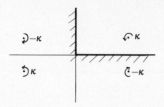

(i.e. leaving out the vortex itself as before)

$$W(z) = (i\kappa/2\pi)\{\ln(z - \bar{z}_0) + \ln(-z - \bar{z}_0) - \ln(-z - z_0)\}.$$

Now

$$(dW/dz)_{z=z_0} = (i\kappa/2\pi)\left(\frac{1}{z_0 - \bar{z}_0} + \frac{1}{z_0 + \bar{z}_0} - \frac{1}{z_0 + z_0}\right),$$

because $\ln(-\zeta) = \ln\zeta + \text{constant}$. Let us put $z_0 = X + iY$ and $\kappa/2\pi = C$; then

$$(dW/dz)_{z_0} = \tfrac{1}{2}\{C/Y + iC/X - iC/(X + iY)\}.$$

But $(dW/dz)_{z_0} = \dot{X} - i\dot{Y}$, from §1(c); so on taking real and imaginary parts, we find that the path of the vortex is given by

$$\begin{cases} 2\dot{X} = C/Y - CY/(X^2 + Y^2), \\ 2\dot{Y} = -C/X + CX/(X^2 + Y^2). \end{cases}$$

Naturally, as the vortex moves, so do its images.

The path can most easily be found by dividing the equations to get

$$\frac{dY}{dX} = \frac{-1/X + X/(X^2 + Y^2)}{1/Y - Y/(X^2 + Y^2)} = -\frac{Y^3}{X^3}$$

which gives

$$1/X^2 + 1/Y^2 = 1/a^2$$

for some constant a. This curve is sketched in fig. XVI.26.

Fig. XVI.26. The path of a line vortex in a corner.

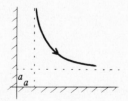

(c) Images in circles: the Milne–Thomson theorem

The next major image theorem concerns images in circular boundaries. We have already seen that images for a singularity at distance c from the centre of a cylinder of radius a are on the line through centre and singularity, at distance a^2/c from the centre; and that sometimes we need an extra singularity at the origin.

To get the correct point for the image of z_0, which must be roughly as shown in fig. XVI.27, we do not want a^2/z_0; for if $z_0 = ce^{i\alpha}$, then

$$a^2/z_0 = (a^2/c)e^{-i\alpha},$$

which is at the wrong angle. What we need is to get an image at the point $a^2/\bar{z}_0$, to achieve both correct distance and angle. So if $f(z)$ has a singularity at z_0, the image potential must be something like

$$\bar{f}(a^2/z),$$

in the same way that an image of z_0 at $\bar{z}_0$ (due to the boundary $y = 0$) required an additional $\bar{f}(z)$.

The theorem, known as the Milne–Thomson circle theorem, is as follows. If $f(z)$ has all its singularities outside the circle $|z| = a$, and we want the flow outside the rigid boundary $|z| = a$ with these singularities, then the appropriate complex potential is

$$w(z) = f(z) + \bar{f}(a^2/z).$$

The proof requires us to show that:
  (i) $\mathscr{I}m\, w(z) = $ constant on $|z| = a$, as then the circle is a streamline;
  (ii) $\bar{f}(a^2/z)$ has no singularities outside $|z| = a$.
For (i), consider the point $z = ae^{i\theta}$ on $|z| = a$. Then

$$f(z) + \bar{f}(a^2/z) = f(ae^{i\theta}) + \bar{f}(ae^{-i\theta}).$$

This is of the form $f(\zeta) + \bar{f}(\bar{\zeta})$, which is real because $\overline{f(\zeta)} = \bar{f}(\bar{\zeta})$ and any complex number plus its conjugate gives a real answer.

For (ii), consider a singularity of $f(z)$ at $z_0 = ce^{i\alpha}$; then $\bar{f}(a^2/z)$ has a singularity when $a^2/z = \bar{z}_0 = ce^{-i\alpha}$, i.e., at $z = (a^2/c)e^{i\alpha}$, which is inside the circle $|z| = a$ and so not in the flow.

Fig. XVI.27. The image point for a circular cylinder.

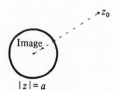

$|z| = a$

Fig. XVI.28. A cylinder with circulation κ in a stream U at angle α.

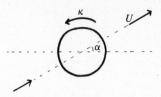

This theorem enables us to write down without any calculation all flows involving the simple boundary $|z| = a$. As a first example, consider a cylinder in a uniform stream. In this case

$$f(z) = Uz$$

for a stream of speed U along the x-axis. By the circle theorem, the complex potential we need is

$$w(z) = Uz + Ua^2/z.$$

This is just the stream plus dipole potential that we have seen before for this flow.

As a slight generalisation let us take a stream U at angle α to the axis, with a circulation κ round the cylinder, as in fig. XVI.28. Deal with the stream first: for this

$$f(z) = Uze^{-i\alpha}.$$

Now introduce the cylinder by the Milne–Thomson circle theorem:

$$\begin{aligned} w(z) &= Uze^{-i\alpha} + U(a^2/z)e^{i\alpha} \\ &= Uze^{-i\alpha} + Ue^{i\alpha}a^2/z, \end{aligned}$$

clearly a simple transformation of the previous formula. Now these two terms are an irrotational stream and dipole, giving no circulation round any curve. To introduce the circulation, add on a term

$$- (i\kappa/2\pi)\ln z;$$

this has no singularity (at any finite point) outside $|z| = a$, and has constant imaginary part on the circle, and gives the correct circulation. Hence it leaves the flow outside the streamline $|z| = a$ with no extra singularity but with a circulation κ. So the final potential is

$$w(z) = Uze^{-i\alpha} + Ue^{i\alpha}a^2/z - (i\kappa/2\pi)\ln z.$$

For a final example on the circle theorem let us take a stream and a cylinder, with a pair of vortices behind the cylinder (see fig. XVI.29), as is seen at Reynolds numbers of about 10–50. The function corresponding to a stream and two vortices is

$$f(z) = Uz + (i\kappa/2\pi)\ln(z - b) - (i\kappa/2\pi)\ln(z - \bar{b}).$$

Fig. XVI.29. A vortex pair in the wake of a cylinder.

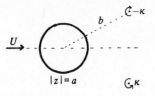

We have here that

$$\bar{f}(a^2/z) = Ua^2/z - (i\kappa/2\pi)\ln(a^2/z - \bar{b}) + (i\kappa/2\pi)\ln(a^2/z - b).$$

But it is easily seen that

$$\ln(a^2/z - \bar{b}) = \ln\{(-\bar{b})(1/z)(z - a^2/\bar{b})\}$$
$$= \ln(-\bar{b}) - \ln z + \ln(z - a^2/\bar{b}).$$

Hence

$$w(z) = f(z) + \bar{f}(a^2/z)$$
$$= U(z + a^2/z) + \frac{i\kappa}{2\pi}\ln\left\{\frac{(z - b)(z - a^2/b)}{(z - \bar{b})(z - a^2/\bar{b})}\right\},$$

on neglecting a constant.

We may ask for what value of κ and b the vortices remain at rest, as they are observed to do. For this, we need to calculate those parts of dw/dz which are *not* due to one of the vortices, at the position of that vortex. Call this

$$\{w'(b)\}^*.$$

Then the equation $\{w'(b)\}^* = 0$ will give information on b and κ in terms of U. Differentiation of the three logarithms separately gives

$$\{w'(b)\}^* = U - Ua^2/b^2 + \left(\frac{i\kappa}{2\pi}\right)\left\{\frac{-\bar{b}}{|b|^2 - a^2} - \frac{1}{b - \bar{b}} + \frac{b}{b^2 - a^2}\right\}.$$

It may be shown from $\{w'(b)\}^* = 0$ that the vortices must lie on the curve

$$2ry = r^2 - a^2$$

where $b = x + iy = re^{i\theta}$. It they lie at distance r on this curve, then

$$\kappa = 2Uy(1 - a^4/r^4).$$

Photographs show that the vortices lie on a curve of this general shape, which is sketched in fig. XVI.30. Typically at $R = 40$ the eddy is centred near $r = 2a$ with $\theta = 15°$. The real flow is reasonably well modelled by the above theory, though the details must be affected by viscosity. Using the model you can show that

$$\kappa = Ua$$

Fig. XVI.30. Possible positions of rest for the upper vortex in fig. XVI.29.

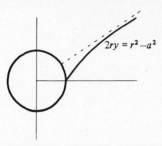

$$2ry = r^2 - a^2$$

approximately (a value which we found above for wake vortices). We could now go on to calculate, with these values for b and κ, the whole flow behind the cylinder. It would be rather a long calculation, and we do not pursue it.

5. Calculation of forces

(a) An integral for the force: Blasius' theorem

So far in this chapter we have concentrated on finding flow patterns, but now we must turn to the other preoccupation of fluid dynamics, finding the associated forces. Since this is a theory for inviscid fluid, the forces are derived from the pressure; and this in its turn is derived from Bernoulli's equation, of which the appropriate form is

$$p + \tfrac{1}{2}\rho v^2 = \text{constant}, \ p_0, \ \text{say}.$$

Now the complex velocity

$$u + iv = \overline{dw/dz},$$

and so

$$p = p_0 - \tfrac{1}{2}\rho |dw/dz|^2.$$

Note that we have neglected any body forces: we are only calculating forces due to flows. Moreover, we have assumed a steady flow field: forces due to unsteady flows, though important, are not discussed here.

We require the force on some surface S in the flow, and since the flow is two-dimensional this is equivalent to requiring

$$\int_C p\mathbf{n}ds,$$

where C is a curve in the flow field which represents the cross-section of S, $\mathbf{n}$ is a unit normal vector to C and ds is an element of length along C. Now in complex variable form dz (as shown in fig. XVI.31) is a length

Fig. XVI.31. Definition sketch for §5(*a*).

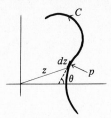

element directed along C, and so

idz

is a length element of C but with direction normal to C, because $i = e^{i\pi/2}$, and so multiplication by i gives a positive rotation of $\frac{1}{2}\pi$ and no change in magnitude. So the inward force element on a piece of C is

$$pidz = (p_0 - \tfrac{1}{2}\rho|dw/dz|^2)idz.$$

We will usually want to have force components parallel to the axes. From the diagram these are

$$\begin{cases} dX = -p\sin\theta\, ds, \\ dY = p\cos\theta\, ds, \end{cases}$$

where θ is the direction of dz, i.e. of the tangent to C. But

$$dz = e^{i\theta}ds$$
$$= (\cos\theta + i\sin\theta)ds$$

and so

$$dX - idY = -ip\overline{dz}.$$

The component force elements are therefore given by the formula

$$dX - idY = (\tfrac{1}{2}i\rho|dw/dz|^2 - ip_0)\,\overline{dz}.$$

Usually the pressure p_0 is irrelevant, as it will integrate to zero round a body; but over any *part* of a body it will give a non-zero force, which the material of the body has to withstand. The more interesting force is that due to the flow (i.e. not due to p_0) on a complete body which has a section given by a *closed* curve C. This force is given by

$$X - iY = \tfrac{1}{2}i\rho\int_C |dw/dz|^2\overline{dz}.$$

We can simplify this expression further when C is a streamline, as it must be if the body is rigid. For then $\psi = $ constant on C, and so $\mathscr{I}m\, w(z) = $

constant on C. That is, $\mathscr{I}m\,dw = 0$ on C, or

$$dw = \overline{dw} \text{ on } C.$$

This is equivalent to

$$(dw/dz)dz = \overline{\{(dw/dz)dz\}} \text{ on } C,$$

or

$$(dw/dz)dz = (\overline{dw/dz})\overline{dz} \text{ on } C.$$

So we can replace $|dw/dz|^2\overline{dz}$ in the integral by $(dw/dz)^2dz$ to get the final formula

$$X - iY = \tfrac{1}{2}i\rho \int_C (dw/dz)^2 dz.$$

In applications the moment of the forces about the origin is often as important as the total force; this is because a body can only be in equilibrium if both the total force *and* the total moment are zero. So we go on to calculate the total moment in a similar fashion.

The force element (dX, dY) acting at the position (x, y) has moment

$$dM = x\,dY - y\,dX$$

about the origin in a positive (anticlockwise) sense. This is related to $dX - idY$ – for which we have a formula – and $z = x + iy$ by

$$dM = \mathscr{R}e\{iz(dX - idY)\}.$$

Hence

$$dM = \mathscr{R}e\{z(p_0 - \tfrac{1}{2}\rho|dw/dz|^2)\overline{dz}\},$$

from the formula above for $dX - idY$. It is left as an exercise to show that

$$\int_C zp_0\overline{dz}$$

is zero round a closed curve C: it is hardly surprising that a uniform pressure gives a zero moment and so no tendency to spin. The total moment formula becomes

$$M = -\tfrac{1}{2}\rho\,\mathscr{R}e \int_C z|dw/dz|^2\overline{dz},$$

and this may be processed into a better looking form as before, by assuming that C is a streamline, to give

$$M = -\tfrac{1}{2}\rho\,\mathscr{R}e \int_C z(dw/dz)^2 dz.$$

This formula, together with

$$X - iY = \tfrac{1}{2}i\rho \int_C (dw/dz)^2 dz,$$

is known as Blasius' theorem for the force and moment on a rigid body fixed in the steady flow given by the complex potential $w(z)$.

(b) Simple examples of Blasius' theorem

Blasius' theorem seems at first sight to present nothing new. It is, after all, equivalent to integrating pressures round a surface, which we have done before. It is only when we take into account two theorems of complex function theory that we see that progress has really been made.

The first theorem (due to Cauchy) is that the two integrals

$$\int_{C_1} f(z)dz$$

and

$$\int_{C_2} f(z)dz$$

are equal provided that $f(z)$ has no singularities between C_1 and C_2. Hence we do not have to calculate the force by integrating round the (possibly awkward) curve that bounds the body: we can instead integrate round some more convenient curve such as the circle illustrated in fig. XVI.32.

The second theorem is that an analytic function has a series expansion

$$f(z) = \sum_{n=-\infty}^{\infty} a_n z^n;$$

this is known as the Laurent series, and is a generalisation of the Taylor series idea. If the integrands in Blasius' theorem are analytic (i.e. smooth, well-behaved) outside the body, then we can express forces and moment in terms of the coefficients a_n in a very simple fashion.

Fig. XVI.32. Contours for Cauchy's theorem.

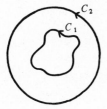

Fig. XVI.33. Circulation κ round a cylinder in a stream.

We will use these two theorems to calculate the force on an arbitrary cylinder, with cross-section C, in a uniform stream U when there is a circulation κ round the cylinder. The situation is sketched in fig. XVI.33. This is directly related to the kind of calculation we did in Chapter XI §9 on D'Alembert's paradox, where it took a lot of trouble to find out about the force; here we can do it for the special case of two dimensions in rather fewer steps.

At large distances the velocity is predominantly that due to a stream and a vortex at the origin (chosen to be inside C),

$$\overline{dw/dz} \sim U + i\kappa/(2\pi z)$$

or

$$dw/dz \sim U - i\kappa/(2\pi z).$$

Now the velocity is analytic outside C, and so the Laurent series for dw/dz must be

$$dw/dz = U - i\kappa/(2\pi z) + Az^{-2} + Bz^{-3} + \cdots$$

Blasius' formula for the force is about $(dw/dz)^2$, and by simple multiplication we see that

$$(dw/dz)^2 = U^2 - i\kappa U/\pi z + \{2AU - (\kappa/2\pi)^2\}z^{-2}$$
$$+ \text{ terms in } z^{-3} \text{ and so on.}$$

Hence we have to calculate terms like

$$\int_C z^{-n}dz$$

for $n \geqslant 0$, in order to calculate the force.

Fig. XVI.34. The equivalent circular contour for use in Blasius' theorem.

The calculations are very easy, when we choose to use a circle instead of C to do the integration round, as we can by Cauchy's theorem. For, on the circle $|z| = a$ (shown in fig. XVI.34), we have

$$\begin{cases} z^{-n} = a^{-n}e^{-in\theta}, \\ dz = ae^{i\theta}id\theta; \end{cases}$$

and so

$$\int_C z^{-n}dz = \int_{\theta=0}^{2\pi} ia^{1-n}e^{(1-n)i\theta}d\theta.$$

This integral is easily seen to have value zero for $n \neq 1$, and value $2\pi i$ when $n = 1$.

We can now use Blasius' theorem

$$X - iY = \tfrac{1}{2}i\rho \int_C \{U^2 - i\kappa U/\pi z + \text{terms in } z^{-n}(n \geqslant 2)\}dz$$

$$= -\tfrac{1}{2}i\rho(i\kappa U/\pi)2\pi i$$

$$= i\rho\kappa U.$$

That is, the force X parallel to the stream is zero, and the Magnus force perpendicular to the stream is

$$Y = -\rho\kappa U.$$

This is a very powerful result. It does not depend on the shape of C at all, which may even have sharp corners. However, its power is somewhat reduced by the requirement that potential flow shall be a good model for the flow, and so in reality C has at least to be streamlined for this result to be obviously realistic.

The moment theorem gives a less powerful result here. In this case we need $z(dw/dz)^2$, and the term in z^{-1} in the integration is then

$$2AU - (\kappa/2\pi)^2,$$

which gives a total moment about the origin

$$-\rho U \, \mathscr{I}m \, A.$$

The value of $\mathscr{I}m \, A$ is not known until you have solved the problem for the flow round C, in general. However, in particular cases it may be obvious that the moment must be zero by symmetry, and so $\mathscr{I}m \, A = 0$. For example, the flow past a plate set against a stream was shown in §2 to have complex potential

$$w(z) = U(z^2 + a^2)^{1/2}$$

with

$$dw/dz = Uz(z^2 + a^2)^{-1/2}.$$

This can be expanded for $|z| > a$ to give the Laurent series

$$dw/dz = U - \tfrac{1}{2}a^2/z^2 + \cdots$$

and the coefficient of z^{-2} is purely real, and so the moment on the plate is zero, as expected.

(c) An alternative method for calculating the integral

The examples in (b) above have been made easier by the fact that dw/dz had no singularity outside C, so that any circle round C was equivalent to C for the integration. When there are such singularities we may (if necessary) proceed slightly differently.

Take as example a line source outside a circular cylinder, as in fig. XVI.35. The complex potential for this can be derived from the circle theorem. In this case

$$f(z) = (m/2\pi)\ln(z - c)$$

for a source of volume flow rate m at position c, which can be taken to be on the real axis without any loss of generality. Then

$$\bar{f}(a^2/z) = (m/2\pi)\ln(a^2/z - c)$$
$$= (m/2\pi)\{\ln(z - a^2/c) - \ln z\}$$

to within a constant: the image system is, as we have seen before, an equal source at the image point a^2/c and an equal sink at the centre.

There is no trouble in calculating $(dw/dz)^2$ from these formulae and

$$w(z) = f(z) + \bar{f}(a^2/z).$$

We derive

$$(dw/dz)^2 = \left(\frac{m}{2\pi}\right)^2 \left\{ \frac{1}{z - c} + \frac{1}{z - a^2/c} - \frac{1}{z} \right\}^2,$$

and so for the force on the cylinder we need the integral

$$\int_C \left\{ \frac{1}{z - c} + \frac{1}{z - a^2/c} - \frac{1}{z} \right\}^2 dz.$$

Fig. XVI.35. Definition sketch for §5(c).

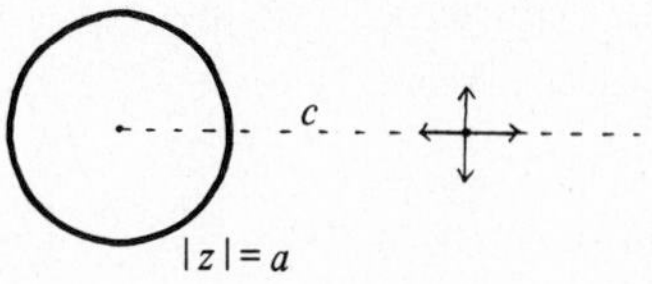

$|z| = a$

In this case we could proceed as before, by calculating the Laurent series of the integrand since C is already a circle. But in general we may replace C by a set of small circles round the singularities inside C (by Cauchy's theorem), and then use expansions near the singularities to integrate round the small circles. We choose this latter course here.

Consider for example the term

$$-2/\{z(z-c)\}$$

which comes into the integral for the force. This has a singularity inside C at $z = 0$, around which its expansion is given by

$$(2/zc)(1-z/c)^{-1} = (2/zc)(1 + z/c + z^2/c^2 + \cdots).$$

The 'residue' is the coefficient of z^{-1}, and the contribution is $2\pi i$ times the residue, which in this case is

$$4\pi i/c.$$

The reason why we need only the term in z^{-1} is, as before, because

$$\int_C z^n dz$$

round a circle is zero for any integer n except -1, and in the case $n = -1$ the value is $2\pi i$.

There are other terms in the integral of $(dw/dz)^2$ which have a non-zero residue. One is

$$2/\{(z-c)(z-a^2/c)\}$$

which has a singularity at $z = a^2/c$. We put $z = a^2/c + \zeta$ to get a suitable expansion near the singularity, and the term becomes

$$-2(c - a^2/c)^{-1}\{1 - c\zeta/(c^2 - a^2)\}^{-1}\zeta^{-1}.$$

The coefficient of ζ^{-1} is clearly $-2c/(c^2 - a^2)$.

The final term that contributes is

$$-2z^{-1}(z - a^2/c)^{-1},$$

which has singularities at both $z = 0$ and $z = a^2/c$. Its expansion near $z = 0$ is

$$2(c/a^2)z^{-1} + \text{terms in } z^{-n}, \quad n \geqslant 2.$$

And its expansion near $z = a^2/c$ has a coefficient for $(z - a^2/c)^{-1}$ equal to $-2c/a^2$.

So finally the residues of $(dw/dz)^2$ inside $|z| = a$ are in total

$$(m/2\pi)^2\{2/c - 2c/(c^2 - a^2)\} = -(\tfrac{1}{2}m^2a^2/\pi^2)c^{-1}(c^2 - a^2)^{-1},$$

Fig. XVI.36. The corresponding set of line charges.

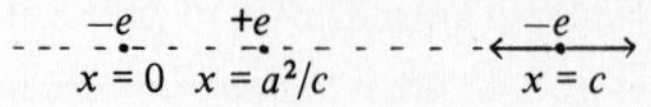

$$-\,-\,-\overset{-e}{\underset{x=0}{\bullet}}\,-\,-\,\overset{+e}{\underset{x=a^2/c}{\bullet}}\,-\,-\,-\,-\,-\,\overset{-e}{\underset{x=c}{\bullet}}\rightarrow$$

and the force is derived from

$$X - iY = \tfrac{1}{2}i\rho 2\pi i(-\tfrac{1}{2}m^2 a^2/\pi^2)c^{-1}(c^2 - a^2)^{-1}.$$

The force is therefore an attraction of amount

$$(\tfrac{1}{2}\rho m^2 a^2/\pi)c^{-1}(c^2 - a^2)^{-1}.$$

This formula for the force is very like the one between line charges in electrostatics. If we have charges $+e$ at $x = c$ and $x = a^2/c$ and $-e$ at $x = 0$, as shown in fig. XVI.36, then the force per length on the charge e at $x = c$ is

$$2(4\pi\varepsilon_0)^{-1}\{-e^2(c - a^2/c)^{-1} + e^2 c^{-1}\}$$
$$= (-\tfrac{1}{2}e^2 a^2/\pi\varepsilon_0)c^{-1}(c^2 - a^2)^{-1}.$$

Thus line sources of volume flow rate m seem to behave in this respect very like line charges e, except that e^2/ε_0 must be replaced by ρm^2.

(d) Example: a vortex and a plane

As a final example on Blasius' theorem, consider, as in fig. XVI.37, a vortex at $z = ia$ above a fixed plane $y = 0$. The flow is found from the first of the image theorems (or from earlier work) to have complex potential

$$w(z) = -(i\kappa/2\pi)\ln(z - ia) + (i\kappa/2\pi)\ln(z + ia).$$

The integral we have to evaluate for the force on the plane is not round a closed curve, but along the line $y = 0$ from $-\infty$ to $+\infty$. This means that we ought to include the pressure p_0 in the calculation, but we shall ask only for the force due to the flow. We need to consider the plane as the upper surface of a (very large) semicircular cylinder to see that in Blasius' theorem the integral needs to be taken in the direction shown in

Fig. XVI.37. Definition diagram for §5(d).

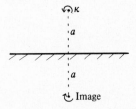

Fig. XVI.38. The plane replaced by the upper surface of a large semicircular cylinder.

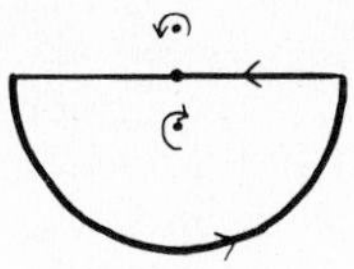

fig. XVI.38 – we always go round the cross-section anticlockwise. Hence the integral we need to calculate is

$$\tfrac{1}{2}i\rho(-\kappa^2/4\pi^2)\int_{C_1}\{(z+ia)^{-1}-(z-ia)^{-1}\}^2 dz$$

where C_1 is the real axis traversed from $+\infty$ to $-\infty$.

We may convert this integral into one round a closed curve by using the fact that the integrand is quite small at large distances. It is not hard to show that

$$dw/dz \sim \text{constant}/z^2$$

in this case, and so

$$(dw/dz)^2 \sim \text{constant}/z^4.$$

So we consider the integral round the (closed) semicircle C shown in fig. XVI.39, of radius R where R is assumed to be large. Then the curved part contributes a term of size

$$(\text{constant}/R^4) \times \pi R$$

to the integral round C, which clearly tends to zero as $R \to \infty$. And the remainder of the integral, from $x = R$ to $x = -R$, gives the integral along C_1 that we need, when $R \to \infty$.

We may calculate the integral round C by the method of residues that we used in (c) above. Notice that in this case there is no circle to integrate round that surrounds C_1 and that does not contain extra singularities, so we must use this method. The integrand

$$\left\{\frac{1}{z+ia}-\frac{1}{z-ia}\right\}^2$$

Fig. XVI.39. The contour for Blasius' theorem in §5(d).

has a singularity at $z = -ia$ inside C. Putting $z + ia = \zeta$ gives

$$\left\{ \frac{1}{\zeta} - \frac{1}{\zeta - 2ia} \right\}^2$$

which has residue (coefficient of ζ^{-1} when we expand for small values of of ζ) $-i/a$. Hence the integral has value

$$\tfrac{1}{2} i \rho (-\kappa^2/4\pi^2)(-i/a) 2\pi i.$$

That is

$$X - iY = -i\rho\kappa^2/(4\pi a);$$

which is to say that the force on the plane is towards the vortex, and of amount

$$\rho/2\pi \times (\text{circulation})^2 \div \text{distance to image}.$$

This is a very similar result to that derived in (*c*).

Exercises

1. Show from the Cauchy–Riemann equations that the curves $\phi = \text{constant}$ and $\psi = \text{constant}$ intersect at right angles. Show also that the curves $\phi = \text{constant}$ can be taken as the streamlines of some flow; identify this flow when $w(z) = A \ln z$ and when $w(z) = A/z$, for real A.

2. Show that

$$\mathscr{R}e \int_C (dw/dz)dz$$

   is the circulation round the closed curve C. Verify that

   $$w(z) = i \ln z$$

   has a non-zero circulation for all curves C round the origin, and that the other $w(z)$ in §1(*b*) give zero circulation.

3. Write down $f(\bar{z}), \bar{f}(z), \bar{f}(\bar{z})$ and $\bar{f}(a^2/z)$ for a real and
   (i) $f(z) = \ln z$,
   (ii) $f(z) = i \ln(z - z_0)$,
   (iii) $f(z) = e^{i\alpha}/(z - z_0)$,
   (iv) $f(z) = \coth(\pi a/z)$.

4. Show that

   $$w(z) = U(a - b)^{-1}\{az - b(z^2 - a^2 + b^2)^{1/2}\},$$

   with the positive square root chosen when z is real and greater than $(a^2 - b^2)^{1/2}$, represents the flow of a stream past the ellipse given by

   $$x = a \cos\theta, \quad y = b \sin\theta.$$

5. Show that

$$w(z) = \sec^2 z$$

represents a flow in the strip

$$y > 0, \quad 0 < x < \tfrac{1}{2}\pi.$$

Sketch the streamlines.

6. Show that the two complex potentials

$$w_1(z) = \cosh^{-1}(z/c),$$
$$w_2(z) = i\pi z^2/(2c^2)$$

each represent flow between the two branches of the hyperbola $x^2 - y^2 = \tfrac{1}{2}c^2$, with $\psi = \tfrac{1}{4}\pi$ on one branch. Do these potentials represent the same flow? Sketch the streamlines of the flow(s).

7. Two (line) vortices each with circulation κ are situated on the real axis at $\pm a$. A third vortex of circulation $-\tfrac{1}{2}\kappa$ is at the origin. Show that the vortices are at rest, and sketch the streamlines of the flow.

8. The circle $|z| = a$ has five vortices each of circulation $2\pi C$ on it at the points $a\exp(2ir\pi/5)$ at time $t = 0$. Show
 (i) by elementary calculation of contributions from other vortices,
 (ii) by showing that the complex potential is

$$w(z) = -iC\ln(z^5 - a^5),$$

that the whole array moves round the circle with angular velocity

$$2C/a^2.$$

9. A stream of speed U parallel to the x-axis moves above the plane $y = 0$ and there is a (line) dipole of strength μ situated at $z = ai$, pointing in a direction at α to the x-axis. Show that if $U = (2\mu\cos\alpha)/a^2$ there is a stagnation point at the origin.

10. A vortex has just been shed from one side of a cylinder in a stream. This is modelled by a vortex of circulation $-\kappa$ at the point $z = ce^{i\alpha}$ outside a cylinder of radius a around which there is a circulation κ, with a uniform stream U at large distances. Write down the complex potential $w(z)$ and calculate the velocity of the vortex. Sketch its path in the case $\kappa = Ua$.

11. Invent and prove a theorem on images in the flanged semicircle boundary sketched in fig. XVI.40

$$y = 0 \quad \text{for } |x| > a,$$
$$|z| = a \quad \text{for } 0 < \theta < \pi.$$

Hence solve again for the flow described in Chapter XI §4.

12. Show that

$$w(z) = U(z^2 + a^2)^{1/2} - (i\kappa/2\pi)\ln\tfrac{1}{2}\{z + (z^2 + a^2)^{1/2}\}$$

is suitable for flow of a uniform stream past a flat plate of width $2a$, with

Fig. XVI.40. A flanged semicircular boundary for Q11.

circulation κ round the plate. Calculate the force and moment on the plate.

13. Calculate the force on the cylinder for Exercise 10 above by using the method of §5(c).

14. A (line) dipole of strength μ and making an angle α with the x-axis is at the point $z = ai$ above the rigid plane $y = 0$. Calculate the force on the plane due to the flow.

References

(a) Useful photographs of the vortex motion associated with a stream flowing past a cylinder can be found in, for example, *Introduction to Fluid Dynamics*, G. K. Batchelor, C.U.P. 1967 and *Physical Fluid Dynamics*, D. J. Tritton, Van Nostrand Reinhold 1977.

(b) Detailed calculations on the stability and drag associated with the Kármán vortex street can be found, for example, in *Hydrodynamics*, W. Lamb, C.U.P. 1945.

(c) Fuller treatments of the complex function method may be found in many texts, for example *Theoretical Hydrodynamics* (2nd edn), L. M. Milne–Thomson, Macmillan 1949.

(d) Any text on complex function theory contains more than is needed in this chapter.

XVII

Conformal mappings and aerofoils

1. An example

The three potentials

(i) $w(z) = Uz$,

(ii) $w(z) = Az^2$

(iii) $w(z) = Bz^{1/2}$,

have been seen in Chapter XVI. They correspond to flows with rather simple boundaries

(i) $\theta = 0$ and $\theta = \pi$,

(ii) $\theta = 0$ and $\theta = \frac{1}{2}\pi$,

(iii) $\theta = 0$ and $\theta = 2\pi$,

with the flows starting on one boundary and ending up along the real axis $\theta = 0$, if U, A, B are all positive. Fig. XVII.1 sketches these flows.

It looks as though the powers in

$$w = Cz^n$$

are related to the corresponding boundaries in a simple way: all the boundaries are

$$\theta = 0 \text{ and } \theta = \pi/n.$$

What is happening here is that we are comparing

$$Uz \text{ and } Cz^n.$$

Fig. XVII.1. The flows for (i) $w = Uz$, (ii) $w = Az^2$, (iii) $w = Bz^{1/2}$.

(i)

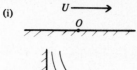

(ii)

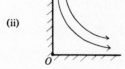

(iii)

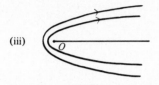

Now $Uz = U(z^{1/n})^n$, and so we are considering the mapping

$$z \mapsto z^{1/n},$$

under which the line $\theta = \pi$ is mapped to $\theta = \pi/n$.

We may guess from the above cases that if we can solve a problem with the boundary $y = 0$, then the mapping

$$z \mapsto z^{1/n}$$

will enable us to solve it for the 'corner' region.

Let us try an example of this transformation. The flow of the wind past the corner of a building is often observed to cause a vortex in which leaves and dust whirl round, much as in fig. XVII.2. This is a flow with boundaries $\theta = 0$ and $\theta = \frac{3}{2}\pi$, which looks as if it might be solved by the transformation

$$z \mapsto z^{2/3}.$$

Certainly, we have no means available otherwise for solving this sort of problem.

Now we know, from the method of images, that the potential for a

Fig. XVII.2. A vortex in a flow past a corner.

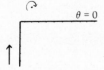

Fig. XVII.3. A vortex in a stream along a wall.

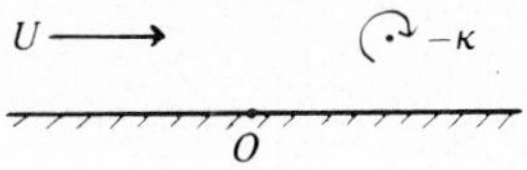

stream and a vortex above the wall $y = 0$ (see fig. XVII.3) is

$$w(z) = Uz + (i\kappa/2\pi)\{\ln(z - z_0) - \ln(z - \bar{z}_0)\}$$

where z_0 is the position of the vortex, and where $0 < \theta < \pi$. Let us try, therefore, the potential

$$W(z) = Uz^{2/3} + (i\kappa/2\pi)\{\ln(z^{2/3} - z_0) - \ln(z^{2/3} - \bar{z}_0)\},$$

where $0 < \theta < \frac{3}{2}\pi$ and see whether it gives:

    (i) a boundary (on which $\psi = $ constant) along $\theta = \frac{3}{2}\pi$ and $\theta = 0$;
    (ii) a vortex at some point outside this boundary;
    (iii) a flow along the wall $\theta = \frac{3}{2}\pi$;
    (iv) no extra singularities in the flow, i.e. for $0 < \theta < \frac{3}{2}\pi$.

We test these one by one.

    (i) If $z = re^{3\pi i/2}$,

$$W(z) = Ur^{2/3}e^{i\pi} + (i\kappa/2\pi)\{\ln(r^{2/3}e^{i\pi} - z_0) - \ln(r^{2/3}e^{i\pi} - \bar{z}_0)\}.$$

Now $e^{i\pi} = -1$, and the logarithm terms are

$$(i\kappa/2\pi)\ln(-r^{2/3} - z_0) + \overline{(i\kappa/2\pi)\ln(-r^{2/3} - z_0)},$$

which is automatically real. Hence on this line $W(z)$ is real, and so it is a streamline. Similarly the line $\theta = 0$ is a streamline for $W(z)$.

    (ii) $W(z)$ has a singularity at

$$z^{2/3} - z_0 = 0,$$

i.e. at

$$z = z_0^{3/2}.$$

Near this point we write

$$z = z_0^{3/2} + \zeta,$$

where ζ is taken to be small. Then

$$\ln(z^{2/3} - z_0) = \ln\{(z_0^{3/2} + \zeta)^{2/3} - z_0\}$$
$$\doteqdot \ln\{2\zeta/(3z_0^{1/2})\}$$

by using the binomial theorem. So *near* the singularity we have

$$W(z) \doteqdot (i\kappa/2\pi)\ln\zeta + \text{constant}$$

which is equivalent to

$$W(z) = (i\kappa/2\pi)\ln(z - z_0^{3/2}).$$

Thus we have a vortex of circulation $-\kappa$ at the point $z_0^{3/2}$ in $W(z)$.

(iii) The velocity from the vortex falls off like $1/z$ whereas the velocity from the term $Uz^{2/3}$ falls off like $z^{-1/3}$ at large distances. So well away from the corner and the vortex we have

$$W(z) \sim Uz^{2/3}$$

which has velocity given by

$$u - iv = 2U/(3z^{1/3}).$$

Now on $\theta = 0$ this is real and so gives a flow along the wall $\theta = 0$. And on $\theta = \frac{3}{2}\pi$ this is

$$\begin{aligned} u - iv &= 2U/\{3(re^{3\pi i/2})^{1/3}\} \\ &= 2U/\{3r^{1/3}e^{i\pi/2}\} \\ &= -2iU/(3r^{1/3}), \end{aligned}$$

which gives a flow up the wall $\theta = \frac{3}{2}\pi$.

(iv) The only other possible singularity in $W(z)$ is where

$$z^{2/3} = \bar{z}_0$$

However, we are taking $z^{2/3}$ to be the function which is real when $\theta = 0$, and so in the flow region $z^{2/3}$ has argument between 0 and π. But $\bar{z}_0$ must have argument between $-\pi$ and 0 (if z_0 is to be in the flow). Hence we cannot have this singularity occurring in the region of the flow.

What we have demonstrated, then, is that

$$W(z) = Uz^{2/3} + (i\kappa/2\pi)\{\ln(z^{2/3} - z_0) - \ln(z^{2/3} - \bar{z}_0)\}$$

represents a flow round a corner of angle $\frac{3}{2}\pi$, with a vortex near the corner. We could now go on to analyse this flow further, for example by asking what values of κ and z_0 give a vortex at rest, in exactly the same way as in the last chapter. But we shall concentrate instead on the more fundamental thing that we have also done, which is to use a simple flow (a vortex in a stream with a wall) to solve a more complicated problem, by using a mapping to get from one problem to another.

Incidentally, we have also shown a method for generating further image theorems: the vortex in the flow round the corner has an 'image' in the corner derived from the image in the wall $y = 0$ that we started with. But we shall not pursue this either.

2. Mappings in general

(a) Transformation of boundaries and regions

We treat mappings in more generality in this section. We assume a flow in the z-plane which has potential

$$w(z)$$

and fixed boundaries C enclosing a domain D; since the boundaries are fixed, $\mathscr{I}m\, w(z) = $ constant on C to give a streamline. Let us take a mapping

$$\zeta = f(z)$$

with inverse

$$z = F(\zeta),$$

where both f and F will be analytic for most points. With this transformation the point z maps to the point

$$\zeta = f(z),$$

and the curve C maps to some new curve C' enclosing a new domain D'. Fig. XVII.4 illustrates this transformation.

In the example in §1 we had

$$\begin{cases} \zeta = z^{2/3}, \\ z = \zeta^{3/2}, \end{cases}$$

and C was $y = 0$ which mapped to the new boundary

$$C' = \{y = 0,\, x \geqslant 0\} \cup \{x = 0,\, y \leqslant 0\};$$

the domain D was $\{y < 0\}$ (or $0 < \theta < \pi$), while D' was $\{x > 0$ and $y < 0\}$ (or $-\frac{1}{2}\pi < \theta < 0$). In this case the functions f and F are not analytic at the origin, and we need a cut plane to define the functions properly; the

Fig. XVII.4. A domain and its contour transformed by $\zeta = f(z)$.

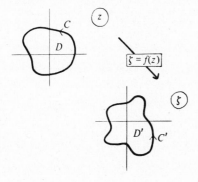

boundaries were such that the cut could go from 0 to ∞ without being in the flow.

(b) Critical points and conformal mappings

Let us take a small element dz at z_0 in the z-plane, and transform it by means of

$$\zeta = f(z).$$

Now if f is differentiable at z_0, which we shall assume to be the case, we must have

$$d\zeta = f'(z_0)dz.$$

It is clear that places where

$$f'(z_0) = 0$$

will be rather special: we shall call them 'critical points' of the transformation. At such a critical point

$$\begin{cases} d\zeta/dz = 0, \\ dz/d\zeta \text{ does not exist (or is infinite)}; \end{cases}$$

that is, critical points are singularities of the inverse transformation $z = F(\zeta)$, points where $F(\zeta)$ is not differentiable and so not analytic.

It is easy to give examples of such critical points. The mapping

$$\zeta = z^2 + z$$

has $f'(z_0) = 0$ when $2z_0 + 1 = 0$, i.e. at $z_0 = -\frac{1}{2}$. The mapping

$$\zeta = z + a^2/z$$

has $f'(z_0) = 0$ when $1 - a^2/z_0^2 = 0$, i.e. at $z_0 = \pm a$. In this latter transformation, it might seem that we ought to have a special name for the point $z = 0$, since ζ looks as though it is not defined there; in fact we do not need to make much special mention of such a point – it is merely mapped to 'the point at infinity'.

If z_0 is not a critical point, then, we have

$$\begin{cases} d\zeta = f'(z_0)dz, \\ f'(z_0) \text{ is not zero}. \end{cases}$$

Now suppose

$$dz = \varepsilon e^{i\alpha}$$

gives the modulus and argument of dz; $f'(z_0)$ may also be put in this form as

$$f'(z_0) = Re^{i\beta},$$

Fig. XVII.5. Transformation of angles at a point by $\zeta = f(z)$.

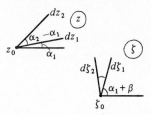

so that

$$d\zeta = \varepsilon \, R e^{i(\alpha + \beta)}.$$

Which means that under this transformation, the element dz at z_0 is

$$\begin{cases} \text{magnified by a factor } R = |f'(z_0)|, \\ \text{rotated by an angle } \beta = \arg f'(z_0). \end{cases}$$

Consider two elements at z_0

$$\begin{cases} dz_1 = \varepsilon_1 \, e^{i\alpha_1} \\ dz_2 = \varepsilon_2 \, e^{i\alpha_2}. \end{cases}$$

These map to $d\zeta_1$ and $d\zeta_2$, where *each* is found by a magnification of $|f'(z_0)|$ and a rotation by $\arg f'(z_0)$ as is shown in fig. XVII.5. However, angle between dz_1 and dz_2 is not changed by the mapping, since

$$\alpha_2 - \alpha_1 \mapsto \alpha_2 + \beta - (\alpha_1 + \beta)$$
$$= \alpha_2 - \alpha_1.$$

This is what is meant by calling a mapping 'conformal': the angles between curves are preserved in the transformation, except at critical points.

Fig. XVII.6. Transformation of a quarter annulus by $\zeta = z^2$.

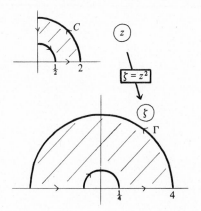

Let us take as an example the mapping of the closed figure shown in fig. XVII.6 by the transformation

$$\zeta = z^2,$$

which is conformal except at the critical point $z = 0$. Notice that a direction is marked on the curve that we map, and the 'inside' of the curve is indicated: it is useful to do this as directions may be altered, and 'inside' sometimes maps to 'outside'. It is not hard to verify that this mapping has the effect shown; for example a point on the quarter circle $z = 2e^{i\theta}, 0 < \theta < \frac{1}{2}\pi$, maps to the point

$$\zeta = 4e^{2i\theta}$$

on the semicircle $0 < 2\theta < \pi$; and a point on the imaginary axis $z = iy$, $\frac{1}{2} < y < 2$, maps to

$$\zeta = -y^2$$

with $-y^2$ lying between -4 and $-\frac{1}{4}$. In this example, lengths have all varied – locally the factor is $2|z_0|$; angles have also changed, having $2\arg z_0$ added to them. But the angles at which curves meet have been preserved: the figure started off with right angled corners and it ended up with them too.

In this example the mapping is not conformal at the origin. An element $dz = \varepsilon e^{i\alpha}$ with one end at the origin (as in fig. XVII.7) transforms to

$$d\zeta = (dz)^2$$
$$= \varepsilon^2 e^{2i\alpha},$$

Fig. XVII.7. Doubling of angles at the origin by $\zeta = z^2$.

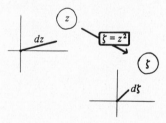

Fig. XVII.8. At a conformal point no corner is created on a smooth curve.

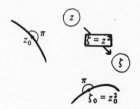

so that the angle here has doubled. It is *not* a universal rule that angles double at critical points: you have to examine each case specially.

It is elementary to verify in this example that the individual parts of the boundary C map into smooth curves. It is also true in general, that a conformal mapping can introduce no extra corners and so a smooth curve maps to a smooth curve. For take z_0 on a smooth part of C as in fig. XVII.8: the two parts of C leave at angle π to each other from z_0, and this angle is preserved by the conformal mapping, and so the image curve is smooth at $\zeta_0 = f(z_0)$. But at critical points corners may indeed be introduced: in §1 the real axis is transformed into two lines at right angles by a transformation which has a critical point at $z = 0$.

(c) *The transformed potential*

Suppose we have a potential $w(z)$ which describes a flow past the boundary C in the z-plane. What do we get by making the transformation

$$\begin{cases} \zeta = f(z), \\ z = F(\zeta)? \end{cases}$$

Firstly, if $w(z)$ is an analytic function in a domain, and if F is also analytic, then the composite function

$$w\{F(\zeta)\}$$

is analytic. Let us call this function

$$W(\zeta).$$

It must therefore represent some flow in the ζ-plane, with velocity components given by

$$U - iV = W'(\zeta);$$

this may be easily expressed in terms of the original potential as

$$U - iV = (dw/dz)(dz/d\zeta)$$
$$= w'(z)F'(\zeta),$$

or, more coherently,

$$U - iV = w'\{F(\zeta)\} F'(\zeta).$$

Secondly, if C is a rigid boundary for the flow in the z-plane with potential $w(z)$, then $\mathscr{I}m\, w = \text{constant}$ on C. That is

$$w(z_c) = \text{real} + i \times \text{constant}$$

when z_c is on C. Now let z_c transform to ζ_Γ, a point on the curve Γ which is the transform of C. Then

$$W(\zeta_\Gamma) = w\{F(\zeta_\Gamma)\}$$

Fig. XVII.9. The vortex above a wall with a stream.

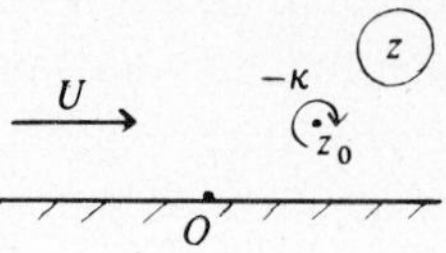

by definition of W, and since

$$\begin{cases} \zeta_\Gamma = f(z_c), \\ z_c = F(\zeta_\Gamma), \end{cases}$$

we must have

$$W(\zeta_\Gamma) = w(z_c)$$
$$= \text{real} + i \times \text{constant}.$$

So Γ is a streamline for $W(\zeta)$, whenever C is a streamline for $w(z)$.

Let us rediscuss the transformation that was done in §1 in the present terms. The flow in the z-plane (illustrated in fig. XVII.9) is a stream above the line $y = 0$ with a vortex at z_0. The potential is

$$w(z) = Uz + (i\kappa/2\pi)\{\ln(z - z_0) - \ln(z - \bar{z}_0)\}$$

which has $y = 0$ as a streamline, on which

$$\mathscr{I}m\, w(z) = 0.$$

The transformation is

$$\begin{cases} \zeta = z^{3/2}, \\ z = \zeta^{2/3}, \end{cases}$$

which are analytic functions in domains that avoid the origin and a cut from 0 to ∞. The new potential is

$$W(\zeta) = w(\zeta^{2/3})$$
$$= U\zeta^{2/3} + (i\kappa/2\pi)\{\ln(\zeta^{2/3} - z_0) - \ln(\zeta^{2/3} - \bar{z}_0)\}$$

as before. The streamline

$$y = 0, x > 0$$

transforms to

$$\zeta = x^{3/2};$$

and the streamline

$$y = 0, x < 0$$

transforms to

$$\zeta = |x|^{3/2} e^{3\pi i/2}.$$

Fig. XVII.10. Fig. XVII.9 transformed by $\zeta = z^{3/2}$.

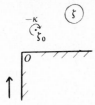

Hence the streamline $y = 0$ transforms to the streamline $\mathscr{I}m\ W(\zeta) = 0$, which is the corner shown in fig. XVII.10. The transformation is *not* conformal at $z = 0$, where $f'(0) = 0$, and angles there are multiplied by $\frac{3}{2}$.

It is clear from our previous work that we might expect separation at the corner in the ζ-plane, with a dividing streamline $\mathscr{I}m\ W = 0$ passing round the vortex, perhaps as in fig. XVII.11. Call this streamline Γ; it must be derived from a streamline C of the original flow which separates from the plane at the origin, and has equation $\mathscr{I}m\ w = 0$. Now in the original flow, streamlines which separate from the plane usually do so at right angles as in fig. XVII.12 (see Chapter V §5(c) and Exercise 9), and so this dividing streamline must leave at $\frac{1}{2}\pi \times \frac{3}{2} = \frac{3}{4}\pi$ in the ζ-plane. This surprising result suggests that this model of flow round a corner of this shape may not be too good in reality; for we would expect the dividing streamline Γ to set off along the imaginary axis in the ζ-plane.

(d) *Transformation of singularities and boundary conditions*

We have already seen a vortex transform into an equal vortex in §1, in one particular transformation. In general, let $w(z)$ contain a term

$$A \ln(z - z_0),$$

Fig. XVII.11. An expected dividing streamline in the ζ-plane.

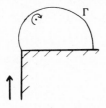

Fig. XVII.12. The dividing streamline in the z-plane.

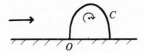

and use the transformation $\zeta = f(z)$ for which z_0 is *not* a critical point. Then we may write the same term as

$$A \ln\{F(\zeta) - F(\zeta_0)\}$$

as part of $W(\zeta)$. But

$$F(\zeta) = F(\zeta_0) + (\zeta - \zeta_0)F'(\zeta_0) + \mathrm{O}(\zeta - \zeta_0)^2,$$

and so near ζ_0 we have the term

$$A \ln\{F'(\zeta_0)(\zeta - \zeta_0)\}$$

approximately, or

$$A \ln(\zeta - \zeta_0) + \text{constant}.$$

This gives us back a term of the same form as we started off with. So it is generally true that

$$\begin{cases} \text{sources transform to sources,} \\ \text{vortices transform to vortices,} \end{cases}$$

away from a critical point.

The situation at a critical point is somewhat different. Take as example a source of volume flow rate m at the origin. It has potential

$$w(z) = (m/2\pi)\ln z.$$

Now apply the transformation

$$\zeta = z^2$$

which has a critical point at the origin. We derive $W(\zeta)$ as usual, by using

$$z = \zeta^{1/2},$$
$$W(\zeta) = (m/2\pi)\ln \zeta^{1/2}$$
$$= (m/4\pi)\ln \zeta.$$

This is still the potential for a source, but the strength is halved. This may be seen to be reasonable by looking at the flow into the z-plane region

Fig. XVII.13. Transformation of a source at the origin by $\zeta = z^2$.

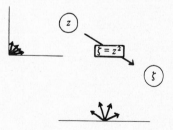

$0 < \theta < \frac{1}{2}\pi$. On putting $\zeta = z^2$, the region opens out to $0 < \arg \zeta < \pi$, as you see in fig. XVII.13, and so the amount of flow which went into an angle $\frac{1}{2}\pi$ now has to fill π: the strength is halved. Again, this is *not* a general result $-\zeta = z^3$ would cut the strength to one third, and so on.

Dipoles are (almost) always altered in strength by a conformal mapping. If

$$w(z) = A/(z - z_0)$$

is mapped by

$$z = F(\zeta)$$

we obtain

$$W(\zeta) = A/\{F(\zeta) - F(\zeta_0)\}$$
$$\doteq \{A/F'(\zeta_0)\}/(\zeta - \zeta_0),$$

by using Taylor's theorem. This is another dipole at the corresponding point ζ_0, but the strength is now modified by the complex number $F'(\zeta_0)$. Thus both strength and direction of the dipole will in general be changed. This should not really be a surprise, if we think of a dipole as a limit of a source and sink separated by an element dz; we have seen that dz is magnified and rotated by $\zeta = f(z)$, and this gives a different strength and direction to the dipole. Fig. XVII.14 illustrates this transformation.

Boundary conditions may sometimes be given in terms of velocity, for example

$$u + iv \sim \text{a uniform stream parallel to the } x\text{-axis.}$$

In the ζ-plane the boundary condition will be different, but simply related to the z-plane condition. The relation comes through

$$dw/dz = (dW/d\zeta)(d\zeta/dz),$$

or

$$u - iv = (U - iV)f'(z).$$

So, for example, if

$$\left. \begin{cases} u \sim u_0 \\ v \sim 0 \end{cases} \right\} \quad \text{as } |z| \to \infty,$$

Fig. XVII.14. Transformation of an approximate dipole by $\zeta = f(z)$.

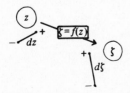

then we use

$$U - iV = (u - iv)/f'(z)$$

to find the asymptotic forms of U and V. They must be

$$\left\{ \begin{matrix} U \sim u_0 \mathscr{R}e\{f'(z)\}^{-1} \\ V \sim -u_0 \mathscr{I}m\{f'(z)\}^{-1} \end{matrix} \right\} \quad \text{as } |z| \to \infty.$$

3. Particular mappings

(*a*) *The underlying problem*

Conformal mappings can be used constructively to solve a considerable number of problems in the two-dimensional flow of an ideal fluid (inviscid, incompressible). The method has, in essence, been shown in §1:

- (i) take a flow in the *z*-plane past boundaries *C*, for which you know the complex potential $w(z)$;
- (ii) apply conformal mappings to the boundaries *C* to get the boundaries Γ in the ζ-plane that you want the flow round;
- (iii) the complex potential you need for the flow round Γ is just $W(\zeta)$, which is $w(z)$ put in terms of ζ.

This seems to be a programme with only one snag in it: finding the function

$$\zeta = f(z)$$

that maps the given *C* to the required Γ; or inversely, the function

$$z = F(\zeta)$$

that maps the boundaries Γ to some boundaries *C* for which you *can* solve the flow problem. This may indeed be difficult or impossible. But practice with simple transformations can often enable you to build up a mapping that does what is needed, as a sequence of easy transformations. It is rather like integrating by substitution – you may have to do several substitutions of standard type before you finally get the answer, and only practice will tell you which substitution to use. So we go on to look at a few basic transformations.

(*b*) *Linear transformations*

The simplest transformation is

$$\zeta = az + b$$

for constants *a* and *b* (they will be complex in general). This has

$$f'(z) = a$$

Fig. XVII.15. Linear transformation of the unit circle.

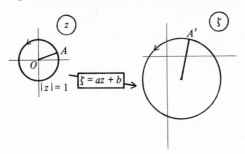

which is constant, so the magnification and rotation are constant over the whole plane. Moreover, $z = 0$ becomes $\zeta = b$, so there is a shift of origin. The transformation is shown in fig. XVII.15. The unit circle in the z-plane transforms to a circle of radius

$$|a|$$

round the point b, and each radius is rotated by an angle

$$\arg a.$$

Note that two particular cases of this transformation are
(i) $\zeta = z + b$, a simple shift of origin,
(ii) $\zeta = e^{i\alpha}z$, a simple rotation of α.
These were used in the last chapter, and are now put in a proper context.

In this transformation the direction in which a curve is traversed is not changed – anticlockwise goes to anticlockwise; and the inside of a closed curve becomes the inside of the transformed curve.

(c) *Powers and inversion*
The mappings

$$\zeta = z^k,$$

which are conformal in regions which avoid the origin and any necessary cuts, have already been seen in §1 for the cases when k is a positive integer or a positive fraction less than 1. There is one other important mapping of this type, given by

$$\zeta = z^{-1},$$

called 'inversion'.

This transformation takes

$$z = re^{i\theta}$$

Fig. XVII.16. Inversion of the unit circle.

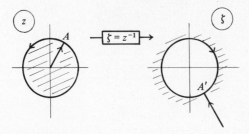

and gives the result

$$\zeta = r^{-1}e^{-i\theta}.$$

This is very like the 'image in a circle' of Chapter XVI §4, except that $0 < \arg z < \pi$ is transformed into $-\pi < \arg \zeta < 0$. But a point outside $|z| = 1$ gives a point inside $|\zeta| = 1$, at the 'image' distance. The point $z = 0$ is transformed into the point at infinity, and vice versa. Fig. XVII.16 shows these changes, which imply that the region $|z| < 1$ is converted into the outside of the unit circle in the ζ-plane. Notice that the sense of traversing the unit circle, and radius to A, is also changed.

The effect of inversion on circles is rather suprising. Take a circle around the point $z = ae^{i\alpha}$. The point P on it can be put as

$$z = ae^{i\alpha} + re^{i\theta}.$$

Then $\zeta = z^{-1}$ and the corresponding points can be calculated. It is easiest to proceed as follows. Let

$$\zeta - ae^{-i\alpha}/(a^2 - r^2) = \zeta - \zeta_0.$$

This is equal to

$$\frac{1}{ae^{i\alpha} + re^{i\theta}} - \frac{ae^{-i\alpha}}{a^2 - r^2}$$

$$= \frac{-r^2 - are^{i(\theta - \alpha)}}{(a^2 - r^2)(ae^{i\alpha} + re^{i\theta})} = \frac{-re^{i\theta}}{a^2 - r^2} \left\{ \frac{re^{-i\theta} + ae^{-i\alpha}}{re^{i\theta} + ae^{i\alpha}} \right\}.$$

And so

$$|\zeta - \zeta_0| = r/|a^2 - r^2|$$
$$= \text{constant},$$

i.e. independent of θ. Hence the circle on which P lies in the z-plane maps into a circle round

$$\zeta_0 = ae^{-i\alpha}/(a^2 - r^2)$$

in the ζ-plane, provided $a \neq r$. Both these circles are drawn in fig. XVII.17.

Fig. XVII.17. Inversion of a general circle.

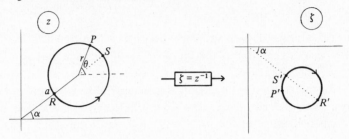

This transformation of circle to circle is not the obvious one; the centre

$$z = ae^{i\alpha}$$

maps to

$$\zeta = a^{-1}e^{-i\alpha}$$

which is *not* the centre of the circle in the ζ-plane. But it is still inside the transformed circle, between the points R' and S' on the diameter through $\zeta = 0$.

The case $a = r$ is rather different, as then the circle in the z-plane passes through $z = 0$ as you see in fig. XVII.18, and this point is transformed to infinity. In this case we may take P to be

$$z = ae^{i\alpha} + ae^{i\theta}$$

so that

$$\zeta = (ae^{i\alpha} + ae^{i\theta})^{-1}$$
$$= \tfrac{1}{2}a^{-1}e^{-i\alpha}\{1 - i\tan\tfrac{1}{2}(\theta - \alpha)\}.$$

This has the form

$$\zeta = z_0 + g(\theta)e^{i\phi}$$

and so represents a line through $z_0 = \tfrac{1}{2}a^{-1}e^{-i\alpha}$ at angle $\phi = \tfrac{1}{2}\pi - \alpha$ to the

Fig. XVII.18. Inversion of a circle through the origin.

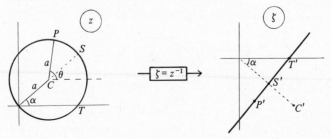

x-axis, with

$$g(\theta) = -\tan \tfrac{1}{2}(\theta - \alpha).$$

In this transformation the circle becomes a line, and the inside of the circle becomes the half plane on one side of the line – notice where the centre C transforms to.

Since the relation of inversion is symmetrical as between ζ and z, we may at once deduce another result: a line in the z-plane that does not pass through the centre of the inversion $z = 0$ transforms to a circle in the ζ-plane. And finally it is easy to show that a line through $z = 0$ transforms to another line through $\zeta = 0$ (and vice versa).

Inversion may be combined with the linear transformation in (b) to obtain the relation

$$\zeta = a/(z - b),$$

which is inversion with centre at $z = b$ and also with a scale change and rotation given by a.

With this transformation we may make changes such as those shown in figs. XVII.19, 20 and 21. Any solution in the z-plane in fig. XVII.19 (see Chapter XI §4 for an example) gives a corresponding solution in a quarter circle. And in fig. XVII.20 $AFDC$ is a circular arc through the centre of the

Fig. XVII.19. Inversion of a region to a quarter circle.

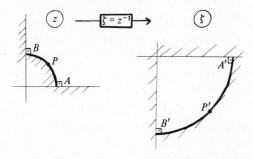

Fig. XVII.20. Inversion of a region to half an annulus.

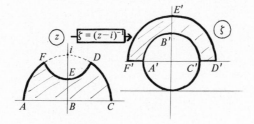

Fig. XVII.21. Inversion of two intersecting circles.

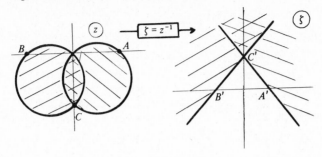

inversion, so it becomes a straight line (through the point $\frac{1}{2}i$ which is the image of $-i$ in the z-plane). The line ABC (through the point at infinity) becomes a circle through the points $\zeta = 0$ and $\zeta = i$. Finally in fig. XVII.21 the circles through the centre of inversion become lines through C'. Angles are preserved, for example the angle $B'\,C'\,A'$ is equal to the angle at which the circles intersect at C.

Note that the 'bilinear transformation'

$$\zeta = \frac{az + b}{cz + d}$$

is an inversion together with other elementary transformations, as it is

$$\zeta = k + l/(z + m)$$

for complex numbers, k, l, m which are related to a, b, c, d.

(d) Other elementary functions

Another group of elementary transformations is based on the exponential function. These are such transformations as

$$\begin{cases} \zeta = e^z, \\ \zeta = \ln z, \\ \zeta = \sin z. \end{cases}$$

Take for example $\zeta = e^z$. This is, for $\zeta = \xi + i\eta$,

$$\begin{cases} \xi = e^x \cos y, \\ \eta = e^x \sin y, \end{cases}$$

and so the infinite strip

$$\begin{cases} -\infty < x < \infty, \\ \quad 0 \leqslant y < 2\pi \end{cases}$$

Fig. XVII.22. Transformation of a half strip to a circle by $\zeta = e^z$.

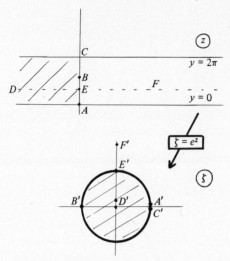

is mapped to the whole ζ-plane. The line $x = x_0$ maps to the circle

$$\xi^2 + \eta^2 = e^{2x_0},$$

as is shown in fig. XVII.22. The line $y = y_0$ maps to the radius at angle $y_0, \frac{1}{2}\pi$ in the diagram.

With this transformation, the whole z-plane is mapped many times over on the ζ-plane; each strip of width 2π in the y-direction gives the whole ζ-plane, so the transformation is only useful when a single strip of the z-plane is being considered.

The transformation

$$\zeta = \ln z$$

needs no extra discussion, as it is just the inverse of the last one. And

$$\zeta = \sin z$$

can be regarded as a combination of previous transformations

$$z \mapsto iz \mapsto e^{iz}$$

followed by a new one which is rather like an inversion

$$e^{iz} \mapsto e^{iz} - (e^{iz})^{-1},$$

and then finally

$$e^{iz} - e^{iz} \mapsto (e^{iz} - e^{-iz})/2i$$
$$= \sin z.$$

We shall go on to consider the transformation

$$z \mapsto z + 1/z$$

in more detail later; meanwhile we deal with sin z directly.

If we have the transformation

$$\zeta = \sin z$$

then this is equivalent to

$$\begin{cases} \xi = \sin x \cosh y, \\ \eta = \cos x \sinh y \end{cases}$$

where $\zeta = \xi + i\eta$ and $z = x + iy$. Clearly the real axis from $x = -\frac{1}{2}\pi$ to $x = \frac{1}{2}\pi$ maps into the real axis from -1 to 1, and any further section of the real axis just gives the same points again. So we restrict attention to the strip $-\frac{1}{2}\pi < x < \frac{1}{2}\pi$ in fig. XVII.23. Next consider the imaginary axis,

$$x = 0.$$

These points map to

$$\begin{cases} \xi = 0, \\ \eta = \sinh y. \end{cases}$$

Take next two points on the line $x = \frac{1}{2}\pi$, at $y = \pm 1$. They both map to

$$\begin{cases} \xi = \cosh 1, \\ \eta = 0. \end{cases}$$

Fig. XVII.23. Transformation of a strip by $\zeta = \sin z$.

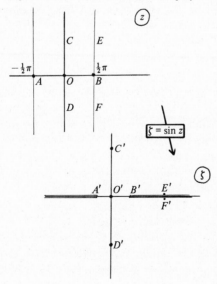

What has happened is that the line *EBF* has folded round *B* to give the (double) line

$$\xi > 1, \eta = 0$$

which is a cut in the ζ-plane. The total result of the transformation is to map the strip $-\frac{1}{2}\pi < x < \frac{1}{2}\pi$ onto the whole ζ-plane, but with cuts from ± 1 to infinity along the real axis. Angles at *A* and *B* have clearly *not* been preserved, and indeed these are points where

$$f'(z) = \cos z$$

is zero. At these points angles like *EBF* have been doubled from π to 2π; and sin *z* is locally a squaring transformation near such points (which doubles angles) because

$$\sin(\tfrac{1}{2}\pi - Z) = \cos Z$$
$$\doteqdot 1 - \tfrac{1}{2}Z^2$$

for small *Z*.

Now a uniform flow down the channel between $x = -\frac{1}{2}\pi$ and $x = \frac{1}{2}\pi$ is given by

$$w(z) = iUz.$$

Hence the solution for the flow through a hole in a plate is given by

$$W(\zeta) = w\{z(\zeta)\}$$
$$= iU \sin^{-1}\zeta$$

where the hole is between $\xi = -1$ and $\xi = +1$. This is a solution that would have been harder to find directly.

(e) *A special transformation*

The transformation

$$\zeta = z + 1/z$$

turns out to be quite important. It is a special case of the transformation

$$\zeta = f(z)$$

where the Laurent series for $f(z)$ around the origin

$$f(z) = \sum_{n=-\infty}^{\infty} a_n z^n$$

has a very simple form with only two non-zero coefficients. This special form firstly makes the planes correspond very closely when $|z|$ is large, as then

$$\zeta = z + o(1)$$

and secondly it gives an inversion near the origin. The correspondence of the planes at large distances is important, because this means that uniform flow at large distances in the z-plane (given by $w(z) \sim U_0 z$) corresponds to uniform flow at large distances in the ζ-plane.

Since the transformation is like an inversion near $z = 0$, let us consider its effect on the circle $|z| = 1$. It is easily seen that

$$z = e^{i\theta} \mapsto \zeta = 2\cos\theta,$$

so that the unit circle $0 \leqslant \theta < 2\pi$ is transformed into the strip of the ξ-axis, $-2 \leqslant \xi \leqslant 2$, traversed twice. Fig. XVII.24 shows the corresponding curves. In fact the region outside the circle $|z| = 1$ is transformed into the whole ζ-plane, cut along the real axis from -2 to 2. The inside of the circle $|z| = 1$ is *also* transformed into the whole ζ-plane, but these points which have $|z| < 1$ are not in the flows which we are going to consider.

This mapping is not conformal at $z = \pm 1$, because

$$f'(z) = 1 - z^{-2},$$

which is zero there. Near $z = +1$ we may write

$$z = 1 + Z$$

where Z is small, and then

$$\begin{aligned}
f(z) &= 1 + Z + (1 + Z)^{-1} \\
&\doteq 1 + Z + (1 - Z + \tfrac{1}{2}Z^2),
\end{aligned}$$

which shows that the transformation is locally a squaring, converting the angle of π on the circle at A into an angle of 2π on the strip at A'.

Fig. XVII.24. Transformation of a circle to a line segment by $\zeta = z + z^{-1}$.

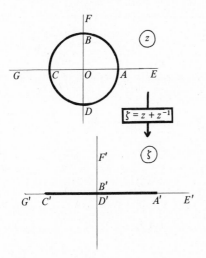

Fig. XVII.25. The circle $|z| = a$ transforms to an ellipse under $\zeta = z + z^{-1}$.

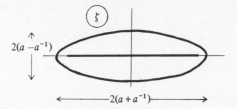

The circle $|z| = a$, $a > 1$, also converts into a simple figure in the ζ-plane. We have

$$\zeta = \xi + i\eta = ae^{i\theta} + a^{-1}e^{-i\theta}$$

and so

$$\begin{cases} \xi = (a + a^{-1})\cos\theta, \\ \eta = (a - a^{-1})\sin\theta. \end{cases}$$

Hence the circle $|z| = a$ has been converted into the ellipse

$$\left(\frac{\xi}{a + a^{-1}}\right)^2 + \left(\frac{\eta}{a - a^{-1}}\right)^2 = 1,$$

which is drawn in fig. XVII.25. For differing values of a this gives a family of confocal ellipses, with the strip $-2 \leqslant \xi \leqslant 2$ as the limiting case.

This transformation immediately provides the potential for many flows to do with ellipses and strips, since most flows round circles can be solved by the Milne–Thomson circle theorem. As an example, take the flow of a uniform stream at angle α to the x-axis past the circle $|z| = a$, indicated in fig. XVII.26. The basic potential is

$$U_0 z e^{-i\alpha},$$

from the last chapter, and by the circle theorem we have

$$w(z) = U_0 z e^{-i\alpha} + U_0 e^{i\alpha} a^2/z.$$

Fig. XVII.26. Flow of a stream at an angle past a cylinder.

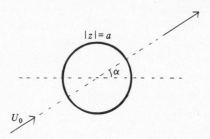

Now make the transformation

$$\zeta = z + 1/z,$$

which changes the circle into an ellipse, and which leaves a uniform stream at angle α at infinity because the planes correspond at infinity. This gives

$$W(\zeta) = U_0 z(\zeta) e^{-i\alpha} + U_0 e^{i\alpha} a^2 / z(\zeta)$$

for the flow past an ellipse; or, if we put $a = 1$, the flow at an angle α past the strip along the ξ-axis.

We may of course solve for z in terms of ζ:

$$z = \tfrac{1}{2}\{\zeta + (\zeta^2 - 4)^{1/2}\},$$

choosing the positive square root so that $z = \zeta$ at large distances. But this is often not necessary as both the force and moment formulae are derived from the velocity, which is the conjugate of

$$dW/d\zeta = (dw/dz)(dz/d\zeta)$$
$$= (dw/dz)(d\zeta/dz)^{-1}.$$

Thus

$$U - iV = (U_0 e^{-i\alpha} - U_0 e^{i\alpha} a^2 / z^2)(1 - 1/z^2)^{-1},$$

and so the velocity at any point ζ can be found from this function of the corresponding point z. Notice that $U - iV$ appears to have a singularity at $z = 0$; but this point is not in the flow and so does not matter. There are also singularities at $z = \pm 1$, and for flow past a strip these *are* points of the flow. This infinite velocity at the ends of the strip is to be expected as we have seen previously that flow round an edge leads to infinite velocity there.

This transformation will be dealt with in more detail below, as will the force and moment on an ellipse or strip in a uniform flow.

4. A sequence of mappings

There are dictionaries of conformal mappings, in which you can find how to transform the most surprising shapes of boundaries into a line or the unit circle. Many of them can be built up by the patient combination of elementary transformations of the types we have considered in §3 above. We consider now an example where we need to use a carefully chosen sequence of transformation:

The flow we choose is shown in fig. XVII.27 and is that along a plane on which a cylinder rests. What we have to achieve is a separation of the points O_-, O_+ just on either side of the point of contact, and then a flattening of the circle through this gap between O_- and O_+. This will give

Fig. XVII.27. Flow past a cylinder resting on a plane.

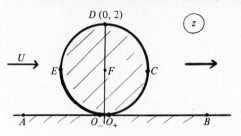

Fig. XVII.28. The first stage of the transformation.

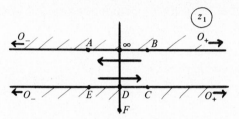

us an easy problem of flow along a plane, provided that the transformations are asymptotically such as to give planes which correspond at infinity.

We start with an inversion

$$z_1 = 1/z$$

as this converts the circle and line through O into a pair of parallel lines $y_1 = 0$ and $y_1 = -\frac{1}{2}$, drawn in fig. XVII.28.

This also separates O_- and O_+ in a helpful fashion, but the flow is now in both directions in this strip.

Now a strip can be converted to a half plane by an exponential transformation, but to do this we need a strip of width π. This needs a magnification

$$z_2 = 2\pi z_1$$

to get such a strip. We follow this with the transformation

$$z_3 = e^{z_2} = e^{2\pi/z},$$

which gives the half plane shown in fig. XVII.29, with O_- now at the origin and O_+ at infinity. The flow is now along the plane, which is good; but the planes do not correspond at infinity, and the original point at infinity

Fig. XVII.29. The second stage of the transformation.

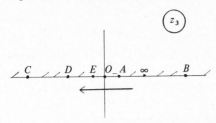

is firmly in the flow. We need another transformation to take ∞ away again, to bring back O_+, and to make the planes correspond at large distances. We do the first with the bilinear transformation (remember that it is a type of inversion)

$$z_4 = \frac{z_3 + 1}{z_3 - 1}.$$

This leaves the x-axis as a whole invariant, but moves the points around as shown in fig. XVII.30. We have clearly done what we wished, by separating O_- and O_+ and squeezing the circle down onto the line. The full transformation is now

$$z_4 = (e^{2\pi/z} + 1)/(e^{2\pi/z} - 1)$$
$$= \coth(\pi/z).$$

Now at large distances $\coth(\pi/z) \sim z/\pi$, so we need a final transformation of scale

$$\zeta = \pi z_4$$

to make the planes correspond at infinity.

We may now choose potential

$$W(\zeta) = U\zeta$$

for a flow along the real axis in the ζ-plane, and the flow round the original

Fig. XVII.30. The final stage of the transformation.

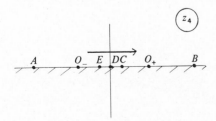

figure must be

$$w(z) = \pi U \coth(\pi/z).$$

This result is surprisingly simple, but would hardly have been guessed. As a solution to a problem of fluid dynamics it suffers from the usual drawbacks. The real flow would separate from the cylinder to leave a wake if the Reynolds number is large; otherwise this model is irrelevant.

5. The Joukowski transformation of an ellipse

(a) The need for a circulation: the Kutta condition

The flows derived from potential theory are often rather unrealistic in practice, because the real flow tends to separate from a boundary, whereas the potential flow sticks to the boundary. As an example of this consider the flow at small angle of incidence α past a thin ellipse. The potential flow, shown in fig. XVII.31, has to have a high velocity to get round the rather sharp front and rear ends, and inevitably the real flow will separate at the rear end, to give a smooth flow off the back, with quite a narrow wake if the ellipse is thin. The modified flow is shown in fig. XVII.32.

Such a shift of the rear stagnation point to the back of the ellipse can be achieved in potential theory by adding on a circulation round the ellipse. Thus we are led to consider the potential

$$w(z) = U_0 z e^{-i\alpha} + U_0 a^2 e^{i\alpha}/z + (i\kappa/2\pi)\ln z$$

Fig. XVII.31. Potential flow past an ellipse.

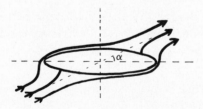

Fig. XVII.32. High Reynolds number flow separates to give a wake behind the ellipse.

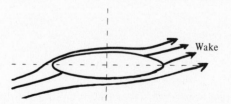

together with the transformation

$$\zeta = z + 1/z,$$

which in the context of such flows is known as a 'Joukowski transformation'. This will give a potential $W(\zeta)$ which describes flow of a uniform stream U_0 at angle α past an ellipse round which there is a circulation $-\kappa$. The rear stagnation point is given by the value of z (near $z = a$) which makes $dw/dz = 0$. That is,

$$U_0 e^{-i\alpha} - U_0 a^2 e^{i\alpha}/z^2 + i\kappa/(2\pi z) = 0.$$

What we shall do is *choose* κ (if we can) so that the separation point is actually at the trailing edge $z = a$. This gives

$$\kappa = 4\pi a U_0 \sin \alpha.$$

This is an example of 'Joukowski's hypothesis' or of a 'Kutta condition'; the circulation κ is chosen so as to give the smooth flow off the trailing edge that is observed in practice at small angles of incidence α, for very thin ellipses.

The choice of κ looks slightly uncertain here. It is not entirely clear that the stagnation point must be exactly at the trailing edge of the ellipse, but clearly it cannot be far from it if the ellipse is thin, i.e. if a is just greater than 1. For $dW/d\zeta$ has a factor

$$(d\zeta/dz)^{-1}$$

in it, which is

$$(1 - 1/z^2)^{-1}$$

and when $z = a \doteq 1$ this can be extremely large; if this factor is not cancelled by a zero of dw/dz at the same point, the velocity round the trailing edge will be too large for the flow to stay attached, and the separation point will move towards the trailing edge.

This thin ellipse is a first example of an 'aerofoil': a two-dimensional shape which automatically generates a circulation (via the Kutta condition and by a process whose dynamics we shall describe in more detail below in §7) when it is put into a uniform stream. And a circulation gives a lift force, as we have found before; so such a shape is a suitable cross-section for a wing.

(b) *The lift force for small angles*

The force and moment on the ellipse in this flow can be calculated by using Blasius' theorem and the methods of Chapter XVI §5. The force is given by

$$X - iY = \tfrac{1}{2}i\rho \int_\Gamma (dW/d\zeta)^2 d\zeta,$$

But this may just as easily be calculated in terms of $w(z)$ as

$$X - iY = \tfrac{1}{2}i\rho \int_C (dw/dz)^2 (dz/d\zeta)^2 (d\zeta/dz)dz,$$

where C is now the circle $|z| = a$. The integral we have to evaluate is

$$\tfrac{1}{2}i\rho \int_{|z|=a} \{U_0 e^{-i\alpha} - U_0 a^2 e^{i\alpha}/z^2 + i\kappa/(2\pi z)\}^2/(1 - 1/z^2)dz.$$

Now there are no singularities outside the circle $|z| = a$, so we may choose instead to integrate round a (large) circle; the term

$$(1 - 1/z^2)^{-1}$$

may then be expanded as

$$1 + 1/z^2 + O(z^{-4}).$$

The only term giving a contribution to the integral will, as before, be the one in z^{-1}, which has coefficient

$$\tfrac{1}{2}i\rho 2 U_0 e^{-i\alpha} i\kappa/2\pi.$$

So finally

$$X - iY = 2\pi i \left(- \rho U_0 e^{-i\alpha}\kappa/2\pi \right)$$
$$= - i\rho U_0 e^{-i\alpha}\kappa.$$

That is,

$$\begin{cases} X = - \rho U_0 \kappa \sin \alpha, \\ Y = \rho U_0 \kappa \cos \alpha. \end{cases}$$

The two components X and Y are along the x and y axes. It is more helpful to think in terms of components along and perpendicular to the stream. Aeroplane flight is arranged by having a wing, which has here

Fig. XVII.33. Force components are taken parallel and perpendicular to the incident stream.

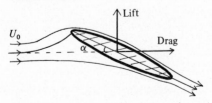

been modelled by an infinite cylinder whose section is a thin ellipse, which is held at a small angle α to the direction of motion through still air. The lift on the aeroplane is thus the force perpendicular to the relative velocity of flow, i.e. perpendicular to the stream in the model, and as shown in fig. XVII.33. This lift force is thus

$$Y \cos \alpha - X \sin \alpha = \rho U_0 \kappa$$

$$= 4\pi a \rho U_0^2 \sin \alpha$$

in this model (for unit length of wing in the third dimension). The drag force is

$$X \cos \alpha - Y \sin \alpha = 0.$$

This model gives quite a good prediction of the lift force, provided the angle α is rather small. But when α exceeds a few degrees the flow separates from the ellipse to give a wide wake (crudely sketched in fig. XVII.34), which gives little or no lift and a large drag. The wing is then said to be 'stalled'.

The prediction of zero drag force in the model flow is rather a poor one. For one thing there will be viscous stresses in the boundary layers, which cannot come into the present theory. For another, there will in reality be a separated wake even when κ is chosen to satisfy the Kutta condition; the wake will have a width which is comparable to the width of the ellipse, as you can see in fig. XVII.35. Now the width of the ellipse is not hard to

Fig. XVII.34. At large angles of incidence the wing stalls.

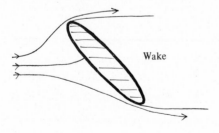

Fig. XVII.35. The wake at the rear of the ellipse.

find: we have transformed the circle

$$|z| = a$$

by

$$\zeta = z + 1/z,$$

and in §3(*e*) we saw that the ellipse so generated has semi-minor axis

$$a - 1/a.$$

Since *a* is near 1 this is approximately

$$2(a - 1).$$

Hence the drag due to the separated wake must be of approximate size

$$\rho U_0^2 (a - 1)$$

for unit length in the third dimension.

The viscous drag can never be eliminated, because it arises from the no-slip condition. But it would seem that the drag due to the separated wake can be eliminated (to a reasonable approximation) by taking $a = 1$. This however is not realistic, as a flat plate of zero thickness, which is what we get when $a = 1$, cannot have any strength. And any plate of non-zero thickness can be modelled adequately by the elliptic section we have just been discussing. The dilemma can be resolved by taking a section for the wing which has non-zero thickness over most of the section, but comes to a point at the trailing edge, as in fig. XVII.36. Then the wake would have, as a first approximation, zero thickness, provided the flow remained attached to the wing surface all the way to the trailing edge. We shall see how to calculate such a flow below.

(c) *The moment acting on the aerofoil*

The moment of the pressure forces on the aerofoil of elliptic section is given by Blasius' formula

$$M = -\tfrac{1}{2}\rho \, \mathscr{Re} \int_\Gamma \zeta (dW/d\zeta)^2 d\zeta.$$

This too can be calculated in the *z*-plane, as

$$M = -\tfrac{1}{2}\rho \, \mathscr{Re} \int_C (z + 1/z)\{w'(z)\}^2/(1 - 1/z^2)dz.$$

Fig. XVII.36. A sharp trailing edge may give a wake of negligible thickness.

As before we look for the term in z^{-1} in the integral, and get

$$M = -\tfrac{1}{2}\mathscr{R}e\{2\pi i(2U_0^2 e^{-2i\alpha} - 2U_0^2 a^2 - \kappa^2/4\pi^2)\}$$
$$= -2\pi\rho U_0^2 \sin 2\alpha.$$

This is the moment per unit length along the wing, and taken about the point $\zeta = 0$. It tends to increase α, so $\alpha = 0$ is an unstable position: this can be seen if you try to 'fly' a long envelope – α increases until the envelope stalls.

The total effects on the air pressures on the wing section, in this model, are given by the force components X and Y, and moment M. In rigid body dynamics we can replace such a system by the forces X and Y acting at some other point, so as to give the same moment about $\zeta = 0$. This clearly requires that (X, Y) acts at a point at distance b along the ellipse from the centre (as shown in fig. XVII.37), where

$$Yb = M$$

or

$$b = a^{-1},$$

on substituting the values we have found for Y, M and κ. That is, the lift may be taken as acting at a distance a^{-1} from the centre, which is approximately midway between the centre and the leading edge, i.e. at a quarter of the 'chord' from the leading edge.

It would appear from this analysis that a wing would fly at constant angle α provided that its weight acted through this 'quarter-chord point'. where the lift may be taken as acting. But the stability of a wing loaded in such a fashion would be marginal, and so aeroplanes of traditional design require extra stability, such as is provided by a tail. Note that these calculations are based on an incompressible fluid, which is only a reasonable approximation for Mach numbers rather less than 1. Supersonic flight must be calculated separately, and different results are found.

Fig. XVII.37. The equivalent force system for an elliptic aerofoil acts at the quarter chord point.

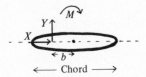

6. The cambered aerofoil

(a) *Sharpening the trailing edge*

We have seen in §3(e) that the mapping

$$\zeta = z + 1/z$$

takes the circle $|z| = 1$ into the strip

$$-2 \leqslant \xi \leqslant 2.$$

If we consider this strip as a wing section, then it has a sharp trailing edge, and should give a wake of zero thickness when the circulation is such as to give smooth flow off the trailing edge. But the circle

$$|z| = a > 1$$

is transformed into an ellipse which has a (useful) non-zero thickness while giving a wake of non-zero thickness as well, which is undesirable. We might hope to get the better points of each curve by using a circle like the one in fig. XVII.38, which touches

$$|z| = a > 1$$

at one end to give thickness, and touches

$$|z| = 1$$

at the other end to give sharpness. Take the circle centred on

$$z = -\delta$$

with radius

$$1 + \delta$$

to fit both these requirements. The circle has equation

$$|z + \delta| = 1 + \delta$$

Fig. XVII.38. A displaced circle for transformation to an aerofoil.

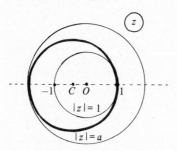

Fig. XVII.39. The symmetric Joukowski aerofoil.

and points on it are given by

$$z = -\delta + (1 + \delta)e^{i\theta}.$$

Since we expect to have a wing whose thickness is small compared to its chord, we shall be taking δ to be small compared to 1.

The cross-sectional shape of the wing we now have (which is sketched in fig. XVII.39) is given by

$$\begin{cases} \zeta = z + 1/z, \\ z = -\delta + (1 + \delta)e^{i\theta}. \end{cases}$$

This may be shown to give

$$\begin{cases} \xi = 2\cos\theta + (-1 + \cos 2\theta)\delta + O(\delta)^2 \\ \eta = 2\delta\sin\theta - \delta\sin 2\theta + O(\delta^2). \end{cases}$$

At $\theta = 0$ we have

$$\xi = 2 \text{ and } \eta = O(\delta^2),$$

which is the trailing edge of zero thickness that we require; at $\theta = \pi$ we have (approximately)

$$\xi = -2 \text{ and } \eta = 0,$$

so that the chord of the wing is still nearly 4 if δ is small.

The upper surface is

$$\eta = 2\delta\sin\theta(1 - \cos\theta)$$

approximately. This has

$$d\eta/d\theta = 0$$

when $\theta = \frac{2}{3}\pi$ or 0.

The maximum thickness is approximately $(3\sqrt{3})\delta$, at the point $\xi = -1$; and there is a 'cusp' of zero angle at $\xi = 2$.

The wing section has to have zero angle between its two sides at the trailing edge because the circle from which it is derived has angle π between any two adjacent parts, and the transformation at the critical point $z = 1$ is locally a squaring transformation (as we saw in §3(e)) which doubles angles; so there is an external angle 2π at the trailing edge, or an internal angle of zero.

This section for a wing is more useful than the ellipse we considered before, since it should give a negligible wake width at high Reynolds numbers, but a cusp is a hard shape to build, so we shall need to add a further refinement to this theory later. Meanwhile, let us note that $z = 1$ is a point in the flow field, and that

$$d\zeta/dz = 0$$

there. This means that

$$dW/d\zeta = w'(z)dz/d\zeta$$

is infinite at $z = 1$ unless $w'(1) = 0$. In this case it is clear that we *must* impose the Kutta condition

$$w'(1) = 0$$

in order to keep the velocity finite. Earlier, it was merely plausible that we needed to have $w'(1) = 0$ in order to get a smooth flow off the trailing edge. In the present case the Kutta condition achieves this smooth flow as well; we shall show this by calculating the component of the velocity perpendicular to the wing section near $\zeta = 2$, and showing it to be zero in the limit as we approach the point $\zeta = 2$. But first let us write down $w(z)$. We need flow round the circle of radius $a = 1 + \delta$, and with centre at $z = -\delta$. We need

$$w(z) = U_0 e^{-i\alpha}(z + \delta) + U_0(1 + \delta)^2 e^{i\alpha}/(z + \delta) + (i\kappa/2\pi)\ln(z + \delta)$$

to allow for both these changes. The Kutta condition is, as we have shown, $w'(1) = 0$, or

$$U_0 e^{-i\alpha} - U_0 e^{i\alpha} + (i\kappa/2\pi)(1 + \delta)^{-1} = 0.$$

So,

$$\kappa = 4\pi U_0(1 + \delta)\sin\alpha$$
$$= 4\pi U_0 a \sin\alpha$$

as before.

We can now calculate the velocity near the trailing edge of this wing section. We need

$$dW/d\zeta$$

near $\zeta = 2$, and this is derived from

$$dw/dz$$

with $z \doteq 1$, together with $d\zeta/dz$. Let us put $z = 1 + \varepsilon$, where ε is a complex number with $|\varepsilon| \ll 1$. Then

$$w'(1 + \varepsilon) = U_0 e^{-i\alpha} - U_0(1 + \delta)^2 e^{i\alpha}(1 + \delta + \varepsilon)^{-2}$$
$$+ (i\kappa/2\pi)(1 + \delta + \varepsilon)^{-1}.$$

Fig. XVII.40. Flow at an angle past the symmetric aerofoil.

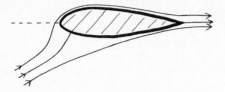

But we have just shown that

$$\kappa = 4\pi U_0 (1 + \delta) \sin \alpha,$$

so this expression can be rewritten as

$$w'(1 + \varepsilon) = U_0 e^{-i\alpha} - U_0 e^{i\alpha} \{1 + \varepsilon/(1 + \delta)\}^{-2}$$
$$+ 2iU_0 \sin \alpha \{1 + \varepsilon/(1 + \delta)\}^{-1}.$$

For small values of ε we can use the binomial theorem to write this as

$$w'(1 + \varepsilon) = 2\varepsilon(1 + \delta)^{-1} U_0 \cos \alpha + O(\varepsilon^2).$$

Now the transformation from z to ζ gives

$$dz/d\zeta = (1 - z^{-2})^{-1}$$

and with $z = 1 + \varepsilon$ we get

$$dz/d\zeta = \tfrac{1}{2}\varepsilon^{-1} \{1 + O(\varepsilon)\}.$$

So finally the velocity near the trailing edge is given by

$$\frac{dW}{d\zeta} = \frac{U_0 \cos \alpha}{1 + \delta} + O(\varepsilon).$$

The first term here is purely real, and so the flow leaves the trailing edge along the real axis, as was expected. The flow is sketched in fig. XVII.40, for an exaggerated angle α.

To round off this piece of work, we ought to calculate the force and moment on this aerofoil. These calculations are of exactly the same form as those done in §5(b) and (c) above, and give only a slightly different result, so they will be left as an exercise.

The aerofoil we have produced is known as the 'symmetrical Joukowski aerofoil' after its discoverer and its shape; we go on to generalise it.

(b) *The cambered aerofoil*

The purpose of a wing is to make the flow of air change direction, and hence provide a force on the wing. It is reasonable to expect a curved wing to do this more effectively than a straight one, provided that the air flow is along the wing all the way round. See fig. XVII.41 for the kind of

Fig. XVII.41. Flow past a cambered aerofoil.

flow we have in mind. And it is indeed found that birds use such a 'cambered' wing section. So we next investigate how to transform a circle in the z-plane into this sort of shape in the ζ-plane.

We continue to use the transformation

$$\zeta = z + 1/z,$$

and we still use a circle in the z-plane, but this time we put the centre of the circle at

$$z = \delta e^{i\phi}$$

where δ is small and positive and

$$\tfrac{1}{2}\pi < \phi < \pi.$$

Such a circle is drawn in fig. XVII.42. So as to continue to have a sharp trailing edge, we make the circle in the z-plane pass through $z = 1$ by choosing the radius of the circle appropriately; we need to have

$$a = (1 - 2\delta \cos \phi + \delta^2)^{1/2}.$$

With these choices we ought to get a wing section with

  (i) a cusp at the trailing edge,
  (ii) a thickness of order δ,
  (iii) a chord of about 4.

The aerofoil shape that we get by transforming this circle is not easily found in detail from the equations we are using. It is easier to look at a graphical solution rather than try some hard analysis; after all, for a given complex number z it is simple to find z^{-1} as $r^{-1}e^{-i\theta}$ by direct measure-

Fig. XVII.42. The displaced circle for a cambered aerofoil.

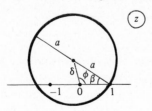

Fig. XVII.43. The geometry of the transformation $z \mapsto z + z^{-1}$.

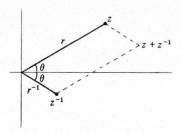

Fig. XVII.44. The cambered aerofoil constructed geometrically; the circle has radius 1.28, centre at distance 0.35 and angle 135° from the origin.

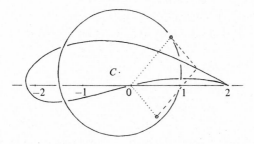

ments of r and θ. Then the addition of z is done by the parallelogram (triangle) rule, or just by adding real and imaginary parts for z and z^{-1}, as in fig. XVII.43. This process gives an adequate idea of the shape of the aerofoil quite quickly. The case illustrated in fig. XVII.44 used

$$\begin{cases} \delta = \tfrac{1}{4}\sqrt{2} = 0.35, \\ \phi = 135°, \end{cases}$$

and such a large value of δ gives rather a thick profile for the aerofoil. But if a smaller value had been used, the cusp region would have been less obvious.

The angle which the cusp makes with the x-axis can be found rather easily. The two circles intersect at angle β at the point $z = 1$, where

$$a\sin\beta = \delta\sin\phi.$$

Hence the displaced circle which we are transforming makes an angle $\tfrac{1}{2}\pi - \beta$ with the x-axis at $z = 1$. Now the transformation

$$\zeta = z + 1/z$$

is locally a squaring transformation, as we have seen before, and this doubles angles at $z = 1$. So the angle $\tfrac{1}{2}\pi - \beta$ is transformed to

$$\pi - 2\beta.$$

That is, the cusp at the trailing edge is at angle 2β with the x-axis at $\zeta = 2$.

What we have gained, then, from the transformation of this displaced circle into a cambered Joukowski aerofoil is this angle 2β; the symmetrical aerofoil in (*a*) above sends the air off along the x-axis, but the cambered aerofoil diverts it an extra 2β. And we might hope that this extra angle 2β would come into the lift force, making the lift proportional to

$$\sin(\alpha + 2\beta)$$

and so giving a better lift for small angles α.

Before we go on to calculate the lift force, you should notice that the angle β comes into the shape of the aerofoil at a point other than the cusp. For the negative x-axis cuts the displaced circle at angle $\frac{1}{2}\pi - \beta$. This angle is preserved by the conformal mapping, and hence the aerofoil profile cuts the negative real axis at angle $\frac{1}{2}\pi - \beta$.

(c) Force and moment for the cambered aerofoil

The flow in the z-plane round the displaced circle is given by the complex potential

$$w(z) = U_0 e^{-i\alpha}z + U_0 e^{i\alpha}a^2/(z - \delta e^{i\phi}) + (i\kappa/2\pi)\ln(z - \delta e^{i\phi}),$$

where κ is chosen so as to satisfy the Kutta condition

$$w'(1) = 0,$$

and where we shall use the Joukowski transformation

$$\zeta = z + z^{-1}.$$

The Kutta condition turns out to be rather easy, because, from the geometry of the situation in fig. XVII.45,

$$1 - \delta e^{i\phi} = ae^{-i\beta}.$$

So we find by an easy calculation that

$$\kappa = 4\pi a U_0 \sin(\alpha + \beta).$$

This means that a larger circulation is needed to make the air come smoothly off the trailing edge, which seems reasonable as the air has to be diverted more.

The force on the aerofoil is calculated exactly as in §5(*b*) above, from

Fig. XVII.45. The geometry of a, δ, β, ϕ.

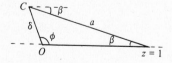

Blasius' formula

$$X - iY = \tfrac{1}{2}i\rho \int_{\Gamma} (dW/d\zeta)^2 d\zeta.$$

Notice that $dW/d\zeta$ is a perfectly well-behaved function of ζ at the trailing edge, just because we have used the Kutta condition; and so there can be no problem in the integration even though the transformation

$$\zeta = z + z^{-1}$$

has a critical point there.

The calculation is done in the z-plane as before:

$$X - iY = \tfrac{1}{2}i\rho \int_{C} (dw/dz)^2 \, (dz/d\zeta)^2 \, (d\zeta/dz)dz,$$

where we choose C to be a circle round the origin and enclosing the displaced circle that we have transformed. This allows us to calculate the integral as before, by expanding terms where necessary and using only the term in z^{-1}. The integrand is

$$\{U_0 e^{-i\alpha} - U_0 e^{i\alpha} a^2/(z - \delta e^{i\phi})^2 + (i\kappa/2\pi)/(z - \delta e^{i\phi})\}^2 \times (1 - z^{-2})^{-1}.$$

We use

$$\begin{cases} (z - \delta e^{i\phi})^{-1} = z^{-1}\{1 + z^{-1}\delta e^{i\phi} + O(z^{-2})\}, \\ (z - \delta e^{i\phi})^{-2} = z^{-2}\{1 + 2z^{-1}\delta e^{i\phi} + O(z^{-2})\}, \\ (1 - z^{-2})^{-1} = 1 + z^{-2} + O(z^{-4}), \end{cases}$$

as the appropriate expansions on C.

We may now see explicitly what we might have noted before, that the terms in $\delta e^{i\phi}$ do *not* come into any coefficients of z^{-1}: the displacement of the circle makes no difference to the forces acting except in so far as the value of κ is changed to maintain a smooth air flow at the trailing edge.

The coefficient of z^{-1} in the integrand is

$$2U_0 e^{-i\alpha}(i\kappa/2\pi),$$

and so the force is given by

$$X - iY = \tfrac{1}{2}i\rho \times 2\pi i \times 2U_0 e^{-i\alpha}(i\kappa/2\pi)$$
$$= - i\rho U_0 e^{-i\alpha}\kappa,$$

exactly as before. And in the same way as before we want the force perpendicular to the incident stream, which is

$$Y\cos\alpha - X\sin\alpha = \rho U_0 \kappa$$
$$= 4\pi a\rho U_0^2 \sin(\alpha + \beta).$$

We have indeed gained extra lift from the camber of the aerofoil, but not as much as the earlier physical argument suggested: the circulation round the wing section is the fundamental quantity, and not the angle that the airflow is diverted through.

The moment of the pressure forces on the wing can also be calculated from Blasius' theorem. This is done exactly as in §5(c) above, but now a slightly modified result arises:

$$M = -\tfrac{1}{2}\rho \mathcal{R}e\{2\pi i(2U_0^2 e^{-2i\alpha} - 2U_0^2 a^2 - \kappa^2/4\pi^2 + i\kappa U_0 \delta e^{i(\phi-\alpha)}/\pi)\}$$
$$= -2\pi\rho U_0^2 \sin 2\alpha + \rho\kappa U_0 \delta \cos(\phi - \alpha)$$
$$= -2\pi\rho U_0^2 \{\sin 2\alpha - 2a\delta \sin(\alpha + \beta)\cos(\phi - \alpha)\}.$$

This is a small modification in the case of thin aerofoils for which δ is small; it is due to a redistribution of the pressure round the wing.

7. Further details on aerofoils

The final modification to this work on the Joukowski aerofoil and transformation is to vary the transformation slightly so that there is no longer a cusp at the trailing edge, which could never be built in practice, but a more wedge shaped trailing edge of small angle. The transformation

$$\zeta = (2 - \varepsilon)\frac{(z+1)^{2-\varepsilon} + (z-1)^{2-\varepsilon}}{(z+1)^{2-\varepsilon} - (z-1)^{2-\varepsilon}}$$

reduces to the ordinary Joukowski transformation when $\varepsilon = 0$, but near $z = 1$ it has a better form. Put $z - 1 = Z$ and consider the transformation for small Z; it is approximately

$$\zeta = (2 - \varepsilon)\frac{2^{2-\varepsilon} + Z^{2-\varepsilon}}{2^{2-\varepsilon} - Z^{2-\varepsilon}}$$
$$= (2 - \varepsilon)(1 + 2^{\varepsilon-1} Z^{2-\varepsilon}).$$

This is no longer a local squaring near $Z = 0$; for ε small and positive it

Fig. XVII.46. The transformation of §7 gives a non-zero angle at the trailing edge.

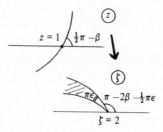

converts an angle of π into an angle of

$$(2 - \varepsilon)\pi = 2\pi - \pi\varepsilon.$$

Hence the trailing edge is now wedge shaped as in fig. XVII.46, with small angle $\pi\varepsilon$, and the upper surface makes angle

$$2\beta + \tfrac{1}{2}\varepsilon\pi$$

with the real axis, approximately (because β is small also).

We shall not go on with further calculations on this modified transformation. It is more profitable to stop at this stage and ask whether the preceding theory is a good model of real flows and whether it is useful in practice.

The Kutta condition on the circulation round a two-dimensional wing section works very well in practice. The initial flow when a wing starts to move through a fluid at rest is close to that obtained from potential theory with zero circulation, as in the first of the sequence of sketches in fig. XVII.47. There is a stagnation point on the upper surface near the trailing edge, and a very high speed round the trailing edge. This is associated with low pressure at the trailing edge A and much higher pressure at the stagnation point B. A secondary flow from B to A is generated in the boundary layer on the aerofoil, leading to separation at A and a vortex behind the trailing edge. This vortex is left behind the wing, and a circulation has to be left round the wing in the opposite sense, by Kelvin's theorem on the constancy of circulation round any circuit moving with the fluid – there

Fig. XVII.47. Separation of a starting vortex from the trailing edge.

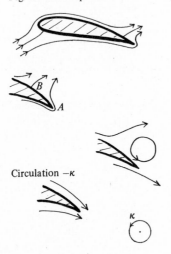

Circulation $-\kappa$

Fig. XVII.48. Kelvin's theorem ensures a circulation round the aerofoil when a vortex is shed.

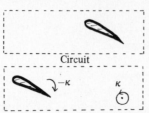

Circuit

is initially none for a large circuit enclosing the wing in fig. XVII.48, and hence there must be none for the same circuit when it encloses wing and vortex.

Not only is the flow pattern given well by the Kutta condition, but the force prediction is also quite good. Measurements for one particular aerofoil which had $\beta = 8°$ are shown in fig. XVII.49. The 'lift coefficient' C_L and the 'drag coefficient' C_D were measured for various values of the angle of incidence α, and the theoretical values are based on the work in §6. These coefficients are defined by

$$\begin{cases} C_L = \text{lift force per length/(chord} \times \tfrac{1}{2}\rho U_0^2), \\ C_D = \text{drag force per length/(chord} \times \tfrac{1}{2}\rho\, U_0^2). \end{cases}$$

From §6(c),

$$\begin{cases} C_L = 8\pi a \sin(\alpha + \beta)/\text{chord}, \\ C_D = 0. \end{cases}$$

For this thin aerofoil, the chord is approximately 4 and $a \doteqdot 1$, and to an adequate approximation

$$C_L \doteqdot 2\pi(\alpha + \beta), \quad C_D = 0.$$

Fig. XVII.49. Lift and drag coefficients C_L and C_D for a particular aerofoil; the full line gives the experimental results of Betz, quoted in Batchelor's text. The dashed line gives the theoretical values.

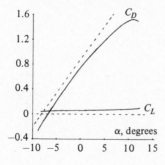

Fig. XVII.50. A stalled aerofoil.

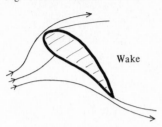

Wake

You see on the graph that the theoretical results are quite good, but viscous forces prevent their being very good. And when α gets too large (or too small, though this is not shown) the theoretical results become poor. For large angles of incidence the flow separates near the front of the aerofoil, much as in fig. XVII.50, and lift decreases and drag increases. The aerofoil is then 'stalled'. One of the advantages of the Joukowski aerofoil over a simple ellipse is that it stalls at a larger angle α, and so can produce relatively high values of the lift coefficient.

There are several reasons why the Joukowski aerofoil is not now used for most aeroplanes. Firstly, profiles which give a higher lift can in fact be found; and better resistance to stalling can be obtained by the use of extra devices such as slots near the leading edge. Secondly, many modern aircraft fly near or above the speed of sound, and so the effects of compressibility can be important; hence different types of design become useful, especially above the speed of sound. And finally, wings can never be two-dimensional; the effects of a finite length of wing must be taken into account, and also the problems of building a wing which is strong enough to support the weight of the body without itself being very heavy.

So the Joukowski aerofoil is part of history, but it provides a good introduction to the basic principles of how aircraft wings provide the lift force essential to flight.

Exercises

1. Flow along a plate occupying the negative real axis has potential

   $w(z) = -Uz.$

   Use a transformation

   $z \mapsto z^k$

   to solve the problem of a flow divided by a right-angled wedge, as in fig. XVII.51.

Fig. XVII.51. Sketch for Q1.

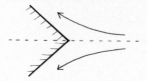

Fig. XVII.52. Sketch for Q2.

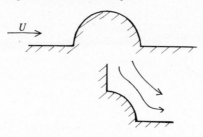

2. Flow along a plane with a semicircular bump on it, sketched in fig. XVII.52, has potential

   $w(z) = U(z + a^2/z)$.

   Use the transformation

   $z \mapsto z^{1/2}$

   to solve for flow in a corner which has a bump in it.

3. Show that in a conformal mapping the flow rate between corresponding streamlines is preserved. Show that stagnation points correspond in general; when do they not correspond?

4. Use the transformation

   $\zeta = z^{1/2}$

   to derive the formula

   $z = a^2 \cot^2 (\pi w/\kappa)$

   for the complex potential w of a line vortex on the negative real axis when the positive real axis is a barrier.

5. Transform the region between the two circles

   $|z| = 2a$ and $|z - a| = a$

   in the z-plane into a strip in the ζ-plane by means of an inversion about a suitably chosen point. Find what the transforms of the two circles are, and to what region the inside of the smaller circle transforms. Write down the complex potential for a uniform flow in the appropriate region of the ζ-plane, and hence solve a flow problem in the z-plane.

Fig. XVII.53. The obstructed channel of Q7.

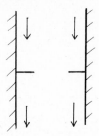

6. Show that the bilinear transformation

$\zeta = e^{i\alpha}(z - a)/(1 - \bar{a}z)$

transforms the unit circle into itself, while moving the point $z = a$ to the origin. What is the effect of the factor $e^{i\alpha}$?

7. Fluid flows down the channel drawn in fig. XVII.53, which has an obstruction in the form of plates fixed to the sides of the channel. Use the transformation

$\zeta = \sin z$

to convert this problem to one of flow through a hole in a plane. Now use another transformation of the same type to map this flow into an easy channel flow. Hence find the potential for the original flow.

8. Use a Joukowski transformation and the Kutta condition to calculate the force and moment for a flat-plate aerofoil of chord l, as sketched in fig. XVII.54.

   The pressure forces on the aerofoil are everywhere normal to the surface. Why is it not true that the total force is perpendicular to the line of the flat plate?

9. Calculate the force and moment for the symmetrical Joukowski aerofoil of §6(a).

10. Points on the cambered Joukowski aerofoil have coordinates (ξ, η) given by

$\xi + i\eta = \zeta = z + z^{-1}$

where

$z = \delta e^{i\phi} + a e^{i\theta}.$

Fig. XVII.54. Flow at an angle past a plate for Q8.

Fig. XVII.55. The circle for transformation in Q12.

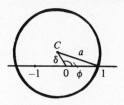

Calculate $\xi(\theta), \eta(\theta)$ to first order in the small number δ. Calculate the extreme values of ξ, and so the chord of this aerofoil. Calculate also the extreme values of η. *Hence* sketch the aerofoil shape, and estimate the maximum thickness.

11. Find a displaced circle which is transformed by a Joukowski transformation to a circular arc. Calculate the circulation induced round such an aerofoil by a uniform stream incident at angle α.

12. The transformation

$$\zeta = f(z)$$

is given to be like a Joukowski transformation in that:
 (i) for large $|z|$, $\zeta \sim z + B_1/z + B_2/z^2 + \cdots$;
 (ii) $f(z)$ has critical points at $z = \pm 1$.
An aerofoil is formed from the circle shown in fig. XVII.55 by using the transformation f. Calculate the force on the aerofoil when a uniform stream is incident at angle α.

13. Show that the transformation of §7 is like a Joukowski transformation in the sense of Q12 above; find the coefficient B_1.

References

(a) The theory of the two-dimensional aerofoil (and much else besides) is clearly set out in full detail in *Theoretical Aerodynamics* (2nd edn), L. M. Milne–Thomson, Macmillan 1952.

(b) There are many texts on Aerodynamics, principally written for Engineers, which give a good account of the effects that must be considered when designing aeroplanes. *Aerodynamics*, L. J. Clancy, Pitman 1975 is such a text which is modern and clear.

(c) The two film loops SFM012 or 28.5040 *Flow separation and vortex shedding*, SFM010 or 38.5038 *Generation of circulation and lift for an airfoil*, form a useful background to the work in § §5–7.

Hints for exercises

Chapter I

1. Resolve in fig. I.1.

2. Sketches need to be three-dimensional versions of fig. I.2: you need either a length dz or a length corresponding to $d\lambda$, and ds will no longer be in the plane of r and θ.

   Otherwise you need to find formulae corresponding to

   $$dx = dr \cos \theta - r \sin \theta \, d\theta,$$

   and calculate ds^2 from

   $$ds^2 = dx^2 + dy^2 + dz^2.$$

3. The formulae for $\hat{\mathbf{r}}, \hat{\boldsymbol{\theta}}, \hat{\boldsymbol{\lambda}}$ are in §2(c) You must differentiate these, and then express the result in terms of these unit vectors.

4. Taylor's theorem in one-dimension is

   $$\phi(x + h) = \phi(x) + h\phi'(x) + \tfrac{1}{2}h^2\phi''(x + \theta h),$$

   when ϕ is smooth enough (twice differentiable with continuous second derivative is certainly enough), and where $0 < \theta < 1$. The last term on the right is less than some constant times h^2 if ϕ'' is continuous on $(x, x + h)$, because ϕ'' is then bounded in that interval.

   Remember that $\mathbf{h}^2 = h_i h_i = h_k h_k$.

5. (i) $\nabla \times (\phi \mathbf{A})$ has ith component

   $$\varepsilon_{ijk}\partial/\partial x_j (\phi \mathbf{A})_k.$$

(ii) When is $\partial^2 u/\partial x \,\partial t = \partial^2 u/\partial t \,\partial x$?

(iii) $\nabla \cdot \mathbf{A} = \partial A_i/\partial x_i$, and here $\mathbf{A} = \phi \nabla \psi$, and $\nabla \psi$ has ith component $\partial \psi/\partial x_i$

(iv), (v) $\varepsilon_{123} = +1$, $\varepsilon_{213} = -1$: look for terms which have these ε in them. Then use $\partial^2/\partial x_1 \partial x_2 = \partial^2/\partial x_2 \partial x_1$ for smooth functions.

(vi) This is $\varepsilon_{ijk}\partial/\partial x_j \{\varepsilon_{klm}\partial A_m/\partial x_l\}$ and use a theorem in §4(*c*).

6. $\nabla \times \nabla = 0$, and write what remains as

$\nabla \times (\mathbf{A} \times \mathbf{B})$,

which works out rather like Q5(vi). Finally put the proper value back for **B**.

7. Use Q5(iii), and the divergence theorem from §5(*b*).

8. For $\nabla \cdot \mathbf{A}$ proceed as in §6(*a*), and use formulae for derivatives of $\hat{\mathbf{r}}, \hat{\boldsymbol{\theta}}, \hat{\boldsymbol{\lambda}}$ from §2(*c*) and Q3. For $\nabla \times \mathbf{A}$ evaluate the determinant in §6(*c*).

9. $\mathbf{A} \cdot \nabla \mathbf{A}$ is composed of terms like

$\mathbf{A} \cdot \nabla (A_r \hat{\mathbf{r}}) = (\mathbf{A} \cdot \nabla A_r)\hat{\mathbf{r}} + A_r(\mathbf{A} \cdot \nabla \hat{\mathbf{r}})$.

See §6(*b*) for $\mathbf{A} \cdot \nabla$ in plane polar coordinates, and §3 for ∇ in spherical polar coordinates.

10. $\nabla^2 \phi$ is $\nabla \cdot \nabla \phi$. $\nabla \cdot \mathbf{A}$ in sphericals is in Q8. $\nabla \phi$ in sphericals is in §3.

Chapter II

(1) (i) Do not use h in forming a dimensionless group, we are trying to find h. What are the dimensions of c, T?

(ii) The time is $T = T_1 + T_2$, where T_1 is the time to fall h_1 (the new estimate of the depth) and T_2 is the time for the sound to travel back up the well. Put both T_1 and T_2 in terms of h_1, and then solve for h_1. Approximate your answer by taking the dimensionless group of (i) to be small enough to use only a few terms of the binomial series.

(2) (i) Compare the forces acting on the stone, and estimate the speed v by using the final (or average) free fall speed. Find a value by thinking of a spherical pebble.

(ii) You need an equation of motion from Newton's law. This can be integrated to give depth y in terms of time t (you need hyperbolic functions). Then the new estimate of depth, h_2, is

$h_2 = y(T)$.

You may approximate this if your dimensionless number in (2) (i) is reasonably small.

Chapter III

1. Can you make a permanent heap of a fluid? Can you do it with salt?

2. See §1, but consider the total momentum in the cube divided by the total mass.

3. Assume that the depression has circular curves of constant pressure (iso-bars), and take two maps separated by about six hours (the time a train takes to go from Bristol to Edinburgh, approximately). Then for each case calculate

$$\frac{\text{pressure at later time} - \text{pressure at earlier time}}{\text{time difference}}$$

an an estimate of $\partial p/\partial t$. Put your answers into standard units at the end.

4. See the example (1) in §2 and (1) in §4(*a*), and add on an equation for the third dimension.

If the *x*, *y* motion is a circle described at uniform speed, and the *z* motion is of constant speed, the result is a helix. This is not hard to draw on the out-side of a cylinder of rolled-up paper.

5. (i) Calculate dy/dx as in example (3) (i) in §2, and remember that *t* is *not* a variable for streamlines, so interpreting the equation is easy. Sketch the result for two different times.
   (ii) Exactly the same, except the interpretation is more bother. Again, use two values for *t*.
   (iii) Use $d\mathbf{r} = dr\,\hat{\mathbf{r}} + rd\theta\,\hat{\boldsymbol{\theta}}$. Then calculate

$$rd\theta/dr.$$

Sketch at least two curves.

6. (i) Again, dy/dx is easiest. But it is also useful to solve for
   (1) $x(s, t)$, $y(s, t)$ for the streamlines
   (2) $x(t)$, $y(t)$ for the particle paths.
   (ii) You can use dy/dx for the streamlines, because *t* is just a parameter. But for the particle paths you must calculate

$$x(t) \text{ and } y(t)$$

and then eliminate *t*.

7. A small square on the boundary is hit by molecules, which reverse a component of their velocity at the collision; many such individual changes of momentum give a total change for the square over a time τ. If you take a suitable size for *a* and for τ then (total momentum change)/(area × time interval) will give a suitable value.

Chapter IV

1. Re-read §1. The fish in a volume *V* increase because of immigration and birth; they decrease because of fishing and death. This gives you an equa-tion to which you can apply the divergence theorem, and the DuBois–Raymond lemma if a continuum theory is sensible.

2. Calculate $\nabla \cdot \mathbf{v}$ as in Chapter I §6(*a*), but using spherical coordinates – see Chapter I Q8. Your calculations obviously do not apply at $r = 0$, **v** is not defined. It is easy to calculate the flow through a *sphere* centred on *O*, be-

cause $\mathbf{v} \parallel d\mathbf{S}$ for such a sphere. Use a suitable sphere round O; the divergence theorem will show that the flow through it equals the flow through S.

If O lies on S you can only fit a hemisphere round O and inside S.

3. (i) $\mathbf{v} \cdot \nabla \mathbf{v}$ is calculated in Chapter I §6(b), with slightly different notation. What is the acceleration of a particle moving at constant speed in a circle?

(ii) Mass conservation is

$v(x)A(x) =$ constant,

where $v(x)$ is the average speed (over the pipe cross-section) at position x. The acceleration is $\mathbf{v} \cdot \nabla \mathbf{v}$, where $\mathbf{v} = v(x)\mathbf{i}$, to a good approximation if the flow is almost in the x-direction.

4. $u = \partial\psi/\partial y$, so integrate to get $\psi(x, y)$. The arbitrary function of x introduced by this is found from $\psi = 0$ on $y = 0$ for all x. Then $v = -\partial\psi/\partial x$.

5. (i) Use a formula for $\nabla \times \nabla \times$ from Chapter I Q5, and show that $\nabla \cdot (\psi\mathbf{k}) = 0$.

(ii) Use a formula for ∇^2 from Chapter I §6(e), and use $\nabla \cdot \mathbf{v} = 0$.

(iii) Work out $\mathbf{v} \cdot \nabla$ in terms of ψ first, then apply it to $\mathbf{v}$ in terms of ψ.

6. Put $\psi = c$ into cartesian coordinates if necessary: it is a conic with a shifted origin. Check the direction of the flow by calculating a velocity component.

7. You must show that $\mathbf{v}$ is zero at large distances and that the velocity of the fluid perpendicular to the cylinder equals the cylinder's velocity in that direction. The origin $r = 0$ is at a position which coincides with the cylinder's axis.

8. $\mathbf{v}$ in terms of $\partial\psi/\partial r, \partial\psi/\partial\theta$ is in §4(b). Try thinking of $\theta = 0$ and $\theta = \frac{1}{2}\pi$ as walls.

9. For the region between two curves, Stokes' theorem has the form

$$\int_{c_1} \mathbf{v} \cdot d\mathbf{l} - \int_{c_2} \mathbf{v} \cdot d\mathbf{l} = \int_S \nabla \times \mathbf{v} \cdot d\mathbf{S}$$

because one curve must be traversed in the opposite direction. Calculate $\nabla \times \mathbf{v}$ for this flow, for $r \neq 0$.

10. Put the streamlines in cartesians to find the behaviour for x large. Solve for r for $\theta = \frac{1}{2}\pi$ on the streamlines

$$\psi = 0, \psi = Ua, \psi = 2Ua, \psi = 3Ua$$

in turn. Calculate the velocity at each of these positions.

11. (i) See §5(c) for $\psi(P)$ due to a source and an image. Here we need $\psi =$ constant on the wall so that it is a streamline. Express the angles in terms of the position of P and the source and its image. The speed along the wall is just a derivative of ψ evaluated on the wall.

(ii) Take the wall to be $z = 0$, and put the source at $z = -c, r = 0$. Use the same procedure as in (i).

12. (i) Treat this as a source and a sink close together.
    (ii) Don't forget the centre of the cylinder.
    (iii) Try the same formula for the image point.

13. You need to show that

    $\partial\Psi/\partial r = rv_z$, $\partial\Psi/\partial z = -rv_r$.

    Try $2\pi\Psi$: the first integrand is

    $v_z(s, z) \times 2\pi s\, ds$,

    i.e. a velocity times the area of the region between two circles of radii

    s and $s + ds$

    with centre on the z-axis. Draw this.

    The other integral is similar, but the area is on a cylinder round the z-axis, with radius a. Draw it. Try the case $a = 0$ to simplify the problem.

14. A source at $z = -a$ (and $r = 0$) has Stokes' stream function

    $$\frac{m}{4\pi}\left\{1 - \frac{z + a}{[(z + a)^2 + r^2]^{1/2}}\right\}$$

    in cylindrical coordinates. Evaluate Ψ on the axis $r = 0$ for $z > a$ and $z < -a$ to find what value of Ψ gives the dividing streamline. Stagnation points on the axis give the length of the Rankine body, see §7(b). Use symmetry to find at what point on the dividing streamline that width is greatest. If $a \to 0$ with $ma = $ constant, you are going to get a dipole at the origin, see §7(c).

15. Evaluate Ψ_1 on $r = 0$ and on $r^2 + z^2 = a^2$. Evaluate v_z on $z = 0$ from Ψ_1.
    For Ψ_2, see §7(c).
    $r^2 + z^2 = a^2$ is a streamline, so what is the velocity component perpendicular to it? Calculate v_θ (spherical coordinates) on $r = a$ for both Ψ_1 and Ψ_2.

Chapter V

1. You should have one principal rate of strain

    $e = -5$ with axis $(0, 1, 0)$.

    This means that the motion is compressive along this line, fluid leaving into directions perpendicular to the line. Find the other principal rates and axes and interpret them similarly. And $\omega = \nabla \times \mathbf{v}$ gives a rotation rate about the direction of ω. See §1.

2. See §1 again. One principal rate of strain is

    $e = 0$ with axis $(y, -x, 0)$,

    i.e. the direction of this zero rate of strain is $\hat{\theta}$, around the pipe's axis. You should find $\nabla \times \mathbf{v}$ from the determinant formula in Chapter I §6. Cylindrical sheets of fluid have to slide over each other, as the speed is higher near the centre.

3. Use the formula
$$\mathbf{i} = \hat{\mathbf{r}} \cos \theta - \hat{\boldsymbol{\theta}} \sin \theta$$
from Chapter I, Exercises, with a similar formula for $\mathbf{j}$.

4. You have ψ in Chapter IV Q4, and you need
$$-\nabla^2 \psi,$$
see §2(*a*). Similarly for Chapter IV Q6.

5. $\mathbf{v}$ in terms of Ψ is in Chapter IV §6(*a*).

   $\nabla \times \mathbf{v}$ for spherical polars is in Chapter I §6(*c*).

6. You need to solve $\nabla^2 \psi = -\omega$, with $\nabla^2 \psi$ given in plane polars in Chapter I §6 (by taking the *z*-dependence out from cylindrical polars). You must then find a particular integral by asking 'what, when you take ∇^2 of it, gives $\sin \theta$?' This suggests that you should take
$$\psi(r, \theta) = f(r)g(\theta)$$
with $f(r)$ satisfying an ordinary differential equation and also $f(a) = 0$.
   For the outer flow you look in Chapter IV §5(*a*) for a flow with $\omega = 0$ and $\psi = 0$ on $r = a$ and the *same* angular dependence $g(\theta)$.
   Finally you adjust the constants so that v_θ has the same value at $r = a$ for both inner and outer flows.

7. The vorticity $\omega = \nabla \times \mathbf{v}$: see Chapter I §6. You can now proceed either geometrically – note that there is *no* vorticity out from the axis of the pipe – or analytically, by solving equations like those for streamlines:
$$d\mathbf{r}/ds = \boldsymbol{\omega}(\mathbf{r}, t).$$

8. The volume dV of a fluid particle must be constant here, and if you put
$$dV = dS\,dz$$
then dS is constant following a particle. Now use Kelvin's theorem for this area dS.
   In Q6 is ω constant for a particle moving round a streamline?

9. The limiting case is $U = 2C/a$. Plotting $\psi = 0$ is hard: find its form *near* $(0, 0)$, and where it cuts $x = 0$. Then sketch in other streamlines.

10. You need an image in each wall, but also the image of these images in the extensions of the walls to $x < 0\ y = 0$ and $y < 0\ x = 0$.
    Calculate the velocity of fluid at the vortex due to the images and use a theorem on the motion of vortex lines. The path of the particle that is the vortex is calculated as in Chapter III §2.

11. The image of a vortex in a cylinder is likely to be another vortex at the image point, with perhaps yet another image at the centre – see Chapter IV

§5(c) for similar images. If there is only one image vortex, you get a circulation round the cylinder. With equal and opposite images you get no circulation. The velocity of the original vortex is due to the image vortex or vortices.

12. You now have extra velocities due to
    (i) the stream,
    (ii) the dipole that is the image of the stream in the cylinder.

13. This is best done by the methods of Chapter XVI. If you want to do it now, you must find the velocities at the vortex due to all other terms arising in Q12, and find points (c, b) [or (r, θ)] such that the total velocity is zero.

14. Take ψ in the form $\psi_1 + \psi_2$ where

    $\psi_1(r)$ is an easy solution of $\nabla^2 \psi_1 = -\Omega$,
    $\nabla^2 \psi_2(r, \theta) = 0$.

    The method of finding ψ_2 is in Chapter XI §4. Reject terms in $\ln r, \theta, r^{-k}$ because they are badly behaved near $r = 0$. Impose $\psi = 0$ on $\theta = \pm \frac{1}{8}\pi$ and also keep *only* the lowest power of r.

15. In Q14 take the easy solution, but with

    $\psi = 0$ on $\theta = \pm \alpha$.

Chapter VI

1. Take a particle of fluid to be at position vector $\mathbf{R}$ in a fixed frame of reference and at $\mathbf{r}$ in the accelerating frame, and consider the apparent acceleration $d^2\mathbf{r}/dt^2$ in terms of $d^2\mathbf{R}/dt^2$. If you have a force $-f\mathbf{i}$, then

    $-\nabla(fx) = -f\mathbf{i}$

    when f is a constant.

2. Take lengths $PA = a$, etc. in §2. Then
    $\mathbf{AB} = (-a, b, 0)$

    and you may calculate $\mathbf{n}\delta A$ from $\mathbf{AB} \times \mathbf{AC}$.

3. Use the result

    $dF_i = \sigma_{ij} n_j dS$

    from the end of §2 (where summation over i is implied by the notation. The normal force is

    $d\mathbf{F} \cdot \mathbf{n}$

    and the tangential force is what is left.

4. Consider $B_i = a_i A_{il}$ where $\mathbf{a}$ is *any* constant vector, and apply the divergence theorem to $\mathbf{B}$.

5. The effective force on the mercury is now (as in §4 and Q1) $-\rho f\mathbf{i} - \rho g\mathbf{k}$. In the first situation the barometer is vertical, in the second it hangs along

the line of

$$f\mathbf{i} + g\mathbf{k}.$$

Atmospheric pressure is unchanged.

6. Take the lowest point of the water to be at height z_0 above the base of the can, and use an equation in §4 (modified to have $z = z_0$ when $r = 0$). The other equation you need is one for conservation of volume of water.

7. The major term is due to the Moon. You need the difference in the potential $\gamma M/r$ at the nearest point to the Moon and at the Earth's centre. Then $g\delta h = \delta\Phi$ gives the change in height where g is Earth's gravity.

8. The outward force at P as sketched in fig. H.1 is $pa\,d\theta$ per length of gutter. Integrate horizontal and vertical components of this, remembering that

$$p = p_0 + \rho gz$$

and that p_0 acts on the outside too. For the moment, you need $BN\,pa\,d\theta$, in terms of θ.

9. At depth z (see fig. H.2) the water pressure is $-\rho gz$ and the force on a strip of area dA is

$$-\rho gz\,dA\mathbf{j},$$

with moment

$$\rho gz^2\,dA$$

about the x-axis. You must integrate these to get the total force and moment

Fig. H.1. A point P on the gutter at a depth z below the centre O. BN is at right angles to OP.

Fig. H.2. An area element DA on the vertical face of the half-full hemisphere.

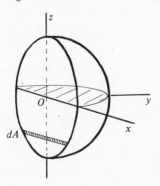

Fig. H.3. A section of the conical container.

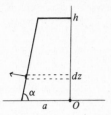

Fig. H.4. The diagram for Q11.

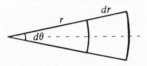

on the vertical face. Then moment/force gives the depth of the centre of pressure.

The second part is better done by noticing that the container as a whole is in equilibrium if supported by a force Mg. The centre of pressure is most easily found by a drawing of three forces which must be in equilibrium.

10. There is an upthrust on the curved sides of the container (see fig. H.3). The total vertical force must be Mg. The volume of the container is

$$\int_0^h \pi(a - z \cot \alpha)^2 dz.$$

11. Neglect z, which is a constant, and resolve forces parallel to the midline (see fig. H.4). The net inward force must equal the mass-acceleration in this motion in a circle.

Chapter VII

1. Look up c_v and c_p for water. Kinetic energy per kilogram is measured by $\frac{1}{2}v^2$, and $\Delta Q/dT$ is a specific heat.

2. Use formulae from §3 for E in terms of T, and the perfect gas law. Find also a formula for S in terms of p and ρ.

   Find the isentropic formula for a perfect gas and use it to eliminate ρ, using the perfect gas law.

3. Calculate dF and replace dE from §2. Hence find $(\partial F/\partial V)_p$ and $(\partial F/\partial T)_V$. Finally calculate $\partial^2 F/\partial V \partial T$ two ways.

4. Use a formula for dS in terms of dE in §2, to calculate $(\partial S/\partial V)_T$.

   Q3 with $pV = f(T)$ shows that both α and β can be expressed purely in terms of f. Hence the new form for β in this question is purely a function of T.

5. Look up Archimedes' theorem in Chapter VI §4, and use this to find where the water level is inside the tin.

   A slow change allows heat flow, and is therefore isothermal. Use this to find the pressure inside the tin.

   Is the tin still floating?

   H is defined by $p_0 = \rho g H$, with ρ the density of water.

6. Use $pV =$ constant for the balloon, and the isothermal atmosphere from §4. Equating two pressures gives a value for z.

   Next use $pV^\gamma =$ constant in a rapidly rising balloon, but the *same* atmosphere.

7. Integrate $dp/dz = -\rho g$ and use the perfect gas law to find $T(z)$. Look at neutral stability in §4.

8. This is an isentropic change, and you need values for T at sea level and 12 km, from §4.

Chapter VIII

1. Read §2(b) and rewrite the argument about momentum transfer into one about kinetic energy (i.e. heat) transfer.

2. In Chapter VII §3 we have $dE = c_v dT$, which is energy per mass (not per volume).

   The equation must express

   'rate of change of heat energy in $V =$
   rate of inflow of heat energy through S
   + rate of heat energy input inside V'

   for any volume V.

   The first term must be d/dt of a volume integral, the second must be like that in §1(iii) but with heat transfer as in Q1, the third must be like §1(ii) but with a heat energy source instead of a momentum source.

   Then proceed as in §1.

3. $D/Dt + \mathbf{\Omega} \times$ in place of D/Dt, and $\mathbf{v} + \mathbf{\Omega} \times \mathbf{r}$ in place of $\mathbf{v}$ in Euler's equation and assume $\mathbf{\Omega}$ is constant.

   Now manipulate $\mathbf{v} \cdot \nabla \mathbf{v}$ by Chapter I §4, and $\mathbf{\Omega} \times (\mathbf{\Omega} \times \mathbf{r})$ as in Chapter VI §1 to get the result.

4. If $2\mathbf{\Omega} \times \mathbf{v} = \nabla p$, then $\mathbf{v}$ is parallel to the curves $p =$ constant on the surface. Is this so in your map? Check up on the sizes of omitted terms, including $\rho_2 \mathbf{g}$ from §4(b). How far up is the presence of the surface likely to be noticed?

5. A length scale l and velocity scale u in the bath give a vorticity of order of magnitude u/l. Estimate u, l and Ω.

6. You must use

   $$dE = \Delta W + \Delta Q$$

and that ρdE is an energy per *volume*. Then the work ΔW is derived from P, and an equation in Chapter V §1 relates dV/V to $\nabla \cdot \mathbf{v}$. Finally ΔQ comes from heat inflow and heat sources as in Q2.

For the other equation you need

$$dE = - PdV + TdS.$$

7. Remember from Chapter IV that

$$\mathbf{v} = (r^{-1}\partial\psi/\partial\theta, \ -\partial\psi/\partial r)$$

in two-dimensional flow in polar coordinates, and that you must have $\nabla \cdot \mathbf{v} = 0$ in a flow which is derived from ψ. The formula for the tensor s_{ij} in terms of e_{ij} is in §2.

8. The formula for $d\mathbf{F}$ on a surface is in Chapter VI §2. You are asked to use just the term

$$2\mu e_{ij}$$

from the stress tensor σ_{ij}. For a cylinder $d\mathbf{S}$ is parallel to $\hat{\mathbf{r}}$, and the resultant force should be parallel to the stream U.

The force will only be correct if the flow is correct: look at photographs in other texts.

Chapter IX

1. Calculate $\nabla \cdot \mathbf{v}, \ \partial \mathbf{v}/\partial t, \ \mathbf{v} \cdot \nabla \mathbf{v}, \ \nabla^2 \mathbf{v}$ for the given velocity $\mathbf{v}$. Hence solve the Navier–Stokes equation and verify that the boundary conditions are satisfied. Neglect gravity. No pressure gradient is needed here.

The stress tensor is

$$\sigma_{ij} = -p\delta_{ij} + 2\mu e_{ij}$$

in this case and e_{ij} can be found from $\mathbf{v}$. The normals to the two walls have direction $(0, \pm 1, 0)$ and stresses on the walls are $\sigma_{ij}n_j$.

2. As in Q1, reduce the Navier–Stokes equation to the form in §1(b). The solution which fits the boundary conditions of zero velocity on both $r = a$ and $r = b$ includes a term in

$$\ln(r/a)$$

The shear stress on the cylindrical walls comes from s_{rz}, whose form is given in §1(b). The force on the inner cylinder ($r = a$, say) parallel to the flow is $2\pi a s_{rz}(a)$. The force from the pressure gradient is proportional to the area $\pi(b^2 - a^2)$ of the flow.

3. The component of gravity causing the motion is $g \sin \alpha$. The other component is balanced by a hydrostatic pressure as in Chapter VIII §4.

The dynamic pressure is constant, because it is constant along the surface, The simplified Navier–Stokes equation has a term in $g \sin \alpha$, and $U(y)$ is quadratic in y. The surface condition is $s_{xy}(h) = 0$, where h is the depth.

The forces parallel to the flow are a component of gravity and a shear force on the slope.

4. Assume there is no imposed pressure gradient. There is again a term in g sin α in the simplified equation, and $U(r)$ is quadratic in r. The force calculations are like those in §1(*b*), Q2 and Q3.

5. The work is as in §1(*c*). For the inner flow you must impose a boundary condition as $r \to 0$, for the outer you need a condition as $r \to \infty$. The couples are calculated as in §1(*c*); one of them is non-zero despite the time-independence of the flow. Calculate the total kinetic energy of the outer flow. How long would it take to set up this flow?

6. The appropriate Navier–Stokes equation is in §2(*a*). Try a solution

$$U(y, t) = F(y)e^{i\omega t}$$

and choose $F(y)$ so that $U(y, t)$ is finite as $y \to \infty$. Remember that

$$V \cos \omega t = \mathcal{R}e(Ve^{i\omega t}).$$

7. Use the solution of §2(*a*) to calculate $\mu(\partial U/\partial y)_{y=0}$. The sign shows the direction of the stress.

   Substitute $U(y, t) = ySF(\eta)$ into the reduced Navier–Stokes equation of §2(*a*) and into the boundary conditions. The equation for $F'(\eta)$ has variables separable. The solution for $F(\eta)$ is best put in the form

$$\int_\eta^\infty -F'(s)ds + D$$

where the integral has a singularity as $\eta \to 0$. Integration by parts gives a form for $F(\eta)$ when η is small.

8. Check that $\nabla \cdot \mathbf{v} = 0$. Use methods from Chapter I to find $\mathbf{v} \cdot \nabla \mathbf{v}$ (it has *four* terms in it) and $\nabla^2 \mathbf{v}$ (*u* does not occur in it, because the velocity $uar^{-1}\hat{\mathbf{r}}$ has $\nabla \times \mathbf{v} = 0$). Show that $\partial p/\partial \theta$ must be zero, as in §1(*c*).

   To interpret your solution, write it in the form

$$C/r - D(r/b)^{R+1}$$

and consider the second term when R is, say 100. Note that $(0.97)^{101} = 0.05$.

9. Use a flow from Chapter III §2 as a model for the initial flow. Pressure gradients are needed to accelerate flows–flow round a curve is accelerated towards the centre of curvature.

   A pressure gradient from B to A will cause a secondary flow from B to A in the boundary layer. Where can this fluid go?

10. Calculate $d/dr(Vr)$ for this flow.

11. If you haven't met the method of separation of variables, leave this exercise till after Chapter XI, where it is explained.

   V must initially equal $-G(a^2 - y^2)/2\mu$, and the separation solution at $t = 0$ is a Fourier series. This initial condition determines the coefficients.

   Which term in the separation series lasts longest?

   How long does vorticity take to diffuse to the centre line from the walls?

Chapter X

1. Vorticity in terms of ψ is in Chapter V §2. You do not need to work with all terms:
   (i) at large distances the last two drop out (and you can put what is left in terms of y),
   (ii) most of the terms satisfy $\nabla^2 \psi = 0$.

2. When you calculate $\boldsymbol{\omega} \cdot \nabla \mathbf{v}$ and $\mathbf{v} \cdot \nabla \boldsymbol{\omega}$ use the methods of Chapter I §6.
   An alternative method is to show that

   $$\nabla \times (\mathbf{v} \times \boldsymbol{\omega}) = 0$$

   by showing that $\mathbf{v} \times \boldsymbol{\omega} = \nabla F(r)$ for some function $F(r)$.

3. Use suffix notation and the equation

   $$\nabla \cdot \mathbf{v} = 0$$

   to rearrange

   $$\rho \nabla \cdot (\mathbf{v} \cdot \nabla \mathbf{v}).$$

   Poisson's equation gives, for example, the gravitational field of a distribution of mass; so it has a reasonably easy solution.

4. Remember, or look up, the theorem that $f(x, y)$ and $g(x, y)$ are functionally related if and only if the Jacobian

   $$\partial(f, g)/\partial(x, y) = 0.$$

5. This is the same as Q4, but you need a different vorticity equation in this flow, and $\mathbf{v}$ and ω in terms of Ψ from Chapter IV §6 and Chapter V §2.

6. The initial vorticity has components ω_θ and ω_z. You can now argue in terms of stretching of vortex lines (see §2(e)); or in terms of the vorticity equation which provides equations for

   $$D\omega_\theta/Dt \quad \text{and} \quad D\omega_z/Dt.$$

7. There is an extra term arising from

   $$\nabla \times (\rho^{-1} \nabla p).$$

   See Chapter I Q5 for the expansion of this product. If $\rho = f(p)$ then $\nabla \rho$ can be put in terms of ∇p by the chain rule.
   When you include a viscous term from the Navier–Stokes equation, you need

   $$`\nabla \times \nabla^2 = \nabla^2 \nabla \times`$$

   which you can prove with tensor methods from Chapter I §4.

8. Re-read the similar problem in §4(b). Assume that the flow down the pipe has constant speed $\mathbf{v}$ and emerges at atmospheric pressure p_0. Assume the flow is (almost) steady because $A \gg a$. You will find a simple equation for dh/dt in terms of h and the initial depth h_0.
   Compare the pressure outside the pipe and in the flow halfway down the pipe.

9. This two-dimensional flow has ω constant, and so by §4(c) there is a special form of Bernoulli's equation which holds everywhere. You need to calculate $\psi(y)$, which you can do from $V(y)$ since this is known both before and after the contraction. You should include gravity, as there is a change of height along a streamline. You can find the pressure drop along the streamline $y = 0$ most easily, and you should show that the pressure is uniform both before and after the contraction.

10. The two-dimensional form is in §4(c). You need to examine $\mathbf{v} \times \boldsymbol{\omega}$ when ω has the form found in §2(d) and $\mathbf{v}$ is derived from Ψ as in Chapter IV §6. This enables you to write $\mathbf{v} \times \boldsymbol{\omega}$ as ∇f where f involves Ψ, ω, r.

11. Ψ gives the speed on the sphere, and hence the pressure through Bernoulli's equation, just as in §5(c). The surface element is now

$$2\pi a^2 \sin \theta \, d\theta$$

because we integrate over a circular strip of radius $a \sin \theta$ and width $a \, d\theta$ at angle θ. Remember to resolve parallel to the stream for a force in this direction.

12. The method is again that in §5(c). The flow far upstream is $V(y)\mathbf{i}$, and the surface streamline goes to infinity along $y = $ constant. The surface pressure is a fairly complicated function of θ, but most of the integrals are zero quite easily. Because of the lack of symmetry you should expect a force in the y-direction.

13. What is the pressure inside the building? Consider the wake at the rear. If the doors are closed, there is a stagnation point on their outside. What is the pressure there? Discuss the local flow and pressures if the doors open slightly.

14. The calculation is harder in two dimensions, but you should at least:
    (i) decide on the model flow as in §6(a);
    (ii) find an equation for the surface, say

    $$r = f(\theta),$$

    and find the pressure on the surface, say

    $$p = g(\theta),$$

    both as in §6(b);
    (iii) find where $p = p_0$ on the surface, and derive an integral for the force parallel to the stream, as in §6(c) and §6(d).

15. You can calculate, on the assumption of no flow up the side tube, the pressure at A in terms of atmospheric pressure, the areas, and the flow rate Q – you need Bernoulli's equation, a mass conservation equation, and an assumption of uniform flow. Then if the pressure inside the lower end of the side tube is less than atmospheric pressure, flow up the tube will take place.

16. This needs both separation of variables (Chapter XI) and Bessel functions

(Chapter XIII) to do it properly. You can come back to it later if necessary.
  Solve the equation for ψ by assuming

$$\psi(r, \theta) = R(r)H(\theta)$$

and separating variables. You must have continuity in θ, and *one* suitable solution is

$$H(\theta) = \sin \theta.$$

The equation for R then comes into a standard form, with suitable solution

$$R(r) = J_1(kr).$$

Choose k to give a simple flow.
  The outer flow is chosen as in the Hill's vortex, Chapter V §3. Choose constants so that v_θ (and hence p) is continuous at $r = a$.

Chapter XI

1. You need to use Kelvin's theorem on circulation here. What is the circulation round *any* circuit before the barrier moves? Where is there any vorticity? Calculate the vorticity in the given rigid body rotation.

2. There must be boundary conditions at the walls $r = a$ and $r = 2a$ and at both sides of the barrier, i.e. at $\theta = 0$ *and* $\theta = 2\pi$. The region V here is simply connected because it is between $\theta = 0$ and $\theta = 2\pi$, so you can use the divergence theorem.

3. Use the divergence theorem. If ϕ has a local maximum, then $\nabla\phi$ locally points towards this maximum – what does this say about

$$\int \nabla\phi \cdot d\mathbf{S}$$

for a small sphere round the maximum?
  Take x-axis along the local velocity, and consider

$$v^2 = (\partial\phi/\partial x)^2.$$

4. The velocity is $\nabla\phi$, and is also $\nabla\psi \times \mathbf{k}$. Try $x^3 - \alpha xy^2$ as a solution of $\nabla^2\phi = 0$. You will see in Chapter XVI why this ϕ and ψ occur together.
  For degree one, x is typical.
  For degree two, xy and $x^2 - y^2$ are typical, but don't forget $r^2 P_2 (\cos \theta)$, which you can put in terms of x, y, z. Use formulae in Chapter I.

5. Read §4 on how separation of variables works. Here you want to use

$$\phi = f(x)g(y).$$

You will find only one sign of separation constant that will satisfy the resulting equation for x *and* the boundary values on $x = \pm a$. The resulting equation for $g(y)$ has to satisfy the condition on $y = 0$.
  The two parts of the solution are *even* and *odd* in x; the boundary condition on $y = a$ is *odd* in x. You need to invert a Fourier series.
  This is a box with a lid hinged at $(0, a)$.
  Calculate d_0 and d_1 in terms of Aa^2.

6. Read §4(*b*). Try zero separation constant and see what you get for $R(r)$ and $H(\theta)$. Try negative constant and show that $H(\theta)$ is not continuous. Try positive constant with the continuity condition to find the values allowed for the constant. Finally add up a lot of suitable solutions.

7. Use the cosine rule to express $|\mathbf{r} - \mathbf{c}|$ in terms of r, c, θ; expand by the binomial theorem. Compare coefficients to find b_n.

8. Discuss circulation and vorticity as in Q1. What is the condition on the wall of the funnel? What is it at the surface, in terms of ϕ? How do you expect ϕ to behave for small r, i.e. near the top of the pipe?

   If you separate variables in r and θ, what function of θ do you expect to get?

9. Read §8. You need a similar ϕ here, but for three dimensions – see §5(*c*). You must calculate $\partial\phi/\partial t$ as in §8 for Bernoulli's equation.

10. Take the cylinder to have velocity (U, V) in cartesian coordinates, and find a potential as in §8. You need $\partial\phi/\partial t$ as in §8, though the formula will be longer. Finally you have to take (U, V) and $(\dot{U}, \dot{V})$ appropriate for a particle moving in a circle (choose a convenient time). The result should look familiar.

11. The potential is as in Q9. You calculate the force as in §8, but remember it is now three dimensions. The added mass is smaller than for a cylinder (because the fluid can get round the sides). The upward force on the sphere follows from Archimedes' theorem, in Chapter VI.

12. The only change from §5 is that we now need $\partial\phi/\partial t$ in Bernoulli's equation, and ϕ depends on t through a. The calculation method is unchanged, but extra terms come into the integration.

13. The equation for Z has an integrating factor, and solves to give Z in terms of H. Since Z is $\frac{1}{2}\dot{H}^2$, $H(t)$ can be found as an integral.

   When c is large and $H < H_0$, you find that Z is an easy function of H, approximately, and that then $\ddot{H} \ll g$. Compare the terms in $\dot\phi$, $\dot{H}^2$ and gH in the equation for the pressure.

14. The image system is in Chapter IV §7. Express the distances from images to any point in the fluid in terms of r, θ and expand the logarithms for $r \gg 2a$.

   The surface pressure comes from Bernoulli's equation. The force integral *can* be evaluated (by parts, $t = \tan\frac{1}{2}\theta$, partial fractions). The result should be like that for a line charge in electrostatics.

   Is the problem of the same type as in §9?

15. What is the surface that you must use at large distance? So what do you use for $d\mathbf{S}$ on this surface? Note that ϕ and $\nabla\phi$ are now larger at ∞ than they were in §9.

   You can still derive the result of §9(*b*). But in §9(*c*) you need to take cartesian components in the integral at large distances. You will find a familiar result.

Chapter XII

1. Aim to get $\partial/\partial t \, \nabla \cdot (\rho \mathbf{v})$ in each equation; you will need the form of Euler's equation that starts with

   $\partial/\partial t (\rho \mathbf{v})$.

   If you include the body force $\rho \mathbf{F}$, you will get a term which shows the interaction of gravity with sound waves.

2. Neglect gravity. Check that $U(y)\mathbf{i}$ satisfies the appropriate equations. Substitute

   $\mathbf{v} + U(y)\mathbf{i}$

   into the equation you found before you linearised in Q1. You must now linearise this equation, and replace

   $\partial/\partial x_j (\rho v_j)$

   by means of a continuity equation that include U. The extra term need not be small.

3. Assume that the potential for $x < 0$ is

   $\phi = \phi_{\text{incident}} + \phi_{\text{reflected}}$,

   where ϕ_{incident} is given and $\phi_{\text{reflected}}$ is a wave travelling in the opposite direction. The reflected wave has the same time dependence as the incident one, so that the boundary condition can hold for all t.

4. What is needed here is a reflected wave which also has the same y-dependence as the incident wave, so that the boundary condition is satisfied for all y.

   You should get a zero reflected wave at a particular angle of incidence.

5. Here you want suitable potentials for

   $\begin{cases} \text{incident and reflected waves for } z > 0, \\ \text{transmitted wave for } z < 0. \end{cases}$

   The boundary conditions are to hold for all x and all t (where x is along the boundary), and they are on normal component of velocity and on pressure at $z = 0$.

   You should find

   $\begin{cases} \text{the wavelength of the transmitted wave,} \\ \text{its angle to the plane,} \\ \text{the transmitted and reflected amplitudes.} \end{cases}$

   For certain conditions the transmitted angle is imaginary; this means that you should have used

   $\phi_{\text{transmitted}} \propto \exp ly$.

   The values for air and water of ρ and c are such that you can approximate the amplitudes.

6. Read §4(*b*). The conditions at $x = 0, l$ are like those for the clarinet. You

also have conditions at $t = 0$ which enable you to find β_n, and then B_n by inverting a Fourier series.

7. Try a solution in terms of $\cos \omega t$ so as to get

$$\partial \phi / \partial x = a\omega \cos \omega t \text{ at } x = 0.$$

You then need a suitable choice of sines and cosines of $\omega x / c$ to get a fit to the boundary condition at $x = l$.

The response is infinite at some values of $\omega l / c$.

8. This is like §5(c) but with a different boundary condition. You are seeking a relation between k, ω and m in the notation of §5(c).

9. This is related to §4(b) and §5(c). Find a suitable combination of sines and cosines to fit all the boundary values, and get ω in terms of three integers from the wave equation.

10. Energy flow rates are shown in §6(d) to be proportional to $|$amplitude of wave$|^2$, and the amplitudes are given in §5(a). No energy is being lost at the wall in this model.

11. You need to extend the theorem of §6(a) to deal with the average of $\mathbf{x} \cdot \mathbf{y}$. The time average densities both depend on y in this example, and energy per length of duct comes from integrating from $y = 0$ to $y = a$. The total energy per length is very simple. The energy flow rate comes from an average of the product of pressure fluctuation and x-component of velocity.

The velocity of flow of the average energy density is not just c, as is apparent from the solution in §5(c).

12. A sound source at $r = 0$ is also a volume source at $r = 0$, so you need no mass flow through a small sphere round $r = 0$.

Put the given conditions in terms of F and G, and integrate to find $F(r)$ and $G(r)$ for all $r \geqslant 0$. Show that F and G must both be zero at large r, from the radiation condition.

Show that the condition at $r = 0$ gives $F(-\xi)$ in terms of $G(\xi)$ for $\xi > 0$, and show that this completes what you need.

13. An image solution is required here that satisfies the wave equation for $z > 0$ and also has zero velocity normal to the plane $z = 0$.

Write the distance from source to P in terms of R, θ by using the cosine rule, then expand both r^{-1} and $\exp(ikr)$ for $R \gg a$, and leave out small terms.

The terms in ka should reduce to a simple form.

14. The image of a dipole in a wall is in Chapter IV, Q12.

This dipole is not along the x-axis, so you need to derive its potential by using an appropriate derivative of the source potential.

After that, proceed as in Q13. But you also need to approximate a sine when R/a is large.

15. Bessel functions are wanted here, so you may want to come back later.

What is needed is

$$\phi = f(r, \theta) \exp i(kz - \omega t)$$

and separation for variables in the wave equation. The separation constant can be zero here. One solution must be a plane wave with $\omega = ck$.

Chapter XIII

1. You need the equation for ϕ from §2(a) and also the boundary condition for ϕ at $z = 0$ from §2(b).

2. There is an extra term $-\gamma/R$ in the pressure boundary condition at $z = 0$. Linearise this in ζ, then eliminate ζ from the velocity boundary condition to get a new condition on ϕ at $z = 0$. Use of the appropriate form for ϕ gives a new relation between ω and k.

3. There is now a term in p_1 in the pressure condition at $z = 0$. How must ϕ depend on x and t so that ϕ and ζ are related as usual through the condition at $z = 0$? Is it surprising that ζ is infinite for a special value of c?

4. Try $\phi = f(z) \cos k(x - ct)$, and determine $f(z)$ to satisfy all necessary conditions. Finally determine ζ from ϕ as in §2. Will ζ be infinite for a special choice of c?

5. You need the surface pressure condition of Q2 and the form of ϕ in §6. These will give a relation between ω and k like both Q2 and §6. Expand a formula for $c^2(k)$ as a Taylor series in k: if the first two derivatives are zero, c^2 will be almost constant. Try values of k to see how far c is constant.

6. You need the methods of §4(a) and (b) applied to the potential of §6(a). Do you expect equal kinetic and potential energy averages? Do your results agree with the energy velocity found in §6(c)?

7. The group velocity is $d\omega/dk$. Evaluate it for large k, for small k, and for the minimum wave speed.

8. The potentials will be as in §5(b). The only change will be in the pressure boundary condition at the interface, where ρ_1, ρ_2 and ζ must now come in. You will find a similar equation for $A(t)$, and

$$\sigma = -i\alpha \pm \beta,$$

where β is real (and so there is instability) if U exceeds some value.

9. At a point in the elimination of B, C, D you need to assume $\omega^2 \neq gk$. Follow the consequences if $\omega^2 = gk$. For a surface wave, you would expect the motion to reduce rapidly with depth.

10. In §7(a) you are given ϕ, from which you can find ζ. Show the lines on which $\zeta = 0$, and decide which case has the largest frequency.

11. Particle paths are calculated in §3(c). In this case you have $\phi(x, y, z, t)$, so there are three equations to solve near x_0, y_0, z_0. You should get solutions

like

$$X(t) = K_1 e^{-i\omega t}$$

with K_1 independent of t.

12. The general solution in §7(b)(ii) has surface motion proportional to $J_n(kr) \cos n\theta$. What determines k and ω? What are the roots of $J_1'(ka) = 0$? Read the work in §7(d).

13. The comparable sound wave example is in Chapter XII §4. The speed of energy flow in the constituent plane waves is $d\omega/dk$, where k is related to K by $\nabla^2\phi = 0$. Show that dK/dk is an easy function of the angle of inclination of the waves.

14. Go back to Chapter IX Q10 to see the method. What are the boundary and initial conditions on $U(r, t)$ when the velocity is taken as $U(r, t)\mathbf{k}$? What velocity do you get when $t \to \infty$? Look for a new function $V(r, t)$ that is zero when $t \to \infty$, and that satisfies a similar equation to $U(r, t)$. Separation of variables for V should give $\exp(-vk^2 t)$ as the time dependence. The coefficients in the series solution are determined as in §8(d). You need not evaluate the integral.

15. (a) See Chapter X Q16
    (b) See Chapter XII Q15

16. Take ϕ_0 independent of θ and show from $\nabla^2\phi = 0$ that it satisfies a Bessel equation. Look up $H_0^{(1)}$ in §8(c). Find how it behaves near $r = 0$ and for large r. Is there a source of material at $r = 0$ (integrate over z)? Is there a radiation of energy at large r? Is this a point source?

Chapter XIV

1. Use Bernoulli's equation and the thermodynamic relations in §1(b). You also need values of a_0 from Chapter XII, and p_0 and R from Chapter VII.

2. Put p in terms of a^2 by eliminating ρ between $a^2 = \gamma p/\rho$ and $p = k\rho^\gamma$.
   Use Bernoulli's equation to get a^2 in terms of ϕ.

3. In Bernoulli's equation put $a^2 = 0$ or $a^2 = v^2$ to get the required speeds.

4. Look back to Chapter VII §2 for dE, and to Chapter X §4 for the derivation of Bernoulli's equation.

5. Set up equations of motion for $r > a$ and $r < a$ in terms of the model flows given: see Chapter V §3 for the Rankine vortex and Chapter X §5 for the pressures when the compressibility is neglected.
   What boundary condition will you use at $r = a$?
   You can get ρ for $r > a$ and also for $r < a$ from the equations you have found.

6. Remember that ∇f is in the x, y plane and perpendicular to the curve $f = $ constant.

7. The equation transforms to $\partial f/\partial \eta = 0$.

8. Read §2(b) and (c). If f is constant on $x^4 - 2t^2 = $ constant, then $f(x, t) = F(x^4 - 2t^2)$. For 'range of influence' read §2(d); the initial condition is given for $x > 0$.

9. Read §3(b). For $|x| > l$ and $t = 0$, $r = s$. For $|x| < l$ and $t = 0$.

$$r = u(x, 0) + a_0/(\gamma - 1)$$

and r and s are constants along their respective characteristics.

   The whole region of flow still occupies length $2l$; the compressive part takes less of this as time goes on.

10. Calculate r and s for the two initial regions. You will find the $+$ characteristics from $x > 0$ steeper than those from $x < 0$, and so a shock forms almost at once. You must also have a rarefaction wave to give a fit between the region totally determined by $x < 0$ and the region with s derived from $x > 0$ and r derived from $x < 0$.

11. From equation (B) in §4(b) you can derive

$$p_2 - p_1 = \rho_1 u_1^2(1 - u_2/u_1)$$

and then divide by

$$p_1 = \rho_1 a_1^2/\gamma.$$

Equation (C) gives a formula for $(a_2^2 - a_1^2)/a_1^2$.

12. The formula for p_2/p_1 in §4(c) shows that

$$M_1^2 = 1 + (\gamma + 1)\varepsilon/2\gamma.$$

The formula for u_2/u_1 is easily approximated, but in the formula for $S_2 - S_1$ you must expand in a series of powers of ε, as the first approximation is 0.

13. Differentiate the equation for dA/dx in §4(d) to get an equation for d^2A/dx^2, and use the alternatives imposed by $dA/dx = 0$ at the throat.

14. Substitute into the equation for dp/dx given in §4(d). You should show that $dp/dx < 0$ at the throat as well.

Chapter XV

1. Take the case when the ship is almost touching the bottom. Use an equation for mass flow and one for pressure on the surface, and eliminate the velocity past the ship.

2. Read §1(b) and (c). You need equations for mass conservation and surface pressure. When you eliminate downstream velocity, you get an equation for downstream depth. Conditions at the narrowest point are found by using $f = 1$ there when a transition occurs.

3. You need to calculate the area of cross-section for depth H upstream, and also for the surface at height $H - d$ when the bottom has moved up by Z. Then the volume conservation equation is velocity $\times$ area = constant.

    Take Z/H and d/H to be small.

4. It is assumed that $u(x, t)$, i.e. u is independent of z. Look back to Chapter XIII §2 to relate w on the surface to derivatives of h, after you have integrated from 0 to h.

5. On OB the value of r comes from the x-axis to the left of O, as does the value of s. Calculate $u - c$ on OB from r and s. Where do values of r and s on DC come from? In the triangle FLG all values come from the uniform region on the x-axis to the right of L.

6. In the equation of §2(a) put

$$h = H[1 + \alpha(x)],$$

    where $\alpha(x)$ is small and so is $u(x)$, so that products of u and α can be neglected.

7. The slopes of the characteristics on the diagram in §2(c) are $(u + c)^{-1}$ for $f(t)$ equal to H_0 and H_1, and these lines pass through t_0 and t_1.

8. Take a frame of reference in which the shallower water is at rest; then the bore advances at speed U_1.

    What is the speed in the (shallow) river in a frame in which the bore is at rest?

9. Use the definitions of F_1 and F_2 in the equations of mass and energy conservation. The equation relating F_1 and F_2 has a factor $F_1^2 - F_2^2$. The final equation for F_2^2 has the form

$$F_2^4 + bF_2^2 - c = 0$$

    where $c > b + 1$.

10. Eliminate H_2 and U_2 from the given equation for T.

11. The two equations to determine U_2 and H_2 are mass conservation and the equation for T (not Bernoulli because there is energy loss). The rate of loss of energy follows as in §3(a). You need to put $U_2/U_1 = 1 + \varepsilon$ and use that α is small.

12. Estimate the terms in the integral for M, after you have substituted for u and w, and keep rejecting terms that are $O(h^3/L^3)$ times those retained. When you need s'' near the end, you estimate it from $s = Q/h$.

13. Substitute numbers into the formulae. The original estimates were in these units. 100 miles per hour is a little too fast, and 100 feet excess height was probably an overestimate.

14. Use the formula near the end of §4(c) to find $s(x)$ in terms of $h(x)$. Then $w(h)$ is given in terms of h and $s(x)$.

15. Formulae for E and M are in §4(d). Use the new values for E_1 and M_1 together with the value of Q to rewrite the equation for h' in the given form.

Chapter XVI

1. What is the gradient of the curve $\phi(x, y) = k$? Consider $W(z) = iw(z)$.

2. Use $dw/dz = qe^{-i\alpha}$ and $dz = dle^{i\theta}$ to replace

$$\int \mathbf{v} \cdot d\mathbf{l}$$

in terms of w and z.

Remember that $\int_C (dw/dz) dz = $ change in w round C.

3. $\bar{f}(z)$ is defined in §4(a), while $f(\bar{z})$ just means replace z by $\bar{z}$. If you are uncertain about coth z, put it in terms of exponentials.

4. Show that $\mathscr{I}m\, w\,(z) = 0$ when

$$z = a \cos\theta + ib \sin\theta$$

and that $w(z) \sim Uz$ as $|z| \to \infty$.

5. As in Q4, a boundary needs to have $\mathscr{I}m\, w(z) = $ constant. Find the velocities on the boundaries, and expand $w(z)$ near the two corners.

6. For $w_1(z)$ and $\psi_1 = \frac{1}{4}\pi$ write

$$z/c = \cosh(\phi + \tfrac{1}{4}\pi i)$$

to find the streamline. Try also $\psi_1 = -\frac{1}{4}\pi$.

Calculate dw/dz near $z = 0$ for the two cases to find the velocity near $z = 0$.

7. To find the motion of one vortex you need the potential for the other two.

Find the stagnation points and the asymptotic form of the potential for all three vortices.

8. (i) Consider the motion of the vortex at $z = a$ under the influence of a symmetrically placed pair of vortices, using that the velocity from each is $(c/\text{distance})$ at right angles to the relative vector.

(ii) The points all satisfy $z^5 = a^5$. This enables you to factorise $z^5 - a^5$, and the logarithm of a product of factors is the sum of the logarithms of the factors.

The velocity of the vortex at $z = a$ comes from the total potential minus the potential of that vortex. Remember that $z^5 - a^5$ has one factor $z - a$.

9. The dipole potential is in §1(d), and the image theorem needed is in §4(a). You then need dw/dz at $z = 0$.

10. This is like the example at the end of §4(c), but there is only one vortex.

To find a path, you need an initial point. Try $c = \frac{3}{2}a$, $\alpha = \frac{1}{4}\pi$ and calculate the velocity there due to all but the vortex which is there; then try a velocity at a suitable later point.

11. Look at §4(*b*): when you have two boundaries you need three images. The proof has two parts as in §4(*c*), but now the boundary is more complicated, and so is $w(z)$.

    The basic flow here is $f(z) = \frac{1}{2}Az^2$.

12. Show that $\mathscr{I}m\, w(z)$ is constant on the plate; you need

    $\ln f(z) = \ln|f(z)| + i \arg f(z)$.

    Show also that $w(z) \sim Uz - (i\kappa/2\pi)\ln z$.

    Calculate $(dw/dz)^2$, and expand it for $|z| > a$. This enables you to calculate the force and moment as in §5(*b*).

13. Expand dw/dz (from Q10) in powers of z near $z = 0$. Then you can find the coefficient of z^{-1} in $(dw/dz)^2$ near $z = 0$.

    Next put

    $z = a^2 e^{i\alpha}/c + \zeta$

    and expand dw/dz near $\zeta = 0$. This gives the appropriate contribution from near $z = a^2 e^{i\alpha}/c$ as the coefficient of ζ^{-1} in the expansion of $(dw/dz)^2$.

14. Look in §4(*a*) for the potential and in §5(*d*) for the method. The integral you need is again $O(z^{-4})$ as $|z| \to \infty$, so you can use a semicircle instead of the real axis as path for the integration.

Chapter XVII

1. You need a power of z which takes $\theta = \pm\pi$ to $\pm\frac{3}{4}\pi$.

2. Check that the boundaries transform correctly, and then use

   $W(\zeta) = w[z(\zeta)]$.

3. A streamline is $\mathscr{I}m\, w(z) = a$, say; and

   $W(\zeta) = w[z(\zeta)]$.

   The flow rate between two streamlines is in Chapter IV §3(*b*).

   A stagnation point is $dw/dz = 0$. Find $dW/d\zeta$ in terms of dw/dz.

4. Take a vortex at $z = ia$ above the half plane barrier $y = 0$, and transform by $\zeta = z^{1/2}$ to get the required potential $W(\zeta)$. Now solve for ζ in terms of W. Finally replace (as a change of notation) ζ and W by z and w.

5. Read §3(*c*) on the effects of inversions. You can send two circles to lines if you invert about a common point.

   Where does the centre of the smaller circle go?

   The flow between the two lines from ∞ to ∞ corresponds to a flow in the z-plane from the point corresponding to $\zeta = \infty$ and back to this point in a different direction. What is the singularity at the common point?

6. On the unit circle $z = e^{i\theta}$. Remember that $|Z| = |\bar{Z}|$ for any Z. Try $z = e^{i\beta}$ where $a = |a|e^{i\beta}$.

7. Take the channel as $x = \pm\frac{1}{2}\pi$ and the plates as $y = 0$, $|x|$ between a and $\frac{1}{2}\pi$. Then $\zeta = \sin z$ gives a flat boundary, which may be stretched to have a gap between $(-1, 0)$ and $(1, 0)$ by multiplying by a constant. Then a $\sin^{-1}$ transformation returns the boundary to a channel with no barriers.

8. To get chord l you need the circle $|z| = \frac{1}{4}l$ and transformation

$$\zeta = z + (\tfrac{1}{4}l)^2/z.$$

The trailing edge is the point corresponding to $z = \frac{1}{4}l$. The calculations follow §5 exactly.

What happens at the leading edge, $z = -\frac{1}{4}l$?

9. The change from the calculations in §5(*b*) and (*c*) is just that $w(z)$ has $z + \delta$ in place of z. Since it is only the coefficient of z^{-1} in each integral that is needed, all you need is the expansion of

$$(z + \delta)^{-1}, (z + \delta)^{-2}$$

in Laurent series; then proceed as before.

10. Work throughout to first order in δ. First estimate a, then approximate ζ. Take real and imaginary parts and check your answers against the results in §6(*a*), which has $\phi = \pi$.

The solutions to $d\xi/d\theta = 0$ must be near $\theta = 0$ and $\theta = \pi$, because they are at those points when $\delta = 0$; so solve approximately.

11. A circular arc is sharp at both ends, so you need a circle through both $z = 1$ and $z = -1$. The camber is provided by taking the centre off the x-axis.

The proof that it is a *circular* arc that results is long.

The circulation follows directly from §6(*c*).

12. Read §6(*c*). Only the first term of the expansion of $dz/d\zeta$ is used in finding the force, and only $d\zeta/dz = 0$ at $z = 1$ is used to find the circulation.

13. Use the binomial theorem for large $|z|$. Use the form of ζ near $z = 1$ to check that $\zeta' = 0$ at $z = 1$.

Answers for exercises

Chapter I

1. $\mathbf{i} = \hat{\mathbf{r}} \cos\theta + \hat{\boldsymbol{\theta}} \sin\theta,$
   $\mathbf{j} = \hat{\mathbf{r}} \sin\theta + \hat{\boldsymbol{\theta}} \cos\theta.$

3. $\partial\hat{\mathbf{r}}/\partial\theta = \hat{\boldsymbol{\theta}}, \qquad \partial\hat{\boldsymbol{\theta}}/\partial\theta = -\hat{\mathbf{r}},$
   $\partial\hat{\boldsymbol{\theta}}/\partial\lambda = \hat{\boldsymbol{\lambda}} \cos\theta, \quad \partial\hat{\boldsymbol{\lambda}}/\partial\lambda = -\hat{\mathbf{r}} \sin\theta - \hat{\boldsymbol{\theta}} \cos\theta.$

8. $\nabla\cdot\mathbf{A} = \partial A_r/\partial r + r^{-1}\partial A_\theta/\partial\theta + (r\sin\theta)^{-1}\partial A_\lambda/\partial\lambda + 2A_r/r + r^{-1}A_\theta \cot\theta.$
   $\nabla\times\mathbf{A} = (r\sin\theta)^{-1}\{\partial/\partial\theta(A_\lambda \sin\theta) - \partial A_\theta/\partial\lambda\}\hat{\mathbf{r}} + (r\sin\theta)^{-1}\{\partial A_r/\partial\lambda$
   $\qquad - \partial/\partial r(A_\lambda \sin\theta)\}\hat{\boldsymbol{\theta}} + r^{-1}\{\partial/\partial r(rA_\theta) - \partial A_r/\partial\theta\}\hat{\boldsymbol{\lambda}}.$

9. $-r^{-1}(A_\theta^2 + A_\lambda^2)\hat{\mathbf{r}} + r^{-1}(A_r A_\theta - A_\lambda^2 \cot\theta)\hat{\boldsymbol{\theta}} + r^{-1}(A_r A_\lambda + A_\theta A_\lambda \cot\theta)\hat{\boldsymbol{\lambda}}.$

10. $\nabla^2\phi = \partial^2\phi/\partial r^2 + r^{-2}\partial^2\phi/\partial\theta^2 + (r\sin\theta)^{-2}\partial^2\phi/\partial\lambda^2 + 2r^{-1}\partial\phi/\partial r$
    $\qquad + r^2 \cot\theta \; \partial\phi/\partial\theta.$

Chapter II

(a) (i) $gT/c, 0.1.$
   (ii) $h_1 \doteqdot h_0(1 - gT/c).$
(b) (i) $\frac{1}{4}\rho v^2 A/mg$, perhaps about 0.2.
   (ii) $h_2 \doteqdot h_0\{1 - \rho g T^2 A/(16m)\}.$
   Perhaps the time taken to generate the sound.

Chapter III

1. No, it piles in heaps, which a fluid cannot for long. About 1 to 5×10^{-4} m. About 100 particles wide.

3. Values are very roughly 2, 5, 20×10^{-2} N m^{-2} s^{-1}.

Fig. A.1. (i) and (ii), streamlines at two times, $t_2 > t_1$; (iii), two streamlines.

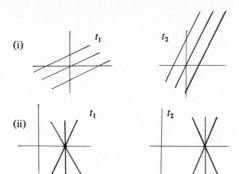

(i)

(ii)

(iii)

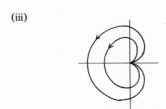

4. The paths are

$$x = y_0 \sin at + x_0 \cos at,$$
$$y = y_0 \cos at - x_0 \sin at,$$
$$z = z_0 + \int_0^t b(\tau)d\tau.$$

Streamlines are

$$x = y_0 \sin as + x_0 \cos as,$$
$$y = y_0 \cos as - x_0 \sin as,$$
$$z = z_0 + sb(t).$$

Rotating flow along a pipe.

5. See fig. A.1.

6. (i) Streamlines $xy = $ constant. Particle paths the same.
   (ii) Streamlines $xy^t = $ constant (i.e. independent of s). Particle paths $y = y_0 \exp\{-(\ln x^2/x_0^2)^{1/2}\}$.

Chapter IV

1. $\partial n/\partial t + \nabla \cdot (n\mathbf{v}) = $ birth rate $-$ death rate $-$ fishing rate, where $\mathbf{v}$ is the total velocity, currents $+$ migration.

2. If S is smooth, $2\pi m$.

3. (i) $-(f^2/r)\hat{\mathbf{r}}$. Yes, central acceleration v^2/r.

Fig. A.2. Two parabolas of the streamline family, and the limiting case along the x-axis.

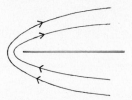

(ii) If Q is the volume flow rate through any section

$Q/A(x)$, $-(Q^2/A^3)dA/dx$.

4. $\psi = U\beta\alpha^{-1}x^{1/3}\tanh(\alpha yx^{-2/3})$.

$v = -\frac{1}{3}U\beta\alpha^{-1}x^{-2/3}\tanh(\alpha yx^{-2/3}) + \frac{2}{3}U\beta yx^{-4/3}\operatorname{sech}^2(\alpha yx^{-2/3})$.

5. (i) $-(\nabla^2\psi)\mathbf{k}$.

(ii) $(\nabla^2\nabla\psi)\times\mathbf{k}$.

(iii) $(\psi_y\psi_{xy} - \psi_x\psi_{yy}, -\psi_y\psi_{xx} + \psi_x\psi_{xy}, 0)$.

6. The streamlines are parabolas, as sketched in fig. A.2. Flow round a half plane barrier.

8. On $\theta = 0$, $\mathbf{v} = 0$. On $\theta = \frac{1}{2}\pi$, $\mathbf{v} = V(1 - \frac{1}{4}\pi^2)\hat{\boldsymbol{\theta}}$.

A scraper moving along a wall.

9. $2n\pi C$, where n is the number of times round.

10. $r/a \doteq 1$, 1.6, 2.4, 3.3 ...

$-v_\theta/U \doteq 2$, 1.4, 1.2, 1.1 ...

11. (i) $x = c$.   (ii) $x = 2^{-1/2}c$.

12. (i) Dipole at $-\alpha$ at the image point.

(ii) Opposite vortex at the image point, perhaps another at the centre.

(iii) Opposite vortex at the image point.

(iv) The actual result is a source at the image point and a line source from there to the centre.

14. $\Psi = \frac{1}{2}Ur^2 + \dfrac{-m}{4\pi}\left\{\dfrac{z-a}{[(z-a)^2 + r^2]^{1/2}} - \dfrac{z+a}{[(z+a)^2 + r^2]^{1/2}}\right\}$.

The dividing streamline is $\Psi = 0$.

The length is the distance between the solution of

$(\pi U/m)(z^2 - a^2)^2 = az$.

The maximum radius is given by the solution of

$(\pi U/m)r^2(a^2 + r^2)^{1/2} = a$.

The body is a sphere of radius $(ma/\pi U)^{1/3}$.

15. See chapter V §3(*b*) for the inner flow. The outer flow is rather like that round a cylinder. $A = -\frac{3}{4}U/a^2$.

16. (*a*) $(2bC/r)\sin\theta$.

(*b*) $-(-2bC)(\partial/\partial y)(\ln r)$.

Fig. A.3. A sketch of some streamlines for Q9.

Chapter V

1. Not easily.
   Vorticity $(0, 5, 0)$. Principal rates of strain are $-5, \frac{1}{2}(5 \pm \sqrt{10})$ with principal axes $(0, 1, 0), (1, 0, \sqrt{10} - 3)$ and $(3 - \sqrt{10}, 0, 1)$. The flow has $\nabla \cdot \mathbf{v} = 0$.

2. $0, br, -br$ with axes $\hat{\boldsymbol{\theta}}, \hat{\mathbf{r}} - \mathbf{k}, \hat{\mathbf{r}} + \mathbf{k}$.
   Vorticity $2br\hat{\boldsymbol{\theta}}$. Like roller bearings.

3. $-2b\hat{\mathbf{r}}\mathbf{k}$.

4. (a) $\omega = -(\partial^2\psi/\partial x^2 + \partial^2\psi/\partial y^2)$ where ψ is in Chapter IV Q4.
   (b) 0.

5. $\omega_\lambda = -(r \sin \theta)^{-1}\{\partial^2\Psi/\partial r^2 + r^{-2} \sin \theta \, \partial/\partial\theta(\operatorname{cosec} \theta \, \partial\Psi/\partial\theta)\}$.

6. $\psi = \frac{1}{8}Ar(r^2 - a^2) \sin \theta \quad$ for $r < a$,
   $\psi = \frac{1}{8}Aa^2r(1 - a^2/r^2) \sin \theta \quad$ for $r > a$.

7. $-w'(r)\hat{\boldsymbol{\theta}} + (u' + u/r)\mathbf{k}$.

9. See fig. A.3.

10. Three vortices; one with the same sign at $(-x_0, -y_0)$, two with the opposite sign at $(-x_0, y_0)$ and $(x_0, -y_0)$.
    The path is $x^{-2} + y^{-2} = x_0^{-2} + y_0^{-2}$.

11. An opposite vortex at the image point and an equal vortex at the centre. Moves round a circle at speed
    $$C(a^2/c)(c^2 - a^2)^{-1}.$$
    An opposite vortex at the image point. The speed is now
    $$Cc/(c^2 - a^2).$$

12. This adds velocities $v_r = U \cos \theta(1 - a^2/r^2)$ and $v_\theta = -U \sin \theta(1 + a^2/r^2)$ to the velocity $v_\theta = -Cr/(r^2 - a^2)$ due to the image vortex.

14. Particular solution $-\frac{1}{4}\Omega r^2$. Final solution $-\frac{1}{4}\Omega r^2(1 - 2^{1/2} \cos 2\theta)$.

15. Angle of region $= \frac{1}{2}\pi$.

Chapter VI

1. $gz + fx = $ constant.

3. $\sigma_{ij}n_j - (\sigma_{kl}n_k n_l) \, n_i$.

Fig. A.4. The weight of the water, the force on the vertical face, and the force on the curved face meet in a point.

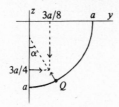

5. (a) No change.
   (b) $p_0 g/(g^2 + f^2)^{1/2}$.
6. $\omega = (g/a)^{1/2}, \frac{1}{2}a$.
7. About $\frac{1}{2}$ m.   About 8 m.
8. $\frac{1}{2}\rho g a^3$; components $\frac{1}{2}\rho g a^2$ horizontally and $\frac{1}{4}\pi\rho g a^2$ vertically.
9. (i) $\frac{2}{3}\rho g a^3$ at a depth $\frac{3}{4}a$.
   (ii) $2Mg/\pi$ horizontally and Mg vertically, where M is the mass of water, through the point Q shown in fig. A.4, where $\alpha = \tan^{-1}(2/\pi)$.

Chapter VII

1. 4.2×10^5 J kg^{-1}, 0.1 K.
2. $E = c_v(p/(R\rho) - T_0)$. $T = T_0(p/p_0)^{(\gamma+1)/\gamma}$.
6. $3\gamma H \ln(a/b)$.
7. $T = T_0(1 - \rho g z/p_0)$, $T_1 = 253$ K.   No.
8. 1044 K   804 K.

Chapter VIII

2. $\partial T/\partial t = (k\nabla^2 T + Q)/\rho c_v$.
4. (i) Non-zero $\mathbf{v}\cdot\nabla\mathbf{v}$ and $\partial\mathbf{v}/\partial t$.
   (ii) Surface effects.
   (iii) Temperature and density gradients.
5. ω has size 'velocity scale/length scale', unlikely to be less than 10^{-2} m s^{-1}/ 1 m; $\Omega = 7 \times 10^{-5}$ s^{-1}.
7. $s_{rr} = -s_{\theta\theta} = 4\mu U a^2 r^{-3} \cos\theta$,
   $s_{r\theta} = s_{\theta r} = 4\mu U a^2 r^{-3} \sin\theta$.
8. 0. The flow used is not realistic.

Chapter IX

1. Normal stress $-p_0$. Shear stress $\pm\mu V/a$.
2. $U(r) = (\frac{1}{4}G/\mu)\{a^2 - r^2 + (b^2 - a^2)\ln(r/a)[\ln(b/a)]^{-1}\}$.
3. $U(y) = \frac{1}{2}(\rho g/\mu)(2hy - y^2)\sin\alpha$.

4. $U(r) = \frac{1}{4}(\rho g/\mu)(a^2 - r^2)\sin \alpha.$

5. (i) 0.

  (ii) $2\mu a^2 \Omega$ per length, on the fluid. The second flow is not realistic as it has infinite energy; the accelerating couple on this steady flow is really needed to accelerate more fluid at large distances.

7. $F'(\eta) = A\eta^{-2}\exp(-\eta^2)$. The boundary conditions are $\eta F \to 0$ as $\eta \to \infty$, $\eta F' + F \to -1$ as $\eta \to 0$.

8. $U'' + r^{-1}(1 - R)U' - r^{-2}(1 + R)\,U = 0.$
   $A = -Va/(b^{R+2} - a^{R+2}), \quad B = -Ab^{R+2}.$
   There is a boundary layer near $r = b$.

10. Stable.

11. $U(\pm a, t) = 0, \quad U(y, 0) = 0. \quad \partial V/\partial t = v\partial^2 V/\partial y^2.$
    $V(\pm a, t) = 0, \quad V(y, 0) = \frac{1}{2}(G/\mu)(y^2 - a^2).$

    About $2a^2/v$; by then vorticity has diffused from the walls to the centre.

Chapter X

2. Yes.

6. The increase of speed along the pipe stretches vortex lines parallel to the axis, and so increases vorticity in this direction. Vortex lines round the axis are compressed, and so this vorticity is reduced.

7. An extra term $\rho^{-2}\nabla\rho \times \nabla p$. An extra term $v\nabla^2\omega$

8. $-(A/a)dh/dt$, where

   $h = \{(H + h_0)^{1/2} - \frac{1}{2}t[(2ga^2)/(A^2 - a^2)]^{1/2}\}^2 - H.$

   About $(2/g)^{1/2}\{(H + h_0)^{1/2} - H^{1/2}\}A/a$. Air enters at the hole.

9. With upstream pressure p_0, the downstream pressure is

   $p_0 - \rho gy - \frac{1}{2}(3Ga/4\rho u)^2.$

10. $\omega\Psi/r + \frac{1}{2}\mathbf{v}^2 + \Phi + p/\rho = $ constant.

11. 0.

12. $F_{\parallel} = 0, F_{\perp} = -2\pi\rho U\tau a^2$ per length.

15. $h < Q^2(B^2 - A^2)/(2gA^2B^2).$

16. k is such that ka is the first zero of J_1, i.e. 3.82. The outer solution is $\psi = Ur\sin\theta - Ua^2r^{-1}\sin\theta$, with $kCJ'_1(ka) = 2U$.

Chapter XI

2. $\partial\phi/\partial r = 0$ at $r = a, r = 2a. \ \partial\phi/\partial\theta = r^2\Omega(t)$ on both sides of the barrier, say $\theta = 0$ and $\theta = 2\pi$.

4. $\phi = x^3 - 3xy^2, \ \psi = 3x^2y - y^3.$

   Degree 0:1.   Degree 1 :x is typical.
   Degree 2 :$2xy = r^2\sin^2\theta\sin 2\lambda$
   $\qquad x^2 - y^2 = r^2\sin^2\theta\cos 2\lambda$
   $\qquad 2z^2 - x^2 - y^2 = 2r^2P_2(\cos\theta).$

5. $d_n = 8Aa^2(-1)^n/\{(2n+1)^3\pi^3 \sinh\tfrac{1}{2}\pi(2n+1)\}$.
Motion in a box when a hinged lid starts to rotate.

7. $b_n = Ac^n$

8. $\nabla^2\phi = 0$ inside. $\partial\phi/\partial\theta = 0$ on the walls $\theta = \tfrac{1}{4}\pi$. $\phi \sim A(t)r^{-1}$ as $r \to 0$ to give a sink. $p = p_0$ on the surface.

9. $dU/dt < 2p_0/\rho a$.

10. $\pi\rho a^2\Omega^2 b$ away from the centre.

11. $(M + \tfrac{2}{3}\pi\rho a^3)dU/dt$. $2g$.

12. $(U + 3u/5)4\pi\rho a^2 u/5$.

13. $dH/dt = -\{2gH_0/(c-1)\}^{1/2}\{(H/H_0) - (H/H)_0^c\}^{1/2}$,
where $c = (A_1^2/A_0^2) - 1$.

14. $\tfrac{1}{3}\pi\rho A^2/a$.

15. $-2\pi\rho U\kappa\mathbf{j}$.

Chapter XII

1. $\nabla^2 p - c^{-2}\partial^2 p/\partial t^2 = \mathbf{F}\cdot\nabla\rho - \partial^2(\rho v_i v_j)/\partial x_i\partial x_j$.

3. $A(1 - \lambda c)/(1 + \lambda c)\exp\{i(-kx - \omega t)\}$.

4. $A(\cos\alpha - \lambda c)/(\cos\alpha + \lambda c)\exp\{i(-kx\cos\alpha + ky\sin\alpha - \omega t)\}$.
The energy is all absorbed.

5. (a) There is a real transmitted wave at angle

$$\beta = \sin^{-1}\{(c_2/c_1)\sin\alpha\}$$

if this is real. Then the reflected and transmitted amplitudes for unit incident wave are

$$(\rho_2 c_2\cos\alpha - \rho_1 c_1\cos\beta)/(\rho_2 c_2\cos\alpha + \rho_1 c_1\cos\beta)$$
$$2\rho_1 c_2\cos\alpha/(\rho_2 c_2\cos\alpha + \rho_1 c_1\cos\beta).$$

(b) Otherwise there is an exponential decay for $z < 0$, with exponent $kz(\sin^2\alpha - c_1^2/c_2^2)^{1/2}$ and the reflected wave has the same strength as the incident wave, but with a phase change.

(c) There is almost total reflection in all cases.

6. $p = 4\alpha\sum A_n\cos\{(2n+1)\pi x/2l\}\cos\{(2n+1)\pi ct/2l\}$,
with $A_n = (-1)^{n+1}/(2n+1)\pi$. $\tfrac{1}{4}c/l$ Hz.

7. $\phi = ac\sec(\omega l/c)\cos\omega t\sin\{\omega(x - l)/c\}$. Not realistic when $\sec\omega l/c$ is large. $a \ll l, a \ll \lambda$.

8. $\omega^2 = k^2 + \tfrac{1}{4}(2n+1)^2\pi^2 c^2/h^2$.

9. $\omega = (c\pi/a)(401/400)^{1/2}$, fundamental $c\pi/(20a)$.

11. $\tfrac{1}{4}\rho_0\omega^2/c^2$. $\tfrac{1}{4}\rho_0\omega k$. kc^2/ω or $\{1 + (m\pi/ak)^2\}^{-1/2}c$.

12. $F(r) = -G(r) = 0$ for $r > a$,
$F(r) = -G(r) = \tfrac{1}{4}\alpha r^2/(\rho_0 c)$ for $0 \leqslant r < a$,
$F(\xi) = -G(-\xi)$ for $\xi < 0$,
$G(\xi)$ not determined for $\xi < 0$.

13. $\phi(P) \sim 2R^{-1} \cos(ka \sin\theta) \exp i(kR - \omega t)$.

15. $\phi = AJ_n(qr) \cos(n\theta + \alpha_n) \exp i(kz - \omega t)$.
where $q^2 = \omega^2/c^2 - k^2$ and $J'_n(qa) = 0$.

Chapter XIII

2. $-\gamma\partial^2\zeta/\partial x^2 + \rho\dot{\phi} + \rho g\zeta = 0$ at $z = 0$. 0.16 m s^{-1}.
The diagram is shown in fig. A.5.

3. $p_1\rho^{-1}/(kc^2 - g)$.

4. (a/kc) (sech $kh)/(1 - c_w^2/c^2)$, where c_w is the normal wave speed $\{(g/k)$ tanh $kh\}^{1/2}$.

5. $\omega^2 = gk$ tanh $kh\,(1 + \gamma k^2/\rho g)$.

6. $\frac{1}{4}\rho k |B|^2 \sinh 2kh$ where B is the amplitude of ϕ. $\frac{1}{4}\rho k |B|^2(kh + \frac{1}{2}\sinh 2kh)$.

7. $c_g^2 = \dfrac{g(1 + 3\gamma k^2/\rho g)}{4k(1 + \gamma k^2/\rho g)}$.

8. Stability if $\rho_1\rho_2 k^2 U^2 < gk(\rho_2^2 - \rho_1^2)$.

9. Surface wave $= \exp(kh) \times$ interface wave.

11. $x = x_0 - (i\pi B/a\omega) \sin \pi x_0/a \cos \pi y_0/a \cosh k(z_0 + h)e^{-i\omega t}$, with similar expressions for y and z. It is part of a line through (x_0, y_0, z_0).

13. $\tan^{-1} \pi/Ka$.

14. $V(r, t) = \sum_i A_i J_0(k_i r) \exp(-\nu k_i^2 t)$, where $J_0(k_i a) = 0$,

$$A_i = \frac{\frac{1}{2}G}{\mu a^2 \{J'_0(k_i a)\}^2} \int_0^a (a^2 - r^2) r J_0(k_i r)\, dr.$$

16. There is an oscillating source of volume along the line $z = 0$. There is outward radiation. Not a point source.

Chapter XIV

1. Your answers should be near

$a = 377$, $T = 353$, $\rho = 1.02$, $M = 0.81$

in suitable units.

2. $a^2/(\gamma - 1) + \frac{1}{2}(\partial\phi/\partial x)^2 + \partial\phi/\partial t = $ constant.

Fig. A.5. The wave velocity c and the group velocity c_g.

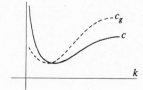

Fig. A.6. The shape of the wave at $t = \frac{2}{3}t_s$.

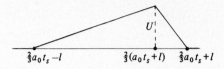

Fig. A.7. Characteristics, shock and expansion region for Q10.

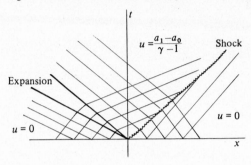

3. About 897 and 366 m s^{-1}.

5. $\rho_0\{1 - (\gamma - 1)\rho_0\Omega^2 a^2/8\gamma p_0\}^{1/(\gamma - 1)}\exp(-\Omega^2 a^2/8RT_1)$.

8. $x^4 - 2t^2 = $ constant. $\exp\left[-(x^4 - 2t^2)^{1/4}\right]$. To the right of the curve $x^4 = 2t^2$.

9. For example, at $t = \frac{2}{3}t_s$ you get the diagram shown in fig. A.6.

10. See fig. A.7.

11. $a_2^2/a_1^2 = 1 + 2(\gamma - 1)(\gamma M_1^2 + 1)(M_1^2 - 1)/(\gamma + 1)^2 M_1^2$.

12. $M_1^2 = 1 + \frac{1}{2}(\gamma + 1)\varepsilon/\gamma$,
$u_2/u_1 = 1 - \varepsilon/\gamma$,
$S_2 - S_1 = c_v(\gamma^2 - 1)\varepsilon^3/12\gamma^2$.

Chapter XV

1. $U = (9g/32)^{1/2}$ m s^{-1}.

2. (a) $(8/3)^{1/2}$ (b) $(243/343)^{1/2}B$ (c) $7H/9$.

3. $A(h) = aH + (H - a)^2 \cot\alpha$ is the area.
$d/Z \doteq \{(gA/aU^2) - 1 - 2(H - a)\cot\alpha/a\}^{-1}$.

5. $-3/2$ on OB; $-12/5$ on DC. FLG, BOA, CDE.

7. Near $x = \{(H_1^{1/2} - 2H^{1/2}/3)t_1 - (H_0^{1/2} - 2H^{1/2}/3)t_0\}/(H_1^{1/2} - H_0^{1/2})$.

8. 2.5 m.

9. $\{[(F_1^4 + 6F_1^2)^2 + 32F_1^2]^{1/2} - (F_1^4 + 6F_1^2)\}/2F_1^2$.

11. $U_2/U_1 \doteq 1 + \alpha F_1^2/(1 - F_1^2)$, $H_2/H_1 = U_1/U_2$.
Rate of loss of energy $= \alpha\rho H_1 U_1^3$.

13. 133 and 139 ft s^{-1}. 172 ft. $F = 1.095$.
    1.33×10^9 cu. ft.

14. $s(x) = UH(1 + \frac{1}{3}h'^2)/h - \frac{1}{6}UH\,h''$.
    $w(h) \doteq -h\,s'(x)$. In both formulae $h(x)$ is the solitary wave shape.

15. $H = H_0 F^{-2/3}$, $U = (gH_0)^{1/2} F^{1/3}$.
    $E = 1.520\ \rho gH_0$, $M = 1.518\ \rho gH_0^2$.
    Cnoidal waves between $h = H_0$ and $9H_0/8$.

Chapter XVI

1. A vortex at the origin.
   A dipole at right angles to that represented by $w(z)$.

3. (i) $\ln \bar{z}$, $\ln z$, $\ln \bar{z}$, $\ln(a^2/z)$.
    (ii) $i\ln(\bar{z} - z_0)$, $-i\ln(z - \bar{z}_0)$, $-i\ln(\bar{z} - \bar{z}_0)$, $-i\ln(a^2/z - \bar{z}_0)$.
    (iii) $e^{i\alpha}/(\bar{z} - z_0)$, $e^{-i\alpha}/(z - \bar{z}_0)$, $e^{-i\alpha}/(\bar{z} - \bar{z}_0)$, $e^{-i\alpha}/(a^2/z - \bar{z}_0)$.
    (iv) $\coth(\pi a/\bar{z})$, $\coth(\pi a/z)$, $\coth(\pi a/\bar{z})$, $\coth(\pi z/a)$.

5. See fig. A.8.

6. No. See fig. A.9.

Fig. A.8. Some velocity directions for Q5. There is a singularity in the right hand corner.

Fig. A.9. The streamlines $\psi = \pm\frac{1}{4}\pi$ are the same hyperbolas, but the flows are different elsewhere.

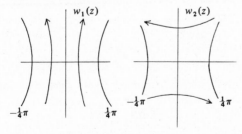

Fig. A.10. A sketch of the streamlines for the three vortices.

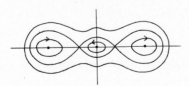

Fig. A.11. A sketch of the path of the vortex in Q10.

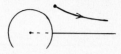

7. See fig. A.10.

10. $w(z) = U(z + a^2/z) + (i\kappa/2\pi)\{\ln(z - ce^{i\alpha}) - \ln(z - a^2e^{i\alpha}/c)\}$.
    $U(1 - a^2e^{2i\alpha}/c^2) + (i\kappa/2\pi)ce^{i\alpha}/(c^2 - a^2)$.
    With initial conditions $c = \frac{3}{2}a$, $\alpha = \frac{1}{4}\pi$, the path is as shown in fig. A.11.

11. If $f(z)$ has all its singularities above the given boundary, then the flow with these singularities and the boundary added has potential
    $$w(z) = f(z) + \bar{f}(z) + \bar{f}(a^2/z) + f(a^2/z).$$

12. $Y = -\rho\kappa U$, $M = 0$.

13. $X - iY = i\kappa\rho U(1 - a^2e^{-2i\alpha}/c^2) + (\rho\kappa^2/2\pi)ce^{-i\alpha}(c^2 - a^2)^{-1}$.

14. $X - iY = \frac{1}{2}i\rho\pi\mu^2/a^3$.

Chapter XVII

1. $W(\zeta) = -U\zeta^{4/3}$, from $k = \frac{3}{4}$.

2. $W(\zeta) = U(\zeta^2 + a^2/\zeta^4)$.

3. At some critical points.

5. Lines parallel to the imaginary axis through
   $\zeta = -\frac{1}{4}a^{-1}$ and $-\frac{1}{2}a^{-1}$.

   The region left of the latter line.
   Flow in the region between the circles with a dipole at $z = 2a$ parallel to the imaginary axis.

6. Rotation of α.

7. $w(z) = iU \sin^{-1}(\sin z/\sin a)$.

8. Lift force $\pi\rho lU^2 \sin\alpha$. Moment $-\frac{1}{8}\pi\rho U^2 l^2 \sin 2\alpha$. A force at the leading edge is not perpendicular to the line of the strip.

9. $X - iY = -i\rho U\kappa e^{-i\alpha}$, where $\kappa = 4\pi U(1 + \delta) \sin\alpha$.
   $M = -2\pi\rho U (1 + \delta) \sin 2\alpha + O(\delta^2)$.

10. $\xi(\theta) = 2\cos\theta + \delta\cos\phi - \delta\cos(\phi - 2\theta)$,
    $\eta(\theta) = \delta\sin\phi - 2\delta\cos\phi\sin\phi - \delta\sin(\phi - 2\theta)$.
    $\xi = \pm 2$ to order δ,
    $\eta = \frac{1}{2}\delta(3\sin\phi \pm 3\sqrt{3}\cos\phi)$.

11. Centre $i(a^2 - 1)^{1/2}$, radius a. The arc has centre
    $-i(2 - a^2)(a^2 - 1)^{-1/2}$, radius $a^2(a^2 - 1)^{-1/2}$.
    $\kappa = 4\pi a U_0 \sin(\alpha + \beta)$, $\sin\beta = a^{-1}(a^2 - 1)^{1/2}$.

12. $-i\rho U_0 e^{-i\alpha}\kappa$, $\kappa = 4\pi a U_0 \sin(\alpha + \beta)$.

13. $B_1 = \frac{1}{3}(3 - \varepsilon)(1 - \varepsilon)$.

Books for reference

This is a collection of the references provided at the ends of the chapters. It is alphabetical by authors, and indicates which chapters the book is mentioned in.

These are *not* the only useful texts available: any book is valuable if it enables you to understand a point more clearly, so hunt around in libraries for yourself.

Adkins, C. J. (1975). *Equilibrium Thermodynamics* (2nd edn). London: McGraw-Hill. Chapter VII.

Batchelor, G. K. (1967). *An Introduction to Fluid Dynamics*. Cambridge: Cambridge University Press. Chapters V, VIII, IX, X, XI, XVI.

Chambers, Ll. G. (1969). *A Course in Vector Analysis*. London: Chapman and Hall. Chapter I.

Chorlton, F. (1967). *Textbook of Fluid Dynamics*. London: Van Nostrand. Chapters XI, XIV.

Chorlton, F. (1976). *Vector and Tensor Methods*. Chichester: Ellis Horwood. Chapter I.

Clancy, L. J. (1975). *Aerodynamics*. London: Pitman. Chapter XVII.

Guggenheim, E. (1960). *Elements of the Kinetic Theory of Gases*. Oxford: Pergamon. Chapter III.

Hunter, S. C. (1976). *Mechanics of Continuous Media*. Chichester: Ellis Horwood. Chapter VIII.

Jaeger, L. G. (1966). *Cartesian Tensors in Engineering Science*. Oxford: Pergamon. Chapter I.

Jeffreys, H. (1931). *Cartesian Tensors*. Cambridge: Cambridge University Press. Chapter I.

Kinsman, B. (1965). *Wind Waves*. Englewood Cliffs, New Jersey: Prentice-Hall. Chapter XIII.

Lamb, H. (1945). *Hydrodynamics* (6th edn). Cambridge: Cambridge University Press. Chapters XI, XIII, XVI.

Lamb, H. (1949). *Statics* (3rd edn). Cambridge: Cambridge University Press. Chapter VI.

Lamb, H. (1960). *The Dynamical Theory of Sound* (2nd edn). New York: Dover. Chapter XII.

Li, W. H. & Lam, S. H. (1964). *Principles of Fluid Mechanics*. Reading, Massachusetts: Addison-Wesley. Chapter XV.

Liepmann, H. W. & Roshko, A. (1957). *Elements of Gas Dynamics*. New York: John Wiley. Chapter XIV.

Lighthill, M. J. (1978). *Waves in Fluids*. Cambridge: Cambridge University Press. Chapters XII, XIII.

Lin, C. C. & Segel, L. A. (1974). *Mathematics Applied to Deterministic Problems in the Natural Sciences*. New York: Macmillan. Chapter IX.

Loeb, L. (1927). *Kinetic Theory of Gases*. London: McGraw-Hill. Chapter III.

Marder, L. (1970). *Vector Analysis*. London: George Allen and Unwin. Chapter I.

Massey, B. S. (1975). *Mechanics of Fluids*. New York: Van Nostrand Reinhold. Chapter XIV.

Milne-Thomson, L. M. (1949). *Theoretical Hydrodynamics* (2nd edn, or 4th edn 1960). London: Macmillan. Chapter IV, V, XI, XIII, XVI.

Milne-Thomson, L. M. (1952). *Theoretical Aerodynamics* (2nd edn, or 4th edn 1966). London: Macmillan. Chapter XVII.

Morse, P. M. & Ingard, K. U. (1968). *Theoretical Acoustics*. New York: McGraw-Hill. Chapter XII.

National Committee for Fluid Mechanics Films (1972). *Illustrated Experiments in Fluid Dynamics*. Cambridge, Massachusetts: MIT Press. Introduction.

Queen, N. M. (1967). *Vector Analysis*. London: McGraw-Hill. Chapter I.

Ramsay, J. A. (1971). *A Guide to Thermodynamics*. London: Chapman and Hall. Chapter VII.

Ramsey, A. S. (1947). *Hydrostatics*. London: Bell. Chapter VI.

Rayleigh, Lord (1945). *Theory of Sound* (2 vols.). New York: Dover. Chapter XII.

Scorer, R. S. (1958). *Natural Aerodynamics*. Oxford: Pergamon. Chapter III.

Silver, R. S. (1971). *An Introduction to Thermodynamics*. Cambridge: Cambridge University Press. Chapter VII.

Trevena, D. H. (1975). *The Liquid Phase*. London: Wykeham Publications. Chapter III.

Tritton, D. J. (1977). *Physical Fluid Dynamics*. Wokingham, Berkshire: Van Nostrand Reinhold. Chapters III, XIII, XVI.

Index

See also the list of contents